Lecture Notes in Electrical Engineering

Volume 596

The book series *Lecture Notes in Electrical Engineering* (LNEE) publishes the latest developments in Electrical Engineering - quickly, informally and in high quality. While original research reported in proceedings and monographs has traditionally formed the core of LNEE, we also encourage authors to submit books devoted to supporting student education and professional training in the various fields and applications areas of electrical engineering. The series cover classical and emerging topics concerning:

- Communication Engineering, Information Theory and Networks
- Electronics Engineering and Microelectronics
- Signal, Image and Speech Processing
- Wireless and Mobile Communication
- Circuits and Systems
- Energy Systems, Power Electronics and Electrical Machines
- Electro-optical Engineering
- Instrumentation Engineering
- Avionics Engineering
- Control Systems
- Internet-of-Things and Cybersecurity
- Biomedical Devices, MEMS and NEMS

For general information about this book series, comments or suggestions, please contact leontina.dicecco@springer.com.

To submit a proposal or request further information, please contact the Publishing Editor in your country:

China

Jasmine Dou, Associate Editor (jasmine.dou@springer.com)

India

Aninda Bose, Senior Editor (aninda.bose@springer.com)

Japan

Takeyuki Yonezawa, Editorial Director (takeyuki.yonezawa@springer.com)

South Korea

Smith (Ahram) Chae, Editor (smith.chae@springer.com)

Southeast Asia

Ramesh Nath Premnath, Editor (ramesh.premnath@springer.com)

USA, Canada:

Michael Luby, Senior Editor (michael.luby@springer.com)

All other Countries:

Leontina Di Cecco, Senior Editor (leontina.dicecco@springer.com)

**** Indexing: The books of this series are submitted to ISI Proceedings, EI-Compendex, SCOPUS, MetaPress, Web of Science and Springerlink ****

More information about this series at http://www.springer.com/series/7818

Vančo Litovski

Electronic Filters

Theory, Numerical Recipes, and Design Practice based on the $\mathcal{RM}$ Software

Springer

Vančo Litovski
Elektronski Fakultet
Niš, Serbia

ISSN 1876-1100 ISSN 1876-1119 (electronic)
Lecture Notes in Electrical Engineering
ISBN 978-981-32-9851-4 ISBN 978-981-32-9852-1 (eBook)
https://doi.org/10.1007/978-981-32-9852-1

This Springer imprint is published by the registered company Springer Nature Singapore Pte Ltd.
The registered company address is: 152 Beach Road, #21-01/04 Gateway East, Singapore 189721, Singapore

To the loving memory of my son
Leonid Litovski

Preface

Filter design as a research subject dates almost a century now. Of course, it was coming as part to the development of the radio and telecommunication industry which became ubiquitous now. Substantial number of solutions was offered in transfer function synthesis as well as in the physical realization which gradually shifted from analogue to a digital one with the ability to be even implemented as software. In that, the interest got shifted mainly towards new technologies while gradually narrowing the scope and the versatility of the approximation solutions. As a consequence, modern filter design is based on reduced set of transfer functions being catalogued and available broadly.

It was our filling that in the historical moment, we are living catalogues and nomograms are not the best way to produce optimal solutions. That is valid especially having in mind the fact that the computers became affordable to anyone and that practically all electrical engineers are fluent in programming languages. We think that the theory and numerical receipts given, anyone may produce any practically feasible transfer functions starting with the proper set of requirements. Similar stands for the circuit or system synthesis.

Having that in mind, this book is to be looked upon as a supportive tool for software development for filter design in both transfer function synthesis (approximation process) and system synthesis (physical design).

It is divided into three parts. The context is created within the first part by introducing the reader into the basic concepts of signals and filters including the $\mathcal{RM}$ software as a supportive tool used for creation of the examples within the book. Then, the transfer function synthesis is addressed which encompasses a wide variety of solutions. It was our intention here not only to describe the solution as such but to give full information for the reader to start software development for the proper approximant by his own. That is supported within every chapter by a paragraph named "Developer's corner" containing advice and transferring experience related to the practical implementation of the concepts just described. In addition, examples are given of relatively high complexity allowing the reader to verify his own solution. In the third part, the system synthesis of analogue and digital systems is visited. Here again, a wide variety of technologies is encompassed allowing the designer to cover most of the frequency spectrum from very low audio frequencies up to tenths of GHz. There is something specific, however, within this part. Namely, in most cases parallel solutions are advised and synthesis concepts are described.

The author is using the opportunity to thank the research staff of the Laboratory for Electronic Design Automation (LEDA) of the Faculty of Electronic Engineering at the University of Niš, Serbia, for their lifelong and everlasting collaboration and support.

Niš, Serbia Vančo Litovski

Contents

Part II Transfer Function Synthesis

Part III Circuit and System Synthesis

About the Author

Prof. Vančo Litovski was born in 1947 in Rakita, South Macedonia, Greece. He graduated from the Faculty of Electronic Engineering in Niš in 1970, and obtained his M.Sc. in 1974, and his Ph.D. in 1977. He was appointed as a teaching assistant at the Faculty of Electronic Engineering in 1970 and became a Full Professor at the same faculty in 1987. He was elected as Visiting Professor (honoris causa) at the University of Southampton in 1999. From 1987 until 1990, he was a consultant to the CEO of Ei, and was Head of the Chair of Electronics at the Faculty of Electronic Engineering in Niš For 12 years. From 2015 to 2017, he was a researcher at the University of Bath. He has taught courses related to analogue electronics, electronic circuit design, and artificial intelligence at the electro-technical faculties in Priština, Skopje, Sarajevo, Banja Luka, and Novi Sad. He received several awards including from the Faculty of Electronic Engineering (Charter in 1980, Charter in 1985, and a Special Recognition in 1995) and the University of Niš (Plaque 1985). Prof. Litovski has published 6 monographs, over 400 articles in international and national journals and at conferences, 25 textbooks, and more than 40 professional reports and studies.

His research interests include electronic and electrical design and design for sustainability, and he led the design of the first custom commercial digital and research-oriented analogue CMOS circuit in Serbia. He has also headed 8 strategic projects financed by the Serbian and Yugoslav governments and the JNA, and has participated in several European projects funded by the governments of Germany, Austria, UK, and Spain, and the EC as well as the Black See Organization of Economic Cooperation (BSEC).

Part I
Basics

Chapter 1
Introduction

> Give a man a fish, and you feed him for a day. Teach a man to fish, and you feed him for a lifetime.

The story of filter design is almost hundred years old now. It follows two paths, one related to transfer function synthesis and the other related to physical implementation synthesis.

While there had to be physical realization for whatever function was developed, due to technological limitations, the branch related to transfer function synthesis (which relied on mathematics) was leading, often facing the "physical realization" obstacle. It grew fast since the first Butterworth's [1] introduction of the maximally flat amplitude characteristic. Solutions were found for equi-ripple amplitude approximation both in the passband and in the stopband of the filter. The problem of linear phase was addressed and solved allowing for subsequent solutions satisfying mixed amplitude and phase requirements. Finally, direct time domain solutions were reported [2].

It was really very difficult to manipulate complicated complex functions in the pre computer era. The introduction of computers in the sixties (and later on) of the last century revolutionized the transfer function synthesis and most of the main problem were now solved with acceptable complexity. Let just mention the problem of frequency transformations [3] proposed by Orchard and Temes and the introduction of iterative design [4].

Later on, with further advent of the computers and digital integrated circuit technology, theory of digital signal processing and digital filter's transfer function synthesis was introduced to allow for an explosion of the telecommunication systems [5].

Nevertheless, after all these years and almost unlimited number of contributions to the subject of transfer function synthesis, even the most advanced contemporary sources are based on a very limited set of solutions (such as Butterworth, Chebyshev, Elliptic, Inverse Chebyshev, and Thomson) and use formulas, nomograms or pre-prepared tables [6–8] for the design. In [9] the authors explicitly state that the transfer function synthesis will be based on existing nomograms. In fact, the real transfer function synthesis became a privilege of a very limited number of

V. Litovski, *Electronic Filters*, Lecture Notes in Electrical Engineering 596,
https://doi.org/10.1007/978-981-32-9852-1_1

scholars who have all the mathematical knowledge, the technological experience, and programming capabilities.

That became an important issue in the progress of the transfer function synthesis research and development frequently leading to non-optimal solutions. Namely, having the wrong feeling that one is fully informed on the subject, research results were published (by world renown publishers) in which explicit ignorance is demonstrated. We will illustrate this claim here with several examples with the intention to draw the attention of the filter design community and to suggest more responsible reviewing. Of course, to avoid negative advertising we will not cite the exact sources. As the first example we will mention the inadequate computation of the phase characteristics. After definition of the transfer function in the form

$$H(\mathrm{j}\omega) = \frac{m_1(\mathrm{j}\omega) + j \cdot n_1(\mathrm{j}\omega)}{m_2(\mathrm{j}\omega) + j \cdot n_2(\mathrm{j}\omega)} \tag{1.1}$$

the authors derive the expression for the phase characteristic as

$$\varphi(\mathrm{j}\omega) = \arctan\left[\frac{n_1 \cdot m_2 - n_1 \cdot m_1}{m_1 \cdot m_2 + n_1 \cdot n_2}\right]. \tag{1.2}$$

Even if we skip the question of a phase being a function of a complex variable and attribute that to negligence, there is a fundamental problem with the method of calculating the phase. Namely, assuming the *m*s and *n*s are real function of a real variable (which must be), the maximum (minimum) asymptotic value of the phase expressed by (1.2) may be $+\pi/2$ (or $-\pi/2$, depending on the sign of the expression in the brackets). That is to be opposed to the right value of

$$\varphi(\omega)_{|\omega\to\infty} = -(k+l) \cdot \pi/2, \tag{1.3}$$

for a filter with k zeros (complex in the right half plane or imaginary on the ω-axis or mix of these two) and l complex poles in the left half plane. So, a tenth order all-pass filter will have asymptotic value of the phase equal to -10π. Any tenth order polynomial filter will, of course, have $-10\pi/2$.

The next example is related to basic knowledge in which no distinction is made between a circuit schematic and the corresponding transfer function. The authors claim the circuit of Fig. 1.1 is the schematic of an elliptic filter.

In fact it is a sixth order ladder filter (using maximum number of inductances and minimum number of capacitances?!) with two transmission zeros at the real part of the ω-axis which, depending on the element values, may perform as so many different transfer functions, among others, as the Inverse Chebyshev, for example. Since it is of even order and has equal terminations, however, it is hard to believe that it is realizable as an elliptic filter. This is not the only example which may be encountered in literature where the topology of a network is associated to a single type of approximant since, probably, that approximant is the only one known to the author.

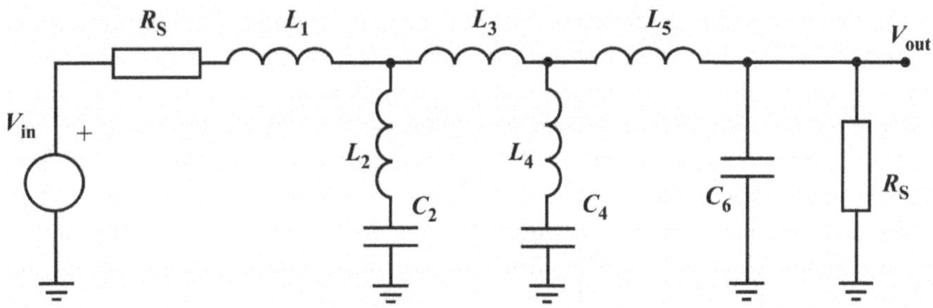

Fig. 1.1 A sixth order LC ladder filter

Finally, one may read about "ultra selective" linear-phase low-pass FIR filters what sounds as a thunder. In fact the authors are reporting a transfer function exhibiting stopband attenuation of 100 dB while the transition region of the filter is 20 times wider than the pass-band width. That means that even the 40th harmonic of a spectral component belonging to the middle of the passband is not filtered. Leaving alone the need for 100 dB of attenuation, a question arises as to: is that a filter at all?

On the physical realization branch, probably with some delay, equally explosive developments could be witnessed. Starting with RLC elements [10], it went first through active RC synthesis [11] which was boosted by the introduction of the integrated operational amplifier [12].

The CMOS technology [13], like the Tesla's alternating currents in the 19th century, became the propellant of the modern society in any respect of the word. It enabled explosion of the information-telecommunication (IT) technologies and did democratize the computing and communication to an unconceivable extent. As for the filter design, first, by introducing the microprocessor the digital filtering era was indicated [14]. Then, by development of powerful programmable devices the digital filtering became affordable. Finally, the custom design of digital integrated circuits allowed for any skilled laboratory to routinely design systems with embedded digital filtering as a subsystem.

Analog CMOS was rising in parallel [15]. In addition to the possibility of embedding the analog filter into a mixed-mode integrated system, fully new ideas were promoted and realized such as switched capacitor (SC) circuits and transconductance-C (Gm-C) circuits which allowed analog signal processing by integrated CMOS circuits at very low and very high frequencies, respectively.

The modern state of the things in the IT may be characterized by immense rise of the awareness of the necessity for every citizen to be capable of writing programming code. One may say that now every electrical and electronic engineer is fluent in one or more programming languages. That, however, does not map into the philosophy of modern filter design. Still, nomograms and tables, generated in the last half of the 20th century, are ruling the trade.

Having that in mind, this book has several main goals.

The first one is to bring forward most of the knowledge and algorithms necessary for successful software development for filter design. In that sense our intention is to give enough information and algorithmic capacity to the reader for he/she, being fluent in modern programming languages and contemporary electronics, to became capable to create his own software for every problem encountered in his/hers everyday professional life.

By acquiring the know-how explicated in this book the reader will be capable to produce approximants (transfer functions of filters) with unique properties accommodated to his/hers problem to a maximal extent. In that sense we want to stress

- The critical monotonic passband amplitude characteristics (CMACs).
- The linear phase filters approximating constant group delay in an equi-ripple manner
- The all-pass and band-pass approximants exhibiting group delay approximating a horizontal and inclined line.
- The amplitude correction algorithms enabling improved both linear phase and selective polynomial prototypes.
- The modified elliptic.
- The low-pass and band-pass group delay correctors.
- The simultaneous group delay and amplitude approximation of band-pass filters.

On the other side our intention was to make the reader capable to create in an existing design environment. To that end we will exemplify all the algorithms and procedures, belonging to transfer function synthesis or to physical system synthesis, on the case of use of the $\mathcal{RM}$ software [16, 17].

In that, one should have in mind that this is not a book on electronics. It is about filter design which means only limited information about the circuits will be given. Namely, one may usually find in filter design books complete information about semiconductor technology, transistor and passive devices design and modelling, and synthesis of low-complexity circuit (maximum to second order cells) leaving the issues of complex synthesis to the reader. Instead, we will try to skip the common knowledge and minimize the theory on synthesis of basic electronic circuits and offer to the potential software developer and filter designer what we think are the best building blocks for proper applications so trying to concentrate on complex real-life solutions and architectures.

The book is divided in three parts.

In the first part we first deal with basics of signals and filtering. Definitions will be given and the basic terminology explained. Attempts will be made not to leave any ambiguities related to the meaning of the terms. As a specialty we introduce for the first time quantification of the selectivity while insisting on design being as selective as possible (the ones saving bands of frequencies are named ecological designs). Next comes the definition of the transfer function and calculations of frequency and time domain responses. Here analog transfer functions are considered only. A specialty in this paragraph is the calculation of the phase response which is frequently misunderstood as exemplified above. We insist that for all phases of design: function approximation, function analysis and physical synthesis, factored form of the transfer function is the only way to represent it. That allows for an integrated approach. Issues related to the transfer function definitions are completed by the description of the frequency transformations in the analog domain. The first part ends with a brief description of the potentials, structure, and functionality of the $\mathcal{RM}$ software for filter design which will be used to exemplify the algorithms and the physical synthesis methods.

The second part is devoted to transfer function synthesis. It encompasses a very wide spectrum of approximants fulfilling requirements in the frequency domain obeying amplitude, group delay or both amplitude and group delay requirements. It is in this book for the first time a complete description and information for synthesis is given for the low-pass selective polynomial filters with critical monotonic amplitude characteristic (CMAC) in the passband. The non-monotonic amplitude characteristic in the passband are studied via the Chebyshev filter while a modification is described which allows for modified amplitude characteristic of the even order Chebyshev filters. These proceedings are followed by comparisons in order to complete the knowledge about polynomial selective filters.

It is a fact that extending the transfer function of a polynomial filter by a finite transmission zero at the ω-axis brings more benefits to the amplitude characteristic than increasing the order of the polynomial filter. That, however, is not advertized in the literature and no books may be found describing the procedure of introducing such zeros. Here we do that in two variants: with distinct and with multiple new transmission zeros. Of course, special cases of these functions are the so called Inverse Chebyshev and the Elliptic filters. Here we introduce the term modified elliptic filters since the procedure we are proposing allows for independent values of the maximum passband and minimum stopband attenuations with number of attenuation zeros equal or smaller than maximum.

Algorithms and examples of synthesis of linear phase low-pass filters exhibiting maximally flat and equi-ripple approximation of constant group delay are discussed next. Synthesis of all-pass low-pass filter is described, too. Algorithms and procedures are given for extending these transfer function with transmission zeros at the ω-axis to improve selectivity while preserving the phase linearity. The rest of this part of the book is related to solutions obeying group delay requirements as the primary ones. First, synthesis of group delay correctors (both low-pass and band-pass) is visited. Then, band-pass all-pass filters approximating a line in maximally flat sense are described. The line here may be horizontal and inclined (rising or falling). Finally, direct synthesis of selective band-pass filters with a group delay approximating a line in maximally flat sense is introduced.

For every approximation type described in this part of book complete information and advice is given enabling the reader to develop his own transfer function synthesis program. Note, for every algorithm and procedure, being inherently iterative, initial solution creation is advised. Detailed examples are accompanying every chapter so allowing the reader to check his own results in every phase of the software action. Readers having no experience in optimization or having difficulties to apprehend the iterative processes recommended in this book, are advised to first read [16].

As for the physical synthesis, to which the third part is devoted, the following technologies are encompassed. Passive LC cascade circuit synthesis; active RC cascade circuit synthesis; parallel active RC circuit synthesis; Gm-C filter synthesis based on LC prototypes; parallel active SC circuit synthesis; and parallel synthesis of IIR digital filters based on bilinear transformation of analog prototypes. For every each of them detailed information is given allowing the reader having a transfer function represented in a factored form to develop his/hers own synthesis program. As in the case of transfer function synthesis, examples are given in every chapter allowing the software developer to check his/hers results in every phase of the transformation of the input data on the transfer function into element values of the schematic.

A special feature of this topic is the parallel synthesis. Namely, the cascade synthesis is not avoided in two basic cases as the passive LC (including the Gm-C filters emanating from them) and the active RC but, since for realization of cascade synthesis laborious procedures are needed which still involve uncertainties, the parallel alternative is offered. It is simple to implement and in some cases advantageous.

The book is supported by 34 video presentations/animations which are on public display [17, 18] (the second was introduced since some of the countries have no access to YouTube). There details on the capacities and ways of implementation of the $\mathcal{RM}$ software are given. We expect that this and the "Developers corner" introduced in every chapter of this book will rise the confidence of the readers for both to take the adventure and develop his/hers own software or to effectively use of existing one.

References

1. Butterworth S (1930) On the theory of filter amplifiers. Exp Wirel Wirel Eng 7(85):536–541
2. Darlington S, Ming N-T, Orchard H, Watanabe H (1968) What, if any, are the important unsolved questions facing filter theorists today? IEEE Trans Circuit Theory 15(4):303–306
3. Orchard H, Temes G (1968) Filter design using transformed variables. IEEE Trans Circuit Theory 15(4):385–408
4. Ishizaki Y, Watanabe H (1968) An Iterative Chebyshev approximation method for network design. IEEE Trans Circuit Theory 15(4):326–336
5. Mitra SK (1998) Digital signal processing: a computer-based approach. McGraw-Hill, New York
6. Poularikas AD (1999) Handbook of formulas and tables for signal processing. CRC Press LLC, Boca Raton
7. Zverev AI (2005) Handbook of filter synthesis. Wiley-Interscience, New York
8. Williams A, Taylor FJ (2006) Electronic filter design handbook. McGraw-Hill, New York
9. Schauman R, Van Valkenburg ME (2001) Design of analog filters. Oxford University Press, New York

10. Darlington S (1999) A history of network synthesis and filter theory for circuits com-posed of resistors, inductors, and capacitors. IEEE Trans Circuits Syst-I: Fundam Theory Appl 46 (1):4–13
11. Sallen RP, Key EL (1995) A practical method of designing RC active filters. IRE Trans Circuit Theory CAS-2(1):74–85
12. Burr-Brown Application Staff (1963) Handbook of operational amplifications. Burr-Brown Corporation
13. Wanlass FM (1963) Low stand-by power complementary field effect circuitry. U.S. Patent No. 3,356,858, filed 18 June 1963
14. Antoniou A (2000) Digital filters. McGraw-Hill Science/Engineering/Math, New York
15. Gray PR, Hurst PJ, Lewis SH, Meyer RG (2009) Analysis and design of analog integrated circuits. Wiley, New York
16. Litovski V, Zwolinski M (1997) VLSI Circuit simulation and optimization. Chapman and Hall, London
17. https://www.youtube.com/channel/UCF_Ipw_YD2gwrRpJDUJJULw/playlists?disable_polymer=1. Last accessed 20 May 2019
18. http://skysupervisor.com/projects/RM%20software/RM%20software.htm. Last accessed 20 May 2019

Chapter 2
Basics of Signals and Filtering

Types of signals are introduced first: continuous time and discrete time. Transposition into the frequency domain via the Fourier transform is used to allow for introduction to the signal processing. By manipulation of the spectrum of an example signal the idea of filtering and filters is introduced. Rationale for the relocation of the spectrum by modulation is given. The term and a measure for selectivity are defined and "green design" is introduced. Categorization of the filters is given based on the way how is the signal processed (continuous time and discrete time); how the signal is represented: analog and digital; where the position of the passband is (low-pass, band-pass, high-pass, all-pass, and band-stop); and what is the technology used for physical implementation (discrete passive LC, discrete and integrated active RC, integrated active SC, integrated active Gm-C, and software realized and integrated digital). Practical issues such as frequency, time, and element values normalization as well as calculator-less decibel (and inverse) calculations are included.

2.1 Introduction

The topic of filtering is rather multidisciplinary in the sense that encompasses serious mathematical apparatus and equally serious technological knowledge. Of course, knowledge of circuit theory and signal processing should not be undermined. That is why preparing a book on filter design is an action full of responsibilities. When starting this book, we decided to try to simplify the explanations as much as possible in order to allow the reader to acquire only the knowledge which is indispensible for the subject. That means that we will try to use examples being as general as possible and in the same time as simple as possible. On the other side an attempt will be made to encompass as much information as possible in order not to bring the designer's horizon too near and so discourage him before starting.

V. Litovski, *Electronic Filters*, Lecture Notes in Electrical Engineering 596,
https://doi.org/10.1007/978-981-32-9852-1_2

Following these consideration this chapter will be an introduction to signal processing and filter design. It will serve as a place for definition the basic terms and concepts.

2.2 Signals

A signal is a function of one or more variables that conveys information about some (usually physical) phenomenon [1]. For the purpose of this book, a signal will be function of a single variable (one-dimensional), the time, t. According to the variation of their time variable and their amplitude, signals can be either continuous-time or discrete-time, analog or discrete amplitude, or digital. This classification relates to the way signals are either processed, stored, or both. According to whether the signals exhibit repetitive behavior or not they are classified as periodic or aperiodic signals [2].

Figure 2.1a depicts a continuous time or an analog signal. Both the time and the voltage variables are allowed to take any value. Analog signals are natural signals and, strictly speaking, all physical signals are analog.

When it come to representation of the signals, e.g. in a computer, problems arise first due to the finite number of significant figures available for representation. This property is referred to as the resolution. No matter how large is the computer word (say 64 bit) only discrete values of both time and the signal may be represented. In between two neighboring values is a difference which is represented by the value of the least significant bit. In addition, storing signal values which are so near to each other (in time and value) may represent a serious request for memory. So, in order to save memory, one discretizes the time, the discretization interval being Δt, and allows for the complete signal to be stored in a limited amount of memory space. Such a signal, depicted in Fig. 2.1b, is still analog since the voltage is allowed to take any value. The process of conversion of a continuous time signal into a discrete time one is referred to as sampling. The final signal is a sampled one. The quantity

$$fs = 1/\Delta t \tag{2.1}$$

is referred to as the sampling frequency and it is expressed in Hz. The reciprocal is usually stated as sampling period $T_s = \Delta t$. For example, $f_s = 25$ kHz means 25,000 samples per second and the sampling period is $T_s = 1/25{,}000$ s = 40 μs.

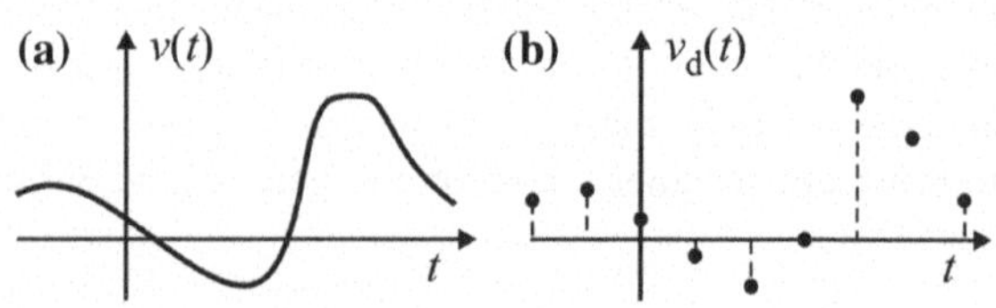

Fig. 2.1 Signals. **a** Continuous and **b** discrete time

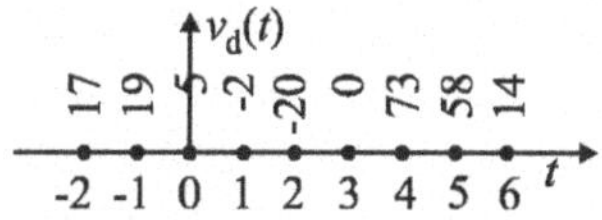

Fig. 2.2 Discretized in time and digitized in value signal

If the instantaneous value of the voltage taken in discrete time intervals is discretized (the term quantized is in use, too), we in fact get a digital signal. The discrete values of the voltage are represented by finite number of figures and, usually, in binary form which allows for storing in a computer. For example, the discretized signal depicted in Fig. 2.1b is converted into a digital one and depicted in Fig. 2.2. Here, to simplify, decimal system is used for representation of the voltage digital values which means that the quantization interval was $\Delta v = 1$ V.

In Fig. 2.2 the discretized time lost its identity. In fact the samples are enumerated and since the sampling time interval Δt is known it may be restored easily. It is obvious that representation of the signal depicted in Fig. 2.2 in a computer is an easy task. Not only that, it will take negligible amount of memory. Both, simplifying the representation and reducing the memory requirements, are what the computer needs. The price for that, however, is lost information in both coordinates. To protect the accuracy one should try to use as small Δv and Δt as possible.

In the next we will very shortly describe the activities frequently referred to as signal processing in all three domains mentioned above.

2.3 The Spectrum of a Signal

Looking to the signal of Fig. 2.1a we see that there is not much to be done with it except to extend its voltage value i.e. to amplify it or to perform sampling and enter the discrete time domain.

In fact, there is much more which is not visible at first sight. The time domain representation we are considering is hiding many properties of the signal which may be modified, corrected, improved etc. To come to these properties we need to change the domain and from the (natural) time domain to switch to (artificial) frequency domain. For that purpose we use the so called Fourier transform which allows for a full insight on the signal and enables implementation of means for processing it.

In the next we will introduce this transform and proceed by use of the means it offers to create context for the need for filtering and ways of doing it.

Given an arbitrary *periodic* continuous time signal $x(t)$ with a period T, according to the Fourier transform it may be represented in a form of an infinite series so as

$$v(t) = c_0 + \sum_{k=1}^{\infty} 2 \cdot \mathrm{Re}\{c_k e^{jk\omega_0 t}\}. \tag{2.2}$$

In the above equation

$$\omega_0 = 2 \cdot \pi / T, \tag{2.3}$$

$\mathrm{j} = \sqrt{-1}$, and c_k, $k = 0, 1, \ldots, \infty$ is a set of constants needed to be determined in order for (2.2) to be valid.

If complex quantities are properly represented by real and imaginary parts a new form of the Fourier series may be created as

$$v(t) = c_0 + 2 \cdot \sum_{k=1}^{\infty} |c_k| \cdot \cos(k\omega_0 t + \theta_k) \tag{2.4}$$

or

$$v(t) = c_0 + \sum_{k=1}^{\infty} [a_k \cdot \cos(k\omega_0 t) + b_k \cdot \sin(k\omega_0 t)] \tag{2.5}$$

where $c_k = |c_k| \cdot e^{\mathrm{j}\theta_k}$, $a_k = \mathrm{Re}\{2 \cdot c_k\}$ and $b_k = -\mathrm{Im}\{2 \cdot c_k\}$. Equation (2.5) is referred to as the *trigonometric form* of the Fourier transform.

Here are the formulas determining c_k:

$$c_k = \frac{1}{T} \int_{t_0}^{t_0 + T} v(t) \cdot e^{\mathrm{j}k\omega_0 t} \cdot \mathrm{d}t \tag{2.6}$$

with

$$c_0 = \frac{1}{T} \int_{t_0}^{t_0 + T} v(t) \cdot \mathrm{d}t \tag{2.7}$$

representing the average value of the signal over the period T. t_0 in (2.6) and (2.7) is arbitrary instant for starting the integration of the periodic signal $v(t)$.

The coefficient c_0 being the average, c_1 is referred to as the first or the main harmonic of the series while c_k for $k > 1$ is referred to as the kth harmonic.

From (2.4) and (2.5) we may see that each complex sinusoid is associated with a particular frequency (which is some integer multiple (k) of the fundamental frequency (ω_0)). So, these coefficients indicate at which frequencies the signal is concentrated. In this way, the Fourier series representation provides a means for

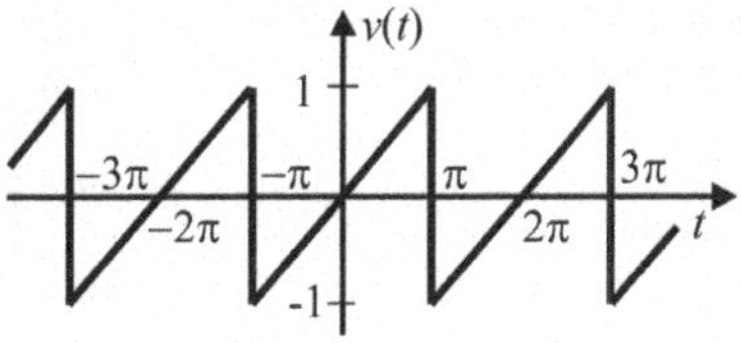

Fig. 2.3 The example saw-tooth signal

measuring the frequency content of a signal. The distribution of the signal over different frequencies is referred to as the *frequency spectrum* of the signal.

Various ways exist to illustrate the frequency spectrum of a signal. Typically, we plot the Fourier series coefficients as a function of frequency. Since, in general, the Fourier series coefficients are complex-valued; we usually display this information using two plots. One plot shows the amplitude of the coefficients as a function of frequency. This is called the amplitude spectrum. The other plot shows the arguments of the coefficients as a function of frequency. In this context, the argument is referred to as the phase, and the plot is called the phase spectrum of the signal.

We will now elaborate an example which will introduce us into the subject of analog signal processing.

Consider the saw-tooth signal, depicted in Fig. 2.3, which is represented by the following set of expressions

$$v(t) = \frac{t}{\pi} \quad \text{for } -\pi < t \leq \pi, \tag{2.8}$$

and

$$v(t + 2\pi \cdot l) = v(t) \quad \text{for } l = 0, 1, \ldots, \infty. \tag{2.9}$$

Since $T = 2\pi$, $\omega_0 = 1$ rad/s. The coefficients of the trigonometric form of the Fourier series related to this signal are

$$a_k = 0 \quad \text{for } k \geq 0 \tag{2.10}$$

and

$$b_k = 2 \cdot \frac{(-1)^{k+1}}{\pi \cdot k} \quad \text{for } k \geq 1. \tag{2.11}$$

After substitution in (2.5) we get

$$v(t) = \sum_{k=1}^{\infty} \left[2 \cdot \frac{(-1)^{k+1}}{\pi \cdot k} \cdot \sin(kt) \right]. \tag{2.12}$$

We will first pay attention to the values of the coefficients b_k. Figure 2.4a depicts the first 20 coefficients and shows that they have alternative signs and decrease

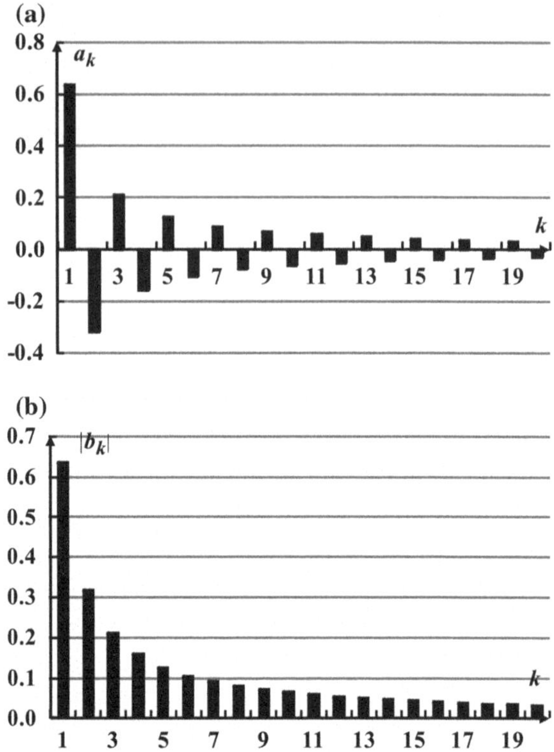

Fig. 2.4 Frequency domain representation of the saw-tooth signal. **a** The b_k coefficients up to the twentieth and **b** modulus of the b_k coefficients—the amplitude spectrum

hyperbolically with increase of k. That means that the higher order coefficients have smaller influence to the overall sum.

Figure 2.4b depicts the amplitude spectrum of the saw-tooth signal with components up to the 20th. Looking to this diagram we come to ideas as to how an analog signal may be processed. Simply, by manipulating the components of the spectrum! Since manipulation may, in some case, mean elimination of a component from the spectrum the term filtering comes forward. We say that that component was eliminated by a filter. Starting with that any manipulation of the spectrum is referred to as filtering and the means for performing filtering is referred to as a filter.

Let us now study the influence of some filters to the time domain presentation of the saw-tooth signal. Consider a filter function defined as

$$f(k) = \begin{cases} 1 & for \quad k \leq m \\ 0 & for \quad k > m \end{cases} \tag{2.13}$$

which is depicted in Fig. 2.5a. We will refer to this type of filtering function as *ideal amplitude low-pas filter*. The value m represents in fact a multiple of ω_0 (i.e. $m \cdot \omega_0$) and is referred to as the cut-off frequency of the low-pass filter.

After multiplying the original signal with the filtering function one gets

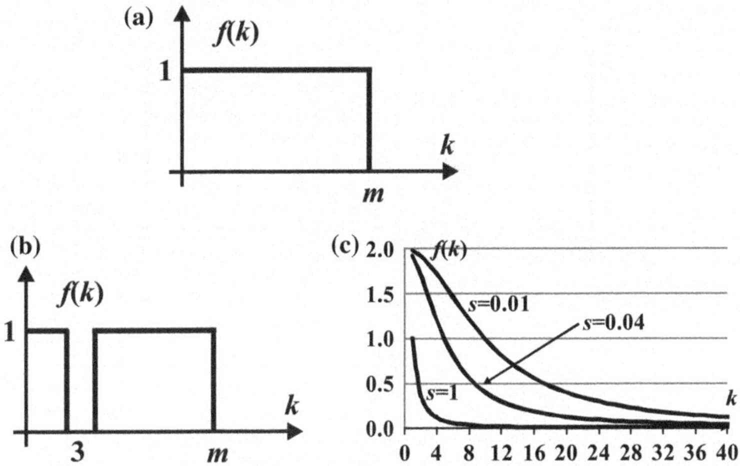

Fig. 2.5 Filtering functions. **a** Ideal low-pass amplitude characteristic, **b** a filtering function eliminating a harmonic component, and **c** amplitude characteristic of a Butterworth filter

$$v_f(t) = v(t) \cdot f(k) = \sum_{k=1}^{m} \left[2 \cdot \frac{(-1)^{k+1}}{\pi \cdot k} \cdot \sin(kt) \right]. \tag{2.14}$$

We refer to this expression as the truncated sum. The result of filtering (seen in the time domain) is produced by performing the summation in (2.14). We will show three results here. Figure 2.6 depicts the obtained waveforms.

Figure 2.6a depicts the time domain representation of the truncated Fourier series, representing the saw-toot signal of Fig. 2.3, for the case $m = 500$. One may see almost perfect matching between the truncated and the original signals. If the cat-off frequency of the filter is reduced to $m = 100$, slight deterioration may be observed from Fig. 2.6b. Finally, by reducing m to the value of 10, as depicted in Fig. 2.6c, significant difference between the original and the "processed" may be seen. To conclude, filtering may limit the frequency spectrum of a signal while producing some distortions. The broader the passband the smaller the distortions.

Figure 2.7a represents the resulting signal after filtering by the function of Fig. 2.5b. Here $m = 10$ was used but, in addition, one component of the spectrum ($k = 3$) was eliminated by notch operation. As can be seen dramatic changes in the final signal are produced.

Finally, the following filtering function was implemented

$$f(k) = \frac{2}{1 + s \cdot k^2}. \tag{2.15}$$

It is depicted in Fig. 2.5c for three values of s: s = 1, $s = 0.04$ and $s = 0.01$. This function is known as the Butterworth filter and will be discussed later in detail.

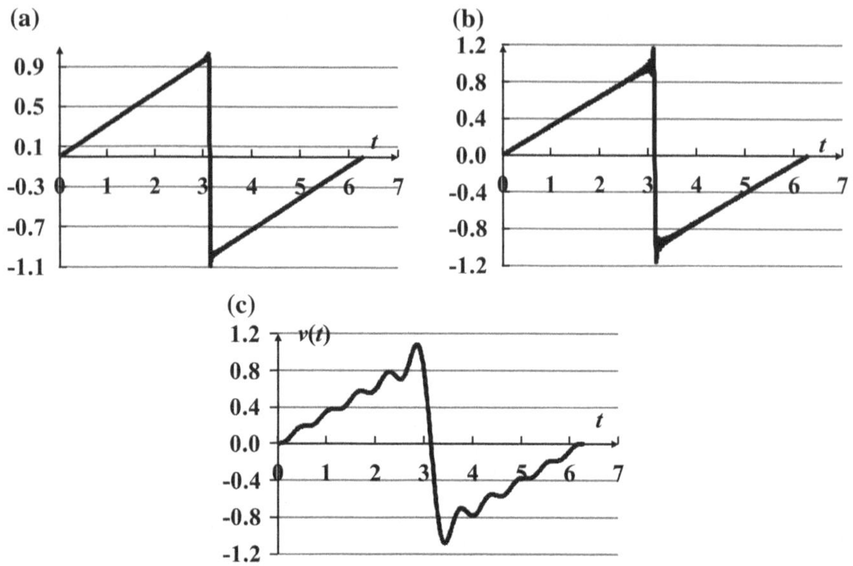

Fig. 2.6 Time domain representation of the truncated saw-tooth Fourier series. **a** $m = 500$, **b** $m = 100$, and **c** $m = 10$

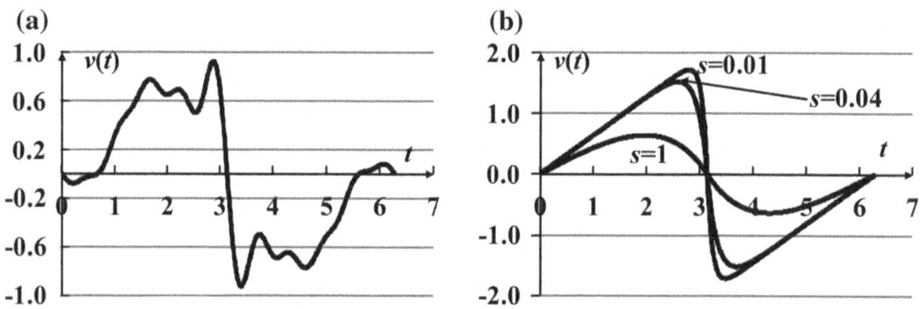

Fig. 2.7 Time domain presentation of a filtered Fourier series of the saw-tooth signal. **a** Filter of Fig. 2.5b and **b** filter of Fig. 2.5c

The results of application of this filtering function are depicted in Fig. 2.7b. By reducing s, in fact, the cut-off frequency $(m = k_s = 1/\sqrt{s})$ is risen and the waveform becomes more and more similar to the original.

To conclude, one important corollary of the above discussion may be drawn. It is the fact that the sinusoid is a function of fundamental importance for representing signals. That is to be added to the Nikola Tesla's sinusoidal alternating current waveform which is the base of power electrical engineering.

Having created rudimentary information on the effects of filtering, in the next, we will try to answer why filtering is necessary. To that end we will first introduce the technique of modulation and specifically, amplitude modulation.

Given a low frequency signal (with an amplitude normalized to unity)

$$S_L(t) = \cos(\omega_L t) \tag{2.16}$$

whose spectrum is a single bar (part of the spectrum of the original signal) on the frequency axis at $f_L = \omega_L/(2 \cdot \pi)$. It may be transposed at a higher frequency if the following is produced

$$S_M(t) = [1 + M \cdot \cos(\omega_L t)] \cdot \sin(\omega_c t) = [1 + M \cdot \cos(\omega_L t)] \cdot C(t) \tag{2.17}$$

where

$$C(t) = \sin(\omega_c t) \tag{2.18}$$

was used as a carrier signal. Here usually $\omega_c \gg \omega_L$. One may say that $S_L(t)$ is impregnated into $C(t)$ to create $S_M(t)$. The new signal $S_M(t)$ is referred to as an amplitude modulated signal with a modulation index $0 \le M \le 1$.

The spectrum of $S_M(t)$ may be recognized from

$$S_M(t) = \sin(\omega_c t) + \frac{M}{2}\{\sin[(\omega_c + \omega_L)t] + \sin[(\omega_c - \omega_L)t]\} \tag{2.19}$$

which is obtained from (2.17). It is depicted in Fig. 2.8a. If it was a complete low frequency signal (complete spectrum) to be used for creation of the amplitude modulated signal, the spectrum would look like the one depicted in Fig. 2.8b where, for simplicity, the low frequency (modulating) signal is supposed to have frequency independent but band limited spectrum.

In that diagram ω_{La} stands for the lowest bound of the spectrum of the low-frequency signal while ω_{Lb} stands for the upper bound. There are practical reasons imposing the need for the low-frequency spectrum to be truncated from both sides. At very low frequencies unwanted electromagnetic interference occurs

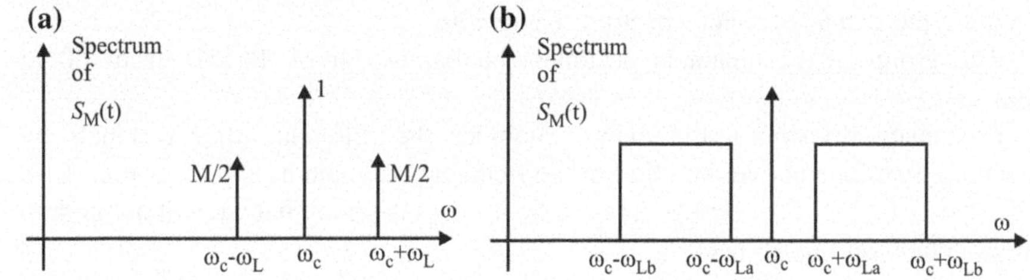

Fig. 2.8 Spectrum of the amplitude modulated signal. **a** Single spectral component and **b** frequency band

creating the so called inter-modulation products which are unwanted. The limitation from the upper side is related to economic reasons since the overall frequency spectrum is limited which may be seen from Fig. 2.8b where $\omega_c + \omega_{Lb}$ may go too high and occupy much of the available frequency band. We will come to this soon. In telephony, for example, the voice spectrum bandwidth is restricted between $f_{La} = 0.3$ kHz and $f_{Lb} = 3.2$ kHz.

Modulation is applied to information signals for a number of reasons, some of which are outlined below.

Many transmission media (cables, for example) have limited pass-bands which means they behave as filters having their own cut-off frequencies at both ends of the band which is transmitted. Modulation methods must therefore be applied to the information signals in order to "frequency transpose" the signals into the range of frequencies that are acceptable for the channel.

Even when the communications channel can support direct transmission of the information-bearing signal, there are often practical reasons why this is undesirable (or impossible). A simple example is the transmission of audio signal (below 10 kHz) via radio wave. The wavelength of the electro-magnetic three-kilohertz signal is 100 km. Since an effective radio antenna is typically as large as half the wavelength of the signal, a three-kilohertz radio wave might require an antenna up to 50 km in length which, of course, is highly unpractical.

The most important reason, however, is related to the fact that in many instances a communications channel is shared by multiple users. In order to prevent mutual interference, each user's information signal is modulated onto an assigned carrier of a specific frequency. When the frequency assignment and subsequent combining is done at a central point, the resulting combination is a frequency-division multiplexed (FDM) signal. Assuming sufficient frequency separation of the carrier frequencies so that the modulated signals do not overlap, recovery of each of the FDM signals is possible at the receiving end. In order to prevent overlap of the signals and to simplify filtering, each of the modulated signals is separated by a guard band, which consists of an unused portion of the available frequency spectrum.

To illustrate, Fig. 2.9 depicts the formation of the so called baseband signal in telephony [3]. Note, since an additional auxiliary monochromatic signal is found at 3850 Hz and due to the fact that the filter has not abrupt change of the gain at cut-off (what was seen in Fig. 2.5c) one use 4 kHz for the width of the filter defining the extended voice spectrum bandwidth.

The group of 12 channels occupies the bandwidth of 48 kHz from 60 to 108 kHz.

Frequency division multiplexing combines the different voice channels by stacking them one above the other in the frequency domain as shown in Fig. 2.10. Of course, the FDM concept is not specific for the telephony but we will not go into more details on this subject.

The next step in FDM is to combine 12 voice channels together. This is the first level of the analog multiplexing hierarchy and it is called group and depicted in

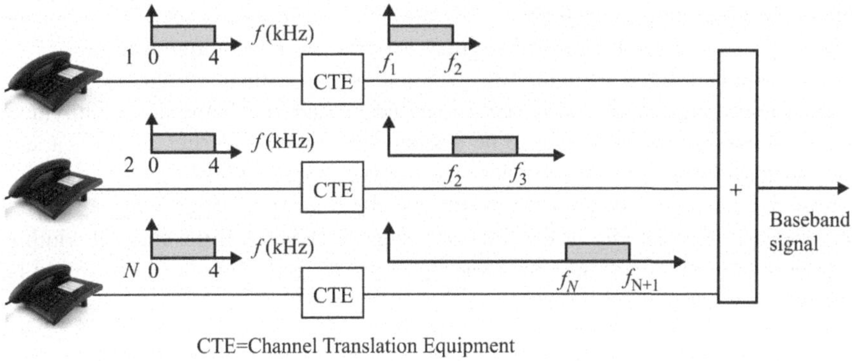

Fig. 2.9 Creation of the baseband signal in telephony

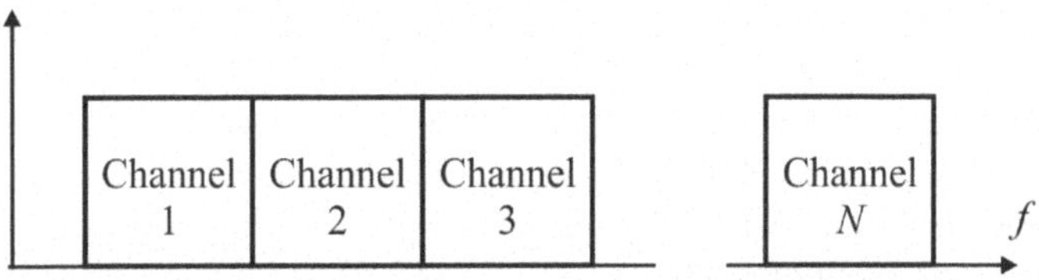

Fig. 2.10 The baseband signal spectrum

Fig. 2.11. These are further used to form a super group as depicted in Fig. 2.12. Master groups are following but there is not the end.

Based on the previous discussion, we think, the reader got the feeling of the role of the modulation to create the groups, super groups and so on. On the other side, what was the main goal, it became noticeable where the filters must be used and how many filters are necessary.

Modern society, as we all know, is based on telecommunication and a wide variety of signals are to be transmitted and received never avoiding proper filtering.

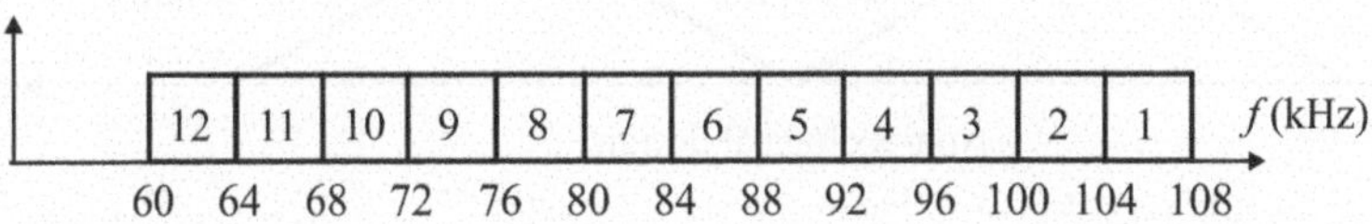

Fig. 2.11 Basic FDM group

Fig. 2.12 FDM super group

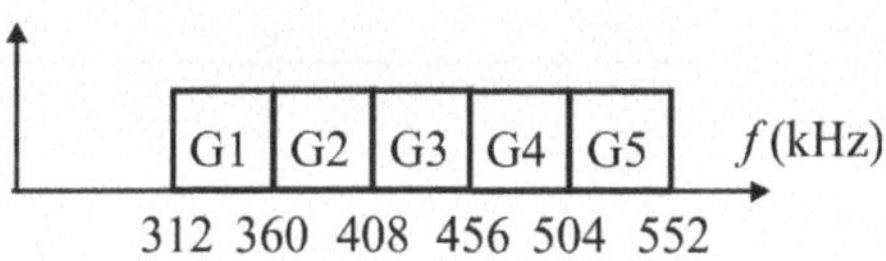

Filters are ubiquitous and effective and rational filter design is seen as one of the most important research and development activities in telecommunication.

Before finishing this paragraph we will address the issue of selectivity. Namely, it was already mentioned that in order to avoid interference between the individual signals, a band of frequencies is to be left between the neighboring channels at any level of the FDM. We will refer here the reserve frequency band as the transition region. The ratio of the channel's width and the transition region will be referred hereafter as the selectivity of the filter defining the channel. If the channel width is denoted by Bw (Hz) and the transition region as Tr (Hz), the selectivity would be

$$S = Bw/Tr. \tag{2.20}$$

The question arises, should the filter be mildly selective (e.g. $S = 1$), very selective (e.g. $S = 2$) or extremely selective (e.g. $S = 10$)? The following considerations, we think, might help when taking decision.

Frequency spectrum is a limited natural resource like the water, soil, and air. There is no renewable resource of frequency, however. Nowhere in the universe! Thus, while designing telecommunication systems and filters as part of them, do save the frequency spectrum. Use as selective filters as possible no matter the price (which comes from the increased physical complexity being realizable by renewable material and energy resources).

If you do so you will further benefit from the reduced overall noise of the system which is directly related to its selectivity. Namely, the noise power of thermal noise (the unavoidable one) is proportional to the noise bandwidth which becomes lower if the amplitude characteristic abruptly changes at the end of the passband.

These considerations are summarized in Fig. 2.13. One may recognize that proper filter design i.e. choice of selective filtering may allow for additional channels to be located on the frequency axis and in that way better (ecologically) use of the frequency spectrum as a natural resource.

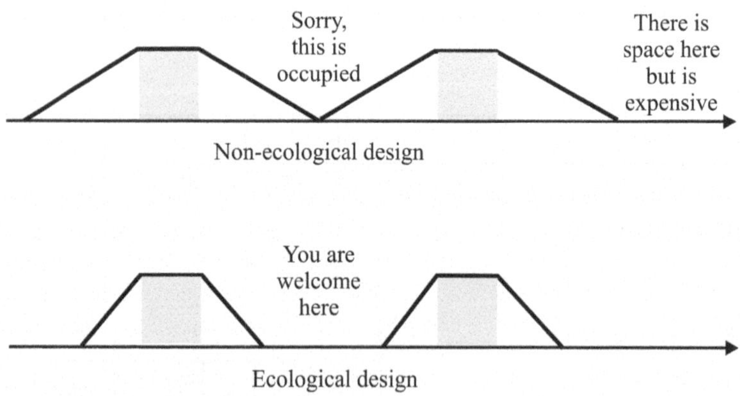

Fig. 2.13 Selectivity and ecological filter design

2.4 Categorization of Filters

There are many aspects of filtering which are used for categorization. These are based on the way how the signal is processed, the way how the filters are produced (technology), the way how they exploit the frequency axis and similar.

In this paragraph we will briefly introduce several attributes which allow distinguishing among different filters.

2.4.1 Categorization Based on the Way How Is the Signal Processed

We already mentioned that there are three types of representation of the signal: analog, discrete, and digital. Accordingly, methods are developed for processing the signal in its own domain so that we have analog filters, discrete time filters, and digital filters.

Using proper mathematical apparatus such as the Fourier transform and its counterpart Discrete Fourier transform a signal, be it analog or discrete, may be represented as a continuous function in the frequency domain.

Analog filters are processing the signal by analog operations which involve complex arithmetic as we will see later on. Here we may say that we achieve maximal accuracy which may be stated as the closeness of the computation results to the corresponding physical system response. From the computational point of view both the time and the signal values are manipulated with the maximal resolution of the proper computer.

To appreciate the accuracy here is how the numbers are represented in binary form within the 32-bit computer

$$(-1)^{x_0}\left\{1+\sum_{i=1}^{23} x_i 2^{-i}\right\}\cdot 2^k \tag{2.21}$$

where x_0 defines the sign bit, $x_i 2^{-i}$ represents the 23-bit mantissa, x_i is a single bit binary number, and 2^k represents the characteristic (it can have $2^8 = 256$ possible values).

If as the minimum increment of a number is considered the one which when added to 1 makes no change, it may be proven; its value will be 2^{-23}. Any number smaller than 2^{-23} will result in no change to the sum. $(2^{-23})_{\text{bin}} = (0.00000011920928955078)_{\text{dec}}$. One may say that in a 32-bit computer we are processing with accuracy of 7 significant figures. This is very important information when the validity of some arithmetic operations is considered. Adding two numbers of the same sign which have similar moduli will produce a number with practically the same accuracy. Subtracting such numbers, however, will produce a number of no accuracy at all. For example, if we subtract 0.16756438e1 and 0.16756324e1 we will get 0.114xxxxxe−4, where x denotes a random figure depending on the previous state of the register containing the number. Now, any number smaller than 10^{-7}, if added to the result of this subtraction, will be "swallowed" by the "noise" represented by the set of x − s. The error may be further amplified if the result of subtraction is multiplied by a large number and used for subsequent computations. Similar effects may be demonstrated when division is considered. That is, despite the fact that high resolution is available, in analog computations care should be paid to preserve numerical accuracy.

We will come back to the general properties of analog filters as compared to the digital ones in a separate section at the end of this paragraph.

A discrete time signal as depicted in Fig. 2.1 may be represented by the following expression

$$v_d(t) = v(t) \cdot \sum_{l=-\infty}^{\infty} \delta(t - l \cdot T_s) \tag{2.22}$$

where δ is the delta function

$$\delta(x) = \begin{cases} 1 & \text{for}\, x = 0 \\ 0 & \text{for}\, x \neq 0. \end{cases} \tag{2.23}$$

For our purposes, however, a slightly modified expression is of interest. In fact we need the value of the signal to stay constant in between the sampling intervals. Such a signal, frequently referred to as sampled-and-held, is depicted in Fig. 2.14. This signal may be expressed as

$$\begin{aligned} v_d(t) = v(t) \cdot \delta(t - l \cdot T_s) \quad &\text{for} \quad l \cdot T_s \leq t < (l+1) \cdot T_s \\ &\text{and} \quad \text{for} \quad -\infty < l < \infty \end{aligned} \tag{2.24}$$

During the sampling period, while the input to the filter is held, transients occur within the circuit which define the value of the output signal in the given time interval. That is how the discrete time filters work: analog operation in restricted time interval.

The electronic system which is processing the frequency spectrum of this signal as such (without quantization of the amplitude) will be referred to here as the discrete time filter. As we will see in the subsequent chapters this processing may

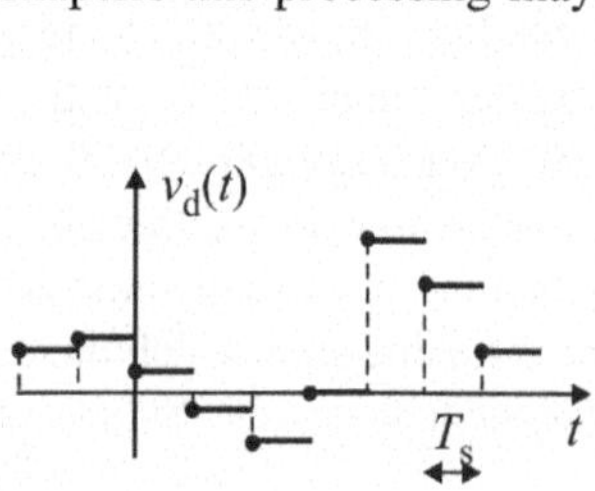

Fig. 2.14 Sampled and held signal

be stated as mixed mode since it is processing an analog signal but in discretized time instants which are digitally controlled.

Digital filters are massively in use nowadays due to the rising computing power and the digital revolution enabled by the CMOS technology. As we mentioned, it is about manipulation of numbers. That, in the simplest explanation, will mean that the samples taken from the analog signal are first quantized. Since binary representation of the final result is expected the accuracy of quantization is related to the value of the least significant figure of the binary number used. Historically, it started by processors based on four bit arithmetic so that four bits were used in so called analog-to-digital converters. Now extensive use of 64-bit processor is encountered. The binary number may be represented as a time series of bits which means that the pulse train representing the quantized value must end (must fit) within the sampling period. Physically, the signal is carried by a single wire. That relates the sampling frequency and the clock frequency of the processor. There will be needed at least as many clocks in the sampling period as the number of bit is. Alternatively, one may use as many wires as the number of bits and process further in parallel. In such a case the sampling frequency and the clock frequency may be (hypothetically) the same.

We may recognize now that digital filtering is in fact performing arithmetic operations on signals represented in binary form as integers. That would be the main difference to analog. A corollary of this claim is that any manipulation of the stream of pulses representing the input signal is considered as digital filtering. For example, there are so called decimation filters which literally decimate the pulse sequence. That will not be the subject of this book since it broadens infinitely the subject of filtering.

Of course, as for the analog, the digital filters need filtering functions as the one represented in Fig. 2.11. These, however, are represented in different circuit form. Namely one use digital (logic) circuits to perform the numerical operations and digital memory cells to store the coefficients of the functions.

Digital filters are separated in two categories. These are the so called Finite Impulse Response (FIR) and Infinite Impulse Response (IIR) filters. We will come back to these two categories later on but here we will mention that in this book only IIR filters will be synthesized. The main reason for that is the simplicity of synthesis which comes from the fact that the IIR filter functions are obtainable from analog prototypes by a simple transformation process.

2.4.2 *Digital Versus Analog Transmission*

Having in mind the difference of the way how the signal is represented and processed one may state some fundamental differences between analog and digital filters in telecommunication applications.

During the course of propagation on the channel, a transmitted pulse becomes gradually distorted due to the non-ideal transmission characteristic of the channel.

Also, various unwanted signals (usually termed interference and noise) will cause further deterioration of the information-carrying digital pulse. However, as there are only two types of signals that are being transmitted by digital transmission, it is possible for us to identify (with a very high probability) a given transmitted pulse at some appropriate intermediate point on the channel and regenerate a clean pulse. In this way, we will be completely eliminating the effects of distortion and noise accumulated till the point of regeneration. Such an operation is not possible if the transmitted signal was analog because there is nothing like a reference waveform that can be regenerated. In addition, when using digital signals we may exploit the advantages such as

- Storing the messages in digital form and forwarding or redirecting them at a later point in time is quite simple.
- Coding the message sequence to take care of the channel noise, encrypting for secure communication can easily be accomplished in the digital domain.

From the technology or circuit point of view the digital technology is not only immune to noise but is also immune to temperature variations since the pulse levels are temperature independent. Furthermore, a digital filter may be implemented as a piece of software as part of the firmware.

Problems may be recognized in sampling and analog-to-digital conversion related to the instability of the oscillator s creating the controlling pulse. In that case variation of the sampling instant occurs, what is referred to as jitter and considered noise.

Digital filtering is not applicable directly to natural signals. For example, in a signal processing or telecommunication systems (like the one discussed in Figs. 2.9, 2.10, 2.11 and 2.12) the natural signal first encounters an analog filter. Similarly, in many applications, the output signal is again analog. That makes the analog filters including the sampled data filters inevitable.

Filters used for transmitting groups and super-groups are again frequently analog. We will see some other specific application during the course of this book.

There is however one special motive for analog filter synthesis. There is a strong theoretical background and software tools which are used for analog filter design in decades now. These results and tools are easily mapped into the digital domain using proper transformations to produce IIR digital filters. In that way the entire analog heritage may be preserved and used effectively.

2.4.3 Position of the Passband and Specifications

The part of the frequency spectrum in which the filter has no attenuation or the attenuation is acceptably small is referred to as the passband. Consequently the stopband is the part of the spectrum in which the attenuation is above a prescribed value. In between is the transition region which defines the selectivity of the filter as

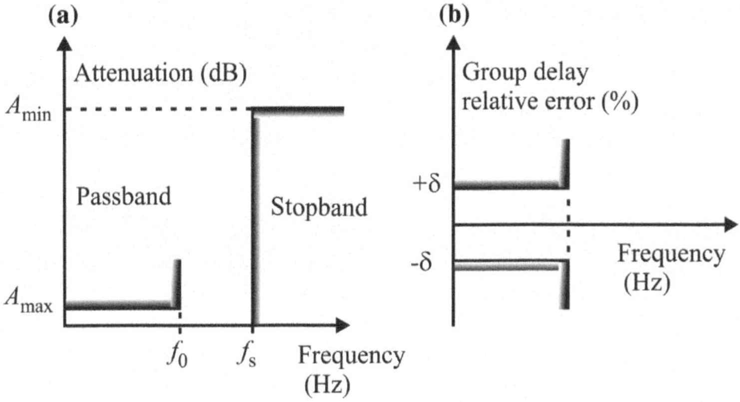

Fig. 2.15 Specifications for a low-pass filter. **a** Attenuation requirements, and **b** group delay requirements

stated by (2.20). The location of the passband is allowed to be anywhere on the frequency axis. Following that a categorization of the filters exists. Low-pass, requirements depicted in Fig. 2.15a, are the ones whose passband starts at the origin and covers the spectrum until the cut-off frequency denoted by f_0. The lower bound of the stopband is denoted by f_s while its upper bound is at infinity. So, the selectivity would be here $S = f_0/(f_s - f_0)$. Two margins are used to define the attenuation characteristic: the maximum passband attenuation a_{max} and the minimum stopband attenuation a_{min}. When requirements on the amplitude characteristics are considered, the set $(f_0, f_s, a_{max}, a_{min})$ is usually a complete one. It happens, however, for the asymptotic slope of the filter's attenuation characteristic is needed not to be lower from a prescribed value. In such a case the minimum asymptotic slope may be a design requirement, too.

Until now we ignored the phase of the spectral components all the time. There are applications, however, insisting phase to be preserved or somehow shaped. In such cases we are interested in the phase characteristic of the filter which represents the argument of the output signal as a function of frequency:

$$\varphi(\omega) = \arg\{V_{out}(j\omega)\} \tag{2.25}$$

where $V_{out}(j\,\omega)$ is the complex-valued output voltage (to be discussed soon).

Since, however, as will be shown in the subsequent chapter, the phase is expressed in a complicated form using transcendental functions; it is the group delay which is more frequently in use. It is defined as

$$\tau_d = -\frac{d\varphi(\omega)}{d\omega}. \tag{2.26}$$

Fig. 2.16 Specifications for a high-pass filter

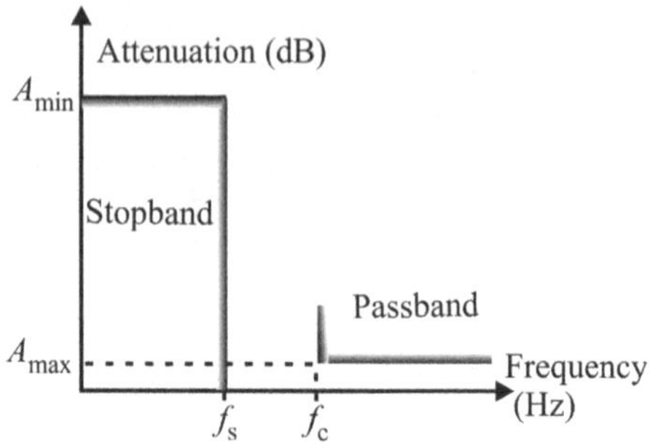

So, when filters are designed, usually linear phase (or constant group delay) is sought which means that all spectral components from the passband will be equally delayed and consequently properly superimposed to each other at the output. This requirement is expressed in Fig. 2.15b where margins for the group delay are set as relative deviation from the nominal (constant) value.

Opposite to the low-pass is the high-pass filter. Requirements for this type of filter are depicted in Fig. 2.16. Here the passband is infinitely broad starting at f_0 and ending at infinity. The frequency f_0 is the lower edge of the passband and is referred to as the cut-off frequency or simply cut-off. The upper edge of the stopband, which starts at the origin, is denoted as f_s. Usually there are no group delay requirements for high-pass filters.

There are filters which are tasked to suppress part of the frequency spectrum coming to the input. These are referred to as band-stop or notch filters. The specifications of the filter's attenuation characteristics are depicted in Fig. 2.17.

In this case there are two pass-bands which in general may have different values of the maximum attenuation (which is not the case in Fig. 2.17). Accordingly, one needs five frequencies to fully define the specifications: the central frequency of the stopband (f_c); the cut-off frequency of the lower pass-band (f_l); the cut-off frequency of the upper passband (f_u), the lower edge of the stopband (f_{sl}), and the upper edge of the stopband (f_{su}).

Fig. 2.17 Specifications for a band-stop filter

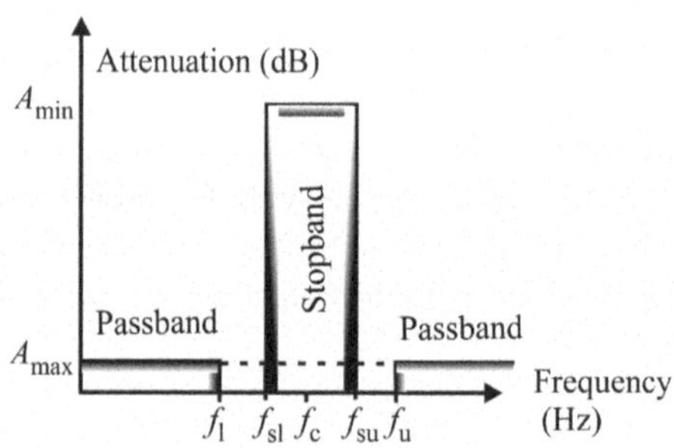

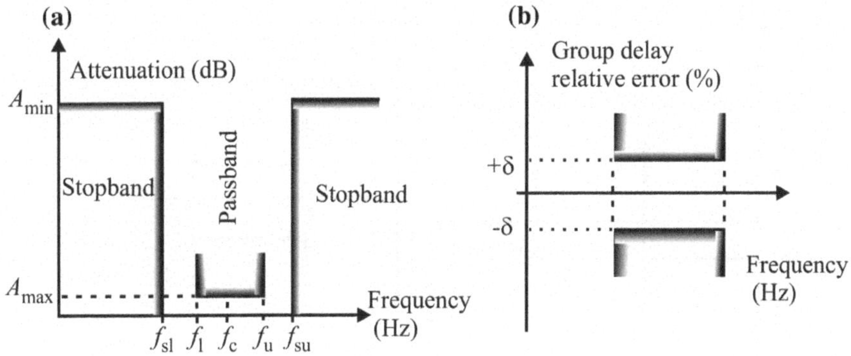

Fig. 2.18 Specifications for a band-pass filter. **a** Attenuation requirements, and **b** group delay requirements

The relative width of the stopband of the filter is defined as

$$B_{s\_r} = (f_u - f_l)/f_c \tag{2.27}$$

Opposite is the band-pass filter. It is tasked to pass part of the incoming spectrum and suppress the rest belonging to the higher and lower frequencies. The proper specification for this type of filters is depicted in Fig. 2.18. The specifics related to the attenuation need, again, five frequencies: the central frequency of the pass-band (f_c); the lower cut-off frequency of the pass-band (f_l); the upper cut-off frequency of the passband (f_u), the lower edge of the upper stopband (f_{su}), and the upper edge of the lower stopband (f_{sl}).

In the case of band-pass filters group delay requirements are imposed very frequently. These are pictured on Fig. 2.18b.

There is a special category of filters which do not exhibit attenuation at any frequency. These are referred to as all-pass filters. In this case the phase or the group delay is in the focus. Usually, a constant group delay is sought, as depicted in Fig. 2.19. Namely, there are applications in which all components of the signal's spectrum (i.e. the signal from the time domain perspective) to be delayed equally. There from comes the second name of this kind of systems: delay lines. There are low-pass all-pass filters which approximate constant group delay as if there is a low-pass filter (depicted in Fig. 2.19a) and band-pass delay line which approximate constant group delay on a frequency band around a central frequency (depicted in Fig. 2.19b).

If two filtering systems are cascaded as depicted in Fig. 2.20, one may create a new system with a much complex attenuation characteristic.

The filter's function in the frequency domain is usually obtained as a ratio of the input and the output voltage:

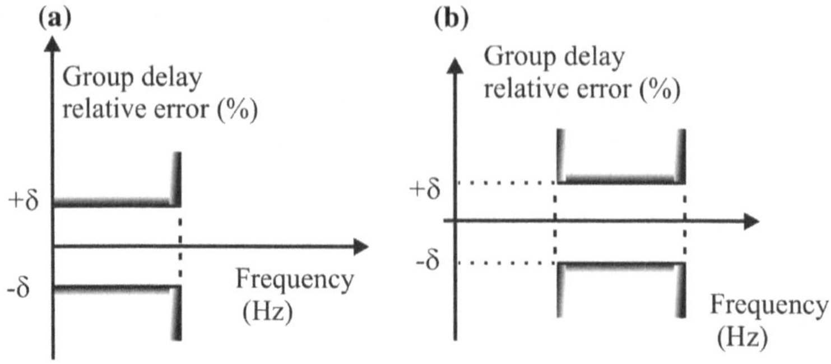

Fig. 2.19 All-pass requirements. **a** Low-pass all-pass and **b** band-pass all-pass

Fig. 2.20 Two stage cascade

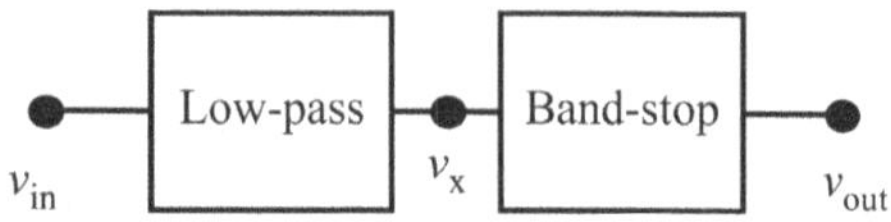

$$G = \frac{v_{out}}{v_{in}}. \tag{2.28}$$

If we introduce the value from the middle of the system we will obtain

$$G = \frac{v_{out}}{v_{in}} = \frac{v_{out}}{v_x} \cdot \frac{v_x}{v_{in}} = G_1 \cdot G_2 \tag{2.29}$$

This expression explains how two cascaded functions create a single one. If, for example, the first one is a low-pass filter and the second is a notch filter whose stop-band interval falls inside the passband of the low-pass filter, one may produce a low-pass filter with two pass-bands as depicted in Fig. 2.5b.

2.4.4 *Filtering and Technologies*

As all electronic systems the filter may be produced in different technologies. There are however phenomena of "non-electrical" nature which are exploited for production of filters, too.

To start with one may categorize the systems realizing the filtering function as a passive (electrical) or active (electronic) circuits.

Passive filters are very rarely (almost never) integrated and are built of discrete components. So they are referred to as discrete filters.

First among passive technology are the RLC filters constructed of capacitors, inductors, transformers and resistors only. These may be stated as classical and the most popular technology implemented in filtering. In addition to this category at higher frequencies (occupying a proper frequency window) one may encounter several technologies based on specific physical effects (mostly resonance) such as

- Cristal filters. These exploit the piezoelectric effect to create mechanical resonators behaving as passive LCR resonators. High stability and low temperature dependence of the resonant frequency of a properly processed quartz crystal contributes to the popularity of this technology.
- SAW (surface acoustic waves) filters. These perform filtering on the surface of a piezoelectric crystal exploiting the resonant properties of the material.
- Waveguide filters. Here we have an example of distributed filtering where transmission lines are included, too. Waveguides are hollow metal tubes inside which an electromagnetic wave may be transmitted. Due to the properties of propagation of the electromagnetic waves usually band-pass filters are produced with this technology. Waveguide filters are used at extremely high frequencies.
- Mechanical filters. Here mechanical (metal) resonators are constructed and coupled to form a cascade behaving as an electrical filter. Transducers are provided to convert an electrical signal to and from a mechanical vibration.

A filter becomes active if electronic components such as transistors, operational amplifiers (OA), operational transconductance amplifiers (OTA), or logic circuits are included. In addition, in the case of sampled data filtering, switches are needed to perform sampling. There are several categories of active filters.

- Active RC filters. These consist of OA, resistors and capacitors. Their introduction allowed the exclusion of inductors at low frequency applications where inductors become bulky, heavy, and expensive. Active RC filters are usually produced as discrete ones but integration is not rare. A serious limitation to the implementation of active RC filters at high frequencies is the frequency characteristic of the OA used.
- Active Gm-C filters. Further reduction of the area of integrated active RC filters is achieved by Gm-C filters consisting of OTAs and capacitors only. Not only area reduction and improved temperature stability is achieved but the frequency span is broadened too, due to better performance of the OTAs. Gm-C filters are integrated in CMOS technology.
- Active SC filters. Sampled filters are implemented using Switches, Capacitors (hence SC), and OA and used for low frequency filtering. In this case a resistor (of an active RC circuit) is substituted by a special connection of a capacitor and regularly commutated switches. SC filters are integrated in CMOS technology. Significant reduction of area is achieved along other advantages.
- Digital filters. Digital filters are all active filters where no passive components are used to create the filter function. Logic circuits are exclusively building blocks of the structure. As mentioned, in a firmware design, a digital filter may be implemented as a piece of software. This category of filtering has become dominant in modern communication. One may build but one does not do digital filters as discrete.

Below a list of frequency range of analog filters is given:

- Discrete active-RC filters: 1 Hz–100 MHz
- On-chip continuous-time active filters: 10 Hz–1 GHz
- Switched-capacitor: 1 Hz–10 MHz
- Discrete LC: 10 Hz–1 GHz
- Distributed: 100 MHz–100 GHz
- SAW: 30 MHz–3 GHz

2.5 Scaling

The values of the signals in telecommunication span over a very large intervals. While at the output of an antenna (input to the telecommunication channel) the voltage is usually in the μV range, at the output of a power amplifier it may rise to hundred of volts. Consequently the gain of a system may span from small values (even smaller than unity when attenuation is needed) to several millions which is usual in radio frequency (RF) systems. On the other side, as already mentioned, the frequency span is starting from mHz and ending with hundreds of GHz. Finally, the element values of the passive components used to build the filters span over extremely wide ranges, too. For example, the resistance may vary from mΩ to several GΩ.

Design engineers prefer to work with numbers near the unity. Such computations lead to smaller numerical errors. Also, designers would like to work with "generalized" dimensionless quantities which will be assigned specific values depending on the design at hand.

All this is achieved by normalization.

To start with, let $V_0 = 1$ V and $\omega_0 = 1$ rad/s are the normalizing voltage and normalizing angular frequency, respectively. The quantity

$$v_\mathrm{n} = v/V_0 \tag{2.30}$$

is the dimensionless normalized voltage. Similarly,

$$\omega_\mathrm{n} = \omega/\omega_0 \tag{2.31}$$

is the dimensionless normalized angular frequency.

To "compress" the span of these two quantities one use logarithmic scale. For the voltage one use the so called Decibel scale as

$$v(\mathrm{dB}) = 20 \cdot \log_{10}(v_\mathrm{n}). \tag{2.32}$$

Table 2.1 represents some examples of the mapping given by (2.32).

Table 2.1 Conversion into decibels

Normalized value	Value in decibels	Normalized value	Value in decibels
1/10,000	−80	10,000	80
1/1000	−60	1000	60
1/100	−40	100	40
1/10	−20	10	20
1/2	−6	2	6
$1/\sqrt{2}$	−3	$\sqrt{2}$	3
1	0	1	0

The value related to $\sqrt{2}$ is added due to the specific importance it has in filter design. Namely, a squared voltage over a unity resistance resistor represents power. If the value of the (normalized) output voltage is $1/\sqrt{2}$ then the power is ½ of the input power (calculated in the same way). This criterion is usually in use to define the cut-off frequency of the filter. It is the frequency at which the power at the load is half of the power obtained at the central frequency of the filter's passband.

If power is concerned, due to the squared voltage, one use

$$p(\text{dB}) = 10 \cdot \log_{10}(p_\text{n}) \tag{2.32}$$

where $p_\text{n} = P/(1\text{ W})$, P is the power to be normalized, and p is the normalized power in Decibels. In that way (using 10 in place of 20) the same value in Decibels will be obtained for the voltage and the power in the same measurement point.

The logarithmic scale used to represent voltage (and current) is nonlinear. One may say that it compresses the large numbers and expands the small ones. That can be easily seen from Table 2.1.

This property is extremely important when frequency dependences are to be drawn. One use "logarithmic scale" for the frequency by putting on the x-axis

$$\omega_{\log} = \log_{10}(\omega/\omega_0). \tag{2.33}$$

There is specific terminology used in this case. It is related to acoustics. Namely, two frequencies related as 2:1 are said to be separated by an octave. For the frequencies related as 10:1 one says that they are separated by decade.

One may use a logarithmic scale for the frequency and linear scale for the voltage to produce a semi-logarithmic (semi-log) frequency diagram. When both quantities are in logarithmic scale one uses a logarithmic-logarithmic (log-log) scale.

When it comes to normalization of the element values one first choose normalizing angular frequency ω_0 (as before) and a normalizing resistance R_0. Then, the normalization of a resistance R goes simply as follows

$$R_n = R/R_0. \tag{2.34}$$

The normalization of a capacitance is a little more complicated. We first normalize the reactance of the capacitance:

$$X_n = \frac{1}{\omega_n C_n} = \frac{X_c}{R_0} = \frac{1}{\omega C R_0} = \frac{1}{\frac{\omega}{\omega_0}\omega_0 C R_0} \tag{2.35a}$$

so that, by comparison, we get

$$C_n = \omega_0 C R_0. \tag{2.35b}$$

For the inductance, in a similar way, we have

$$X_n = \omega_n L_n = \frac{X_L}{R_0} = \frac{\omega L}{R_0} = \frac{\omega}{\omega_0}\frac{\omega_0 L}{R_0}, \tag{2.36a}$$

which leads to

$$L_n = \frac{\omega_0 L}{R_0} \tag{2.36b}$$

At the end of the design one goes for denormalization. This means that starting with the normalized values obtained in the filter synthesis process one will calculate the real values of the capacitances and the inductances as follows

$$C = \frac{C_n}{\omega_0 R_0} \tag{2.37a}$$

and

$$L = \frac{L_n R_0}{\omega_0}. \tag{2.37b}$$

2.6 Developer's Corner

Complex function analysis including complex numbers manipulation is an inherent necessity of filter design. The problem is that it is not performed by using complex arithmetic. Instead always two quantities are manipulated: the real and the imaginary part. That means that the software developer will have to develop his own routines for calculation of the filter's functions and their properties. In Chap. 3, some advice will be given and some misunderstandings will be highlighted.

Table 2.2 Decibels to be converted into gain

A(dB)	55	56	57	58

Conversion of ordinary numbers into decibels and vice versa seems complicated and, one would say, needs at least a calculator to be used. We will explain here on a simple example that, by the help of Table 2.1, one can convert from integer Decibels into ordinary numbers without a calculator.

The example task is to find the values of A from Table 2.2 out of the values of A (dB).

To perform the back-conversion we decompose the Decibels into values depicted in Table 2.1. So, for example, $A(\mathrm{dB}) = 55$ dB = 40 + 6 + 6 + 3 dB. Since the sum of logarithms represents a logarithm of a product, using Table 2.1 we will get $A = 100 \cdot 2 \cdot 2 \cdot 2^{0.5} \approx 564$ times. Similarly, if $A(\mathrm{dB}) = 56$ dB = 80 − 6 − 6 − 6 − 6 dB is to be converted, we have $A = 100 \cdot 100/(2 \cdot 2 \cdot 2 \cdot 2) = 625$ times. Further, if $A(\mathrm{dB}) = 57$ dB = 60 − 3 dB = (20 + 20 + 20 − 3) dB, one gets $A = 10 \cdot 10 \cdot 10/2^{0.5} \approx 709$ times. Finally, if $A(\mathrm{dB}) = 58$ dB = (40 + 6 + 6 + 6) dB, one gets $A = 100 \cdot 2 \cdot 2 \cdot 2 = 800$ times. All these are summarized in Table 2.3.

Since the four values encompass the "discretization" within 3 dB, the method described above may be used for any back-conversion of integer decibels.

By inspection of Table 2.3 one may conclude that 1 dB makes a serious difference which is larger with larger values of the gain.

For direct conversion without a calculator one has to memorize the logarithms of 2, 3, 5, and 7 which may be used for approximative conversion. These logarithms are given in Table 2.4.

Using this, for example, one has $\log_{10}(24) = \log_{10}2 + \log_{10}2 + \log_{10}2 + \log_{10}3 = 3 \cdot 0.3 + 0.48 = 1.38$. Of course, since not all prime numbers are encompassed in Table 2.4, the conversion based on Table 2.4, on first sight, will be not as successful as the back-conversion. If large numbers are to be converted, however, one may always approximate to the nearest value tractable by Table 2.4. In that one may profit from the property mentioned above that 1 dB corresponds to a large change of the gain which means that a small change of the gain will not seriously affect the decibels. For example, 271 is a prime number which may be approximated as $271 \approx 270 \approx 2 \cdot 5 \cdot 27 \approx 2 \cdot 5 \cdot 3 \cdot 3 \cdot 3$. There from we have $\log_{10}(271) \approx 0.3 + 0.7 + 0.48 + 0.48 + 0.48 = 2.44$. This is to be compared with the exact value of 2.43136. So, instead of the exact 48.6 dB one would get approximate 48.8 dB. One may expect even smaller difference if larger gains are to be converted into Decibels in this way.

Table 2.3 Back-conversion results

A(dB)	55	56	57	58
A	565.68	625	707.1	800

Table 2.4 Logarithms of the first four prime numbers

x	2	3	5	7
$\log_{10}(x)$	0.3	0.48	0.7	0.85

References

1. Adams MD (2013) Continuous-time signals and systems. The University of Victoria, Victoria, BC, Canada
2. Chaparro LF (2001) Signals and systems using MATLAB®. Elsevier, Amsterdam
3. Catala P (2002) FDM hierarchy. In: Gibson JD (editor-in-chief) The communications handbook. CRC Press, Boca Raton

Chapter 3
Transfer Function and Frequency and Time Domain Response

It is a custom for the filtering function to be expressed in the frequency domain as a rational function of polynomials of s, the complex frequency variable. Since the natural signals are found in the time domain, the so called Laplace Transform is used to create a representation in the complex frequency domain (from now on: in the s-domain). That allows for the differential equations representing the system to be transformed into algebraic ones and consequently to extract the so called transfer functions which are their s-domain substitute. Ones in the s-domain one may optimize the transfer function in order to control the properties of the physical system which will be created later on. To perform optimization one needs to calculate the properties of the transfer function. This chapter is devoted to the transfer function as such and to the methods allowing expressing its properties in the frequency domain. Procedures are given for calculation the so called amplitude, phase, and group delay characteristics of the system. Furthermore, for two commonly accepted time-domain signals the procedure will be described to go back into the time domain. That will be based on The Inverse Laplace transform and the calculus of residues.

3.1 Introduction

According to basic circuit theory, using the constitutive equations of the circuit elements and the Kirchhoff's laws, we create a set of differential equations governing the signals within the circuit [1]. Since the solution is a function of time we call the action as time domain analysis. Unfortunately, it is very difficult and even impossible to solve such systems of equation manually even for relatively simple linear circuits. That means we need an alternative way to analyze electrical and electronic circuits.

V. Litovski, *Electronic Filters*, Lecture Notes in Electrical Engineering 596,
https://doi.org/10.1007/978-981-32-9852-1_3

In this chapter we will introduce the Laplace transform as the powerful means to create a mathematical representation or, better to say, encapsulation of the properties of linear electrical and electronic circuits. That will be referred to as a transfer function.

Given the transfer function however, one needs means to extract the properties of the system to which it is associated. Hence, we will elaborate the procedures for time and frequency domain characterization of electronic systems for which the transfer function was synthesized elsewhere.

3.2 The Laplace Transform and the Transfer Function

To make the story short, based on Fourier's and Laplace's transforms we transpose the analysis into the frequency domain. The Fourier transform (as mentioned in the previous chapter) manipulates real quantities to represent the time domain signal into a series of sinusoids being of different amplitudes and having different phase angles. The Laplace transform is going further. It represents the real signal coming from the time domain by a complex one containing the amplitude and the phase captured by its complexity. Here we will proceed with the Laplace transform and its application to create transfer functions. The later are containing the intrinsic properties of the linear circuit and, when proper transform of the input signal is available, allows for analysis in both the frequency and time domain.

The unilateral (or one-sided) Laplace transform is defined by the integral

$$F(s) = \mathcal{L}\{f(t)\} = \int_{0^-}^{\infty} f(t) \cdot e^{-s \cdot t} dt. \tag{3.1}$$

In this expression the lower limit is written as $t = 0^-$ because we wish to allow $f(t)$ to include an impulse function at $t = 0$. It is assumed that the value of $f(t)$ is zero for $t < 0$. If the complex frequency is represented as $s = \sigma + \mathrm{j} \cdot \omega$, where $j = \sqrt{-1}$ is the imaginary unit, the value of σ should be such as to ensure the integral (3.1) to converges.

The expression (3.1) transposes a function given in the time domain into a new form in the complex frequency domain. We use capital letters to denote that the voltages and currents are transformed.

The opposite transformation i.e. the inverse Laplace transform, may be expressed as

$$f(t) = \mathcal{L}^{-1}\{F(s)\} = \frac{1}{2\pi \cdot \mathrm{j}} \int_{c-\mathrm{j}\infty}^{c+j\infty} F(s) \cdot e^{s \cdot t} ds, \tag{3.2}$$

Table 3.1 Laplace transform pairs

Time domain $f(t)$	s-domain $F(s)$
$\delta(t) = \begin{cases} \infty & \text{for } t = 0 \\ 0 & \text{othervise} \end{cases}$	1
$u(t) = \frac{\partial}{\partial t}\{\max(t, 0)\}$ for $t > 0$ or $u(t) = \int_{-\infty}^{t} \delta(r)dr$	$1/s$
$\frac{t^{n-1} \cdot e^{a \cdot t}}{(n-1)!}$	$\frac{1}{(s-a)^n}$
$\frac{2 \cdot t^{n-1} \cdot e^{a \cdot t}}{(n-1)!}[A \cdot \cos(bt) - B \cdot \sin(bt)]$	$\frac{A+\mathrm{j} \cdot B}{(s-a-\mathrm{j} \cdot b)^n} + \frac{A-\mathrm{j} \cdot B}{(s-a+\mathrm{j} \cdot b)^n}$

Here c is a constant chosen so that to preserve convergence. The final result, of course, does not depend on the value of c.

Table 3.1 contains a short list of Laplace transform pairs. $\delta(t)$ in this table stands for the unit impulse, while $u(t)$ is the unit step. Putting specific values to the constants in this table we can obtain a much wider set of transform pairs. For example, when $n = 1$, for $f(t) = \mathrm{e}^{a \cdot t}$, we get $F(s) = 1/(s - a)$. Similarly when $n = 1$, $a = 0$, $A = 1$ and $B = 0$, for $f(t) = 2 \cdot \cos(\omega_0 t)$ we obtain

$$F(s) = \frac{1}{s - \mathrm{j} \cdot \omega_0} + \frac{1}{s + \mathrm{j} \cdot \omega_0} = \frac{1}{s^2 + \omega_0^2}.$$

More detailed data on s-domain functions with multiple poles may be found in [2] while [3] contains an extensive list of Laplace transform pairs. The special case of filters having all transmission zeros at infinity is considered in detail in [4]. Much more complex transformations may be performed easily if the theorems presented in Table 3.2 are used in conjunction with Table 3.1. Some of them are of crucial importance for the application of this concept to circuit analysis.

To introduce the Laplace transform into circuit analysis, we first transform the constitutive equations of the basic circuit elements. For the resistive branch in the time domain we have

$$i_{\mathrm{G}}(t) = G \cdot v_{\mathrm{G}}(t) \tag{3.3a}$$

which maps into

$$I_{\mathrm{G}}(s) = G \cdot V_{\mathrm{G}}(s). \tag{3.3b}$$

For the capacitive branch described in the time domain by

$$i_{\mathrm{C}}(t) = C \cdot \frac{dv_{\mathrm{C}}(t)}{dt}, \tag{3.4a}$$

Table 3.2 Laplace transform theorems

Property	f(t)	F(s)
Linearity	$a \cdot f_1(t) + b \cdot f_2(t)$	$a \cdot F_1(s) + b \cdot F_2(s)$
Time domain first derivative	$\frac{df(t)}{dt}$	$sF(s) - f(0^-)$
Time domain second derivative	$\frac{d^2f(t)}{dt^2}$	$s^2F(s) - s \cdot f(0^-) - \frac{df(t)}{dt}\vert_{t=0^-}$
General derivative	$\frac{d^nf(t)}{dt^n}$	$s^nF(s) - \sum_{k=1}^{n} s^{n-k} \frac{d^{(k-1)}f(t)}{dt^{(k-1)}}\vert_{t=0^-}$
Integration	$\int_{0^-}^{t} f(\tau)d\tau$	$\frac{F(s)}{s}$
s-domain differentiation	$t^nf(t)$	$(-1)^n \frac{d^nF(s)}{ds^n}$
Convolution	$\int_{0^-}^{t} f_1(\tau) \cdot f_2(t-\tau) \cdot d\tau$	$F_1(s) \cdot F_2(s)$
Initial value	$\lim_{t \to 0} f(t)$	$\lim_{s \to \infty} s \cdot F(s)$
Final value	$\lim_{t \to \infty} f(t)$	$\lim_{s \to \infty} s \cdot F(s)$

in the s-domain we have

$$I_C(s) = s \cdot C \cdot V_C(s) - C \cdot v_C(0^-). \tag{3.4b}$$

Finally, for the inductive branch described in the time domain by

$$v_L(t) = L \cdot \frac{di_L(t)}{dt} \tag{3.5a}$$

in the s domain we have

$$I_L(s) = \frac{1}{s \cdot L} \cdot V_L(s) + \frac{i_L(0^-)}{s}. \tag{3.5b}$$

The network representations of (3.3a, 3.3b)–(3.5a, 3.5b) are depicted in Fig. 3.1.

Now, we will apply the above results to the equation formulation of the circuit of Fig. 3.2. To that end we first create a mapped circuit as depicted in Fig. 3.3. Then, we substitute directly (3.3b), (3.4b) and (3.5b) (when appropriate) into the KCL system of equations for the circuit. The so obtained system of equations describing the circuit in the s-domain is given by (3.6).

$$\begin{bmatrix} G_1 + sC & -sC & 0 \\ -sC & sC + \frac{1}{sL} & -\frac{1}{sL} \\ 0 & -\frac{1}{sL} & \frac{1}{sL} + G_2 \end{bmatrix} \cdot \begin{bmatrix} V_1(s) \\ V_2(s) \\ V_3(s) \end{bmatrix} = \begin{bmatrix} I(s) \\ 0 \\ 0 \end{bmatrix} + \begin{bmatrix} Cv_C(0^-) \\ -Cv_C(0^-) - \frac{i_L(0^-)}{s} \\ \frac{i_L(0^-)}{s} \end{bmatrix}. \tag{3.6}$$

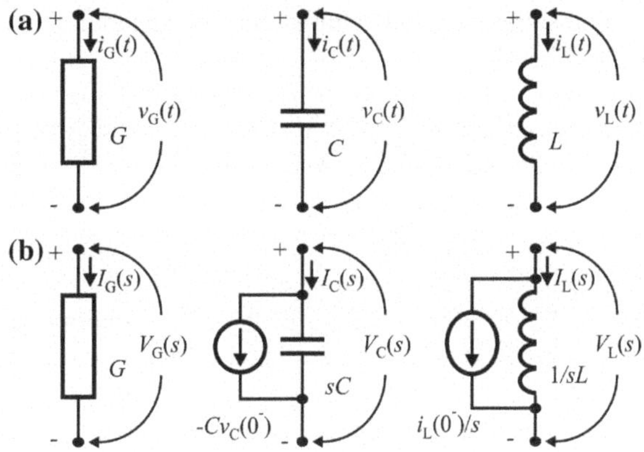

Fig. 3.1 **a** Original (time domain) and **b** mapped (s-domain) branches

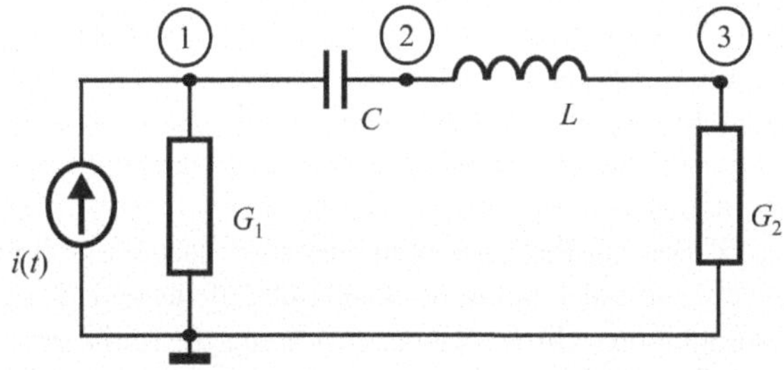

Fig. 3.2 A dynamic linear circuit

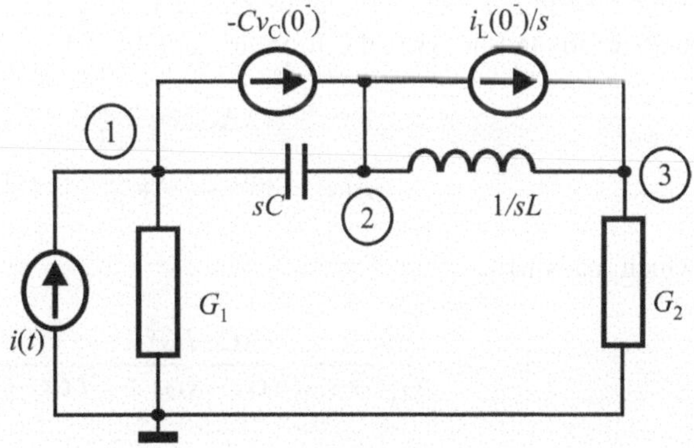

Fig. 3.3 Mapped circuit. It is assumed that in the original circuit $v_C = v_1 - v_2$, and i_L is flowing from node 2 toward node 3

For simplicity, but without loss of generality, we will solve only for V_3. By Cramer's rule:

$$V_3(s) = \frac{\Delta_{13}(s)}{\Delta(s)} I(s) + \frac{\Delta_{13}(s)}{\Delta(s)} \cdot C v_C(0^-) + \frac{\Delta_{23}(s)}{\Delta(s)} \left[-C v_C(0^-) - \frac{i_L(0^-)}{s} \right] + \frac{\Delta_{33}(s)}{\Delta(s)} \cdot \frac{i_L(0^-)}{s}. \tag{3.7}$$

where Δ_{nm}(s) is the cofactor of the corresponding (n, m) element of the matrix in (3.6), and Δ(s) is the determinant of the system. An important property of the solution can be found by examining (3.6) and (3.7). The RHS vector is partitioned so that the first part represents the excitation, while the rest is due to the initial conditions. It is important to note that the complete solution of the system can be obtained using the superposition principle by solving (3.6) twice. The first solution is obtained for the case where only the excitation is present in the RHS vector, and is referred to as the zero-state response. The term implies that the initial state of the circuit is assumed to have zero energy, i.e. all voltages and currents at $t = 0^-$ are equal to zero. On the other hand, if only the initial conditions are present in the right hand side, we obtain the zero-input response, implying that there is no excitation (at the input).

Of course, the zero input response is of less importance because it refers to specific initial conditions. Generally, if needed, it can be obtained by additional computation, which means considering the initial-state source as if it were an input signal.

Conversely, the zero-state response, expressed in symbolic form, may be used to extract the general properties of the circuit by introducing the concept of the network function. To see this, let us reduce (3.6) to the zero-state part. The determinants evaluated in symbolic form are as follows:

$$\Delta_{13}(s) = \frac{sC}{sL}$$
$$\Delta(s) = \frac{G_1 G_2 + sC(G_1 + G_2) + s^2 LCG_1 G_2}{sL} \tag{3.8}$$

which leads to

$$V_3(s) = \frac{sC \cdot I(s)}{G_1 G_2 + sC(G_1 + G_2) + s^2 LCG_1 G_2} = Z(s) \cdot I(s), \tag{3.9}$$

where

$$Z(s) = \frac{V_3(s)}{I(s)} = \frac{sC}{G_1 G_2 + sC(G_1 + G_2) + s^2 LCG_1 G_2} \tag{3.10}$$

is the transfer impedance in the s-domain.

Before proceeding with the analysis of this result let mention that we could get different node or branch responses and combine them with other node or branch quantities, pronouncing them as excitations, to produce different *transfer functions*. For example, the quotient of the branch currents $I_{G2}(s)$ and $I(s)$, i.e. $A_i(s) = I_{G2}(s)/I(s)$, may be considered as current gain transfer function. In filter theory, due to the passive LC-filters heritage, a voltage transfer is most frequently in use.

Let us now consider the result expressed by (3.10). First of all, we can conclude that the solution is the product of a rational function of s and the transform of the excitation $I(s)$. In other words, according to Table 3.2, the solution is the convolution of these two functions. $Z(s)$ neither depends on the excitation nor on the initial conditions. It simply represents the properties of the circuit. Furthermore, the response to a new excitation with a new function $i(t)$, and correspondingly $I(s)$, will be obtained by multiplying $Z(s)$ by the new excitation. Thus, no further analysis of the circuit is needed for the different excitation, as would be necessary in the time domain. With appropriate function evaluation, this can also be applied to the computation to the zero-input response.

The polynomial in the denominator of $Z(s)$ is also the numerator of the system determinant of (3.6). The roots of the polynomial are the poles of $Z(s)$ and are also eigenvalues of the system matrix. These are commonly referred to as the natural frequencies of the circuit.

3.3 Transfer Function Analysis and Response Calculation

Filter design is usually performed in two phases: transfer function synthesis and system synthesis. The transfer function synthesized, before proceeding to system synthesis, one needs a thorough analysis of the properties of the function in frequency and in time domain. To those analyses is devoted the rest of this paragraph. While this issue is most frequently avoided in the literature we consider it is important to be discussed in some detail what, we expect, will become apparent soon.

The polynomials obtained by circuit analysis performed symbolically and, in some cases, by transfer function approximation procedures, are represented in a form of sums so that

$$H(s) = A_0 \cdot \frac{\sum_{i=0}^{i=m} a_i s^i}{\sum_{i=0}^{i=n} b_i s^i}, \tag{3.11}$$

where $H(s)$ is a general representative of any transfer function and A_0 is a constant. To get the properties of the circuit, however, one needs the poles and zeros of the transfer function. In fact, only the amplitude characteristic (will be defined later on) may be correctly computed from the form containing ratio of sums. We will come back to this issue but here we will stress that in case when real frequencies are substituted in the above expression, and the order of the filter is high, one gets extremely large numbers which may lead to serious errors during calculations. For that, and many other reasons, in the next, we will manipulate exclusively with factored form of the polynomials. To develop good filter design software, as may be understood, one will unavoidably need a good routine to solve higher order polynomials. Add to that the fact that some synthesis algorithms are dealing with squared moduli leading to risen order of the polynomials to be solved.

So, the transfer function of a filter may be represented in the following way

$$H(s) = A_0 \cdot \frac{\prod_{i=1}^{m}(s - z_i)}{\prod_{i=1}^{n}(s - p_i)} \tag{3.12}$$

The quantities z_i and p_i are the zeros and the poles respectively. These are frequently referred to as characteristic frequencies of the filter. m and n are the orders of the numerator and denominator polynomials, respectively, as they are in (3.11). For stability reasons $n \geq m$ must be preserved. n is referred to as the order of the filter. A_0 is adjusted so that the function to have a prescribed value of the gain at the central frequency i.e. it is the solution of the equation $|H(j\omega_0)| = G$, where ω_0 is the central angular frequency while G is the proper nominal gain. In the next $G = 1$ will be used.

For low-pass prototype filters normalized so that the gain at the origin is equal to unity one use

$$A_0 = \frac{\prod_{i=1}^{i=n}(-p_i)}{\prod_{i=1}^{i=m}(-z_i)}. \tag{3.13a}$$

Note, some even-order amplitude characteristics exhibit minimum of the gain at the origin. For these cases the value of A_0 must be found after calculating the maximum value of the gain in the passband.

For odd order high-pass filter one has

$$A_0 = 1, \tag{3.13b}$$

which should be corrected as described above for even-order filters if necessary.

For band-pass filter we use

$$A_0 = \frac{\prod_{i=1}^{i=n}(j \cdot 1 - p_i)}{\prod_{i=1}^{i=m}(j \cdot 1 - z_i)}, \tag{3.13c}$$

which will be corrected properly in the case when half of the order of the filter is even (if necessary).

Finally, for band-stop uses (3.13a) or (3.13b) depending on which of the pass-bands is normalized.

In the case when the poles are grouped in conjugate pairs, the transfer function may be written in the following form

$$H(s) = \begin{cases} A_0 \cdot \dfrac{\prod_{i=1}^{\lfloor m/2 \rfloor} (a_{0i} + a_{1i}s + s^2)}{\prod_{i=1}^{\lfloor n/2 \rfloor} (b_{0i} + b_{1i}s + s^2)} & \text{for} \quad n, m - \text{even} \\ A_0 \cdot \dfrac{a_{00} + s}{b_{00} + s} \dfrac{\prod_{i=1}^{i=\lfloor m/2 \rfloor} (a_{0i} + a_{1i}s + s^2)}{\prod_{i=1}^{i=\lfloor n/2 \rfloor} (b_{0i} + b_{1i}s + s^2)} & \text{for} \quad n, m - \text{odd} \end{cases} \tag{3.14}$$

with $a_{0i} = |z_i|^2$, $a_{1i} = -2 \cdot \mathrm{re}\{z_i\}$, $b_{0i} = |p_i|^2$, $b_{1i} = -2 \cdot \mathrm{re}\{p_i\}$, for the complex zeros and poles and $a_{00} = -\mathrm{re}\{z_i\}$, $b_{00} = -\mathrm{re}\{p_i\}$ for the real zeros and poles. A_0 is adjusted as above. $\lfloor . \rfloor$ (the floor function) stands for the minimum integer.

3.3.1 Amplitude Characteristic and Attenuation

The amplitude characteristic is defined as the modulus of the transfer function computed at the imaginary axis of the complex frequency plane. Hence it is a function of the angular frequency. By definition it is obtained as

$$|H(\omega^2)| = \sqrt{H(s) \cdot H(-s)_{|s=j\omega}} \tag{3.15}$$

or

$$|H(\omega^2)| = |A_0| \sqrt{\frac{\prod_1^m (\omega^2 + z_i^2)}{\prod_1^n (\omega^2 + p_i^2)}}. \tag{3.16}$$

Substituting $p_i = \alpha_i + j \cdot \beta_i$ and $z_i = \gamma_i + j \cdot \delta_i$ one gets the programmable form of the expression for the amplitude characteristic

$$|H(\omega^2)| = |A_0| \sqrt{\frac{\prod_1^m [\omega^2 + \gamma_i^2 - \delta_i^2 + 2j \cdot \gamma_i \delta_i]}{\prod_1^n [\omega^2 + \alpha_i^2 - \beta_i^2 + 2j \cdot \alpha_i \beta_i]}} \tag{3.17a}$$

or

$$|H(\omega^2)| = |A_0| \cdot \frac{\prod_1^m \sqrt{4(\omega^2 + \gamma_i^2 - \delta_i^2)^2 + (2 \cdot \gamma_i \delta_i)^2}}{\prod_1^n \sqrt{4(\omega^2 + \alpha_i^2 - \beta_i^2)^2 + (2 \cdot \alpha_i \beta_i)^2}} \tag{3.17b}$$

The attenuation of the filter is, for historical reason, one of the most important quantities to characterize a filter. Namely, the first filters were passive and one could only attenuate the signal by a filter. It is defined as

$$a(\omega) = 20 \cdot \log\{1/|H(\omega^2)|\}. \tag{3.18}$$

The value is normally positive. It will become negative if the amplitude characteristic becomes larger of unity.

3.3.2 *Phase Characteristic*

Being a complex quantity the transfer function may be represented in a polar form

$$H(s)_{|s=j\omega} = |H(\mathrm{j}\omega)| \cdot e^{\mathrm{j}\cdot\Phi(\omega)} \tag{3.19}$$

where $\Phi(\omega)$ represents the phase characteristic. In general, it is obtained as

$$\Phi(\omega) = \frac{1}{2\mathrm{j}} \cdot \ln\left\{\frac{H(\mathrm{j}\omega)}{H(-\mathrm{j}\omega)}\right\}. \tag{3.20}$$

That expression however is not convenient for transfer function analysis. If (3.12) is given, the phase characteristic is obtained from the following expression

$$\Phi(\omega) = \sum_{1}^{m} \operatorname{arctg}\frac{\omega - \delta_i}{-\gamma_i} - \sum_{1}^{n} \operatorname{arctg}\frac{\omega - \beta_i}{-\alpha_i}. \tag{3.21}$$

Before proceeding with the peculiarities related to the evaluation of this expression we will stress that if all zeroes are in the right half plane or on the ω-axis i.e. if $\gamma_i \geq 0 \ \forall\, i = 1, 2, 3, \ldots, m$ one has

$$\lim_{\omega \to \infty} \Phi(\omega) = -m \cdot \frac{\pi}{2} - n \cdot \frac{\pi}{2} = -(m+n) \cdot \frac{\pi}{2}. \tag{3.22}$$

This expression took into account the fact that for the system to obey the stability criterion the condition $\alpha_i < 0, \ \forall\, i = 1, 2, 3, \ldots, n$ must be satisfied.

For example, if $m = 0$, which corresponds to a transfer function having all its zeros at infinity, the asymptotic value of the phase is $-n \cdot \pi/2$. It is not possible to get these results from the polynomial form of the transfer function given by (3.11). If that were to be used, as it was done in [5], the asymptotic value would be $\pm\pi/2$ only which is, of course, wrong.

The evaluation of expression (3.21) seems straightforward but it is not. Namely, while implementing it, care must be taken for every addend to find the proper quadrant to which points the phase vector related to the corresponding pole or zero. The angles must be measured strictly counterclockwise. For example, for a zero in the left half of the complex frequency plane i.e. if $\gamma_i < 0$, when $\omega - \delta_i < 0$, as

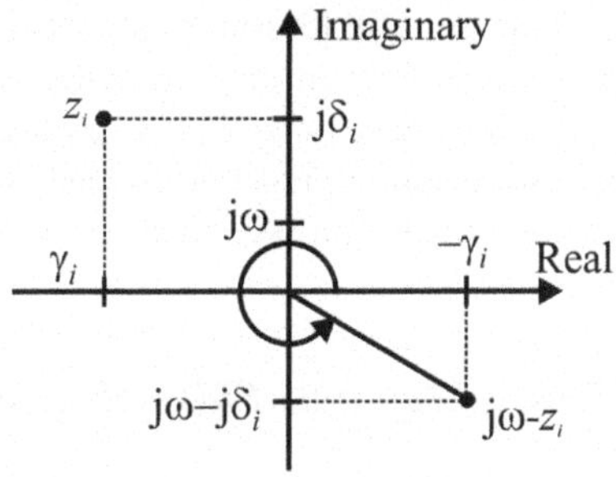

Fig. 3.4 Calculation a fraction of the phase at frequency ω

illustrated in Fig. 3.4, the contribution to the overall phase angle will be $2\pi - \arctan\{|(\omega - \delta_i)/\gamma_i|\}$.

In a special case when the function has a zero at the imaginary axis (i.e. $\gamma_i = 0$) its contribution will be $\pm\pi/2$ depending on the sign of the difference: $\omega - \delta_i$. That means that at frequencies when the value of $\omega - \delta_i$ passes through zero (changes sign), the phase will "jump" for π.

3.3.3 Group Delay

The group delay is a physical quantity expressing the time for which the envelop of the signal is delayed. It is defined and calculated as

$$\tau(\omega) = -\frac{d\Phi(\omega)}{d\omega} = \sum_{1}^{m} \frac{-\gamma_i^2/\gamma_i}{(\omega - \delta_i)^2 + \gamma_i^2} - \sum_{1}^{n} \frac{-\alpha_i^2/\alpha_i}{(\omega - \beta_i)^2 + \alpha_i^2}. \tag{3.23}$$

It is easy to recognize that zeros at the imaginary axis (i.e. $\gamma_i = 0$) will not contribute to the group delay since after the "jump" the phase continues with uninterrupted value of its derivative.

Here, again, the use of (3.11) would lead to a wrong value of the group delay.

Since the group delay is a derivative of the phase, its integral should be equal to the phase. So, the area under the $\tau(\omega)$ curve is

$$P_\tau = \int_0^\infty \tau(\omega) \cdot d\omega = -\Phi(\infty) + \Phi(0). \tag{3.24}$$

Note, due to the symmetry of the complex poles and zeros, $\Phi(0) = 0$.

For polynomial filters $P_\tau = n \cdot \pi/2$. In the case when complex right half plane zeros are present (i.e. for non-minimum phase systems) one has $P_\tau = (m + n) \cdot \pi/2$.

This is useful for rough estimation of the value of the group delay at the origin of a low-pass filter in the case when the passband width is normalized to unity. For example a sixth order filter with two right-half-plane complex zeros, at the origin, will have approximately (but lower than) $\tau(0) = (2 + 6) \cdot \pi/2 = 4 \cdot \pi$. A fifth order all-pass filter will have just below $\tau(0) \approx (5 + 5) \cdot \pi/2 = 5 \cdot \pi$.

3.3.4 Step and Pulse Responses

Before proceeding with recommendation on the most effective and accurate ways of calculation of the circuit's time domain response, based on a given transfer function and excitation, we will shortly revisit Table 3.2.

Namely, one of the theorem stated there is the convolution theorem which we repeat here for convenience: Given the functions $f(t)$ and $g(t)$, and their respective Laplace transforms $F(s)$ and $G(s)$, so that $f(t) = \mathcal{L}^{-1}\{F(s)\}$ and $g(t) = \mathcal{L}^{-1}\{G(s)\}$, than a Laplace transform of $(f * g)(t) = \int_{0^-}^{t} f(\tau) \cdot g(t-\tau) \cdot d\tau$ is $F(s) \cdot G(s)$. In that way, if $G(s)$ is the Laplace transform of the excitation function, $g(\tau)$ is its time domain original. That stands for the $F(s)$ and $f(t)$ pair. Having in mind that the product $F(s) \cdot G(s)$ is easily obtained in the s-domain, it may be used for performing effective inverse Laplace transform in order to find the time domain response to a given time domain excitation.

When discussing the calculation in the time domain we will suppose, as we did for the frequency domain, that the transfer function was created by the approximation procedure or, what should be the same, by transfer function synthesis. The same is applicable to the functions obtained by circuit analysis but that would need some additional steps.

To allow the calculation of the time domain response to a Dirac pulse, we will first rewrite the expression for the transfer function in the following form:

$$H(s) = \frac{N(s)}{D(s)} = \frac{N(s)}{\prod_{i=1}^{k} (s - p_i)^{m_i}}. \tag{3.25}$$

This expression allows for the poles with higher multiplicity (than one) to be represented properly. Namely, if $m_i > 1$, the pole p_i is multiple. Its multiplicity is m_i. In the above k is the number of distinct poles. The order of the denominator is

$$n = \sum_{i=1}^{k} m_i. \tag{3.26}$$

For the case when all poles are simple $n = k$.

Now, according to the discussion on the inverse Laplace transform and Table 3.1, the response to a Dirac excitation is obtained by

$$f(t) = \frac{1}{2\pi \cdot \mathrm{j}} \int_{c} T(s) \cdot e^{s \cdot t} ds \tag{3.27}$$

where c is a suitable contour in the complex s-plane.

To begin with the practical implementation of (3.27) one is to express the transfer function as a sum of partial fractions as follows

$$\begin{aligned} F(s) = {} & \frac{RA_{m_1}}{(s - s_1)^{m_1}} + \frac{RA_{m_1 - 1}}{(s - s_1)^{m_1 - 1}} + \cdots + \frac{RA_1}{(s - s_1)} \\ & + \frac{RB_{m_2}}{(s - s_2)^{m_2}} + \cdots + \frac{RB_1}{(s - s_2)} \\ & + \frac{RR_{m_k}}{(s - s_k)^{m_k}} + \cdots \frac{RR_1}{(s - s_k)}. \end{aligned} \tag{3.28}$$

In the above expression one uses the following formulas for the residues R:

$$RA_{m_1} = (s - s_1)^{m_1} \cdot F(s)_{|s=s_1} \tag{3.29m1a}$$

$$RA_{m_1 - 1} = \frac{d}{ds}[(s - s_1)^{m_1} \cdot F(s)]_{|s=s_1} \tag{3.29m1b}$$

$$\vdots$$

$$RA_1 = \frac{1}{(m_1 - 1)!} \frac{d^{m_1 - 1}}{ds^{m_1 - 1}}[(s - s_1)^{m_1} \cdot F(s)]_{|s=s_1} \tag{3.29m1z}$$

$$RB_{m_2} = (s - s_2)^{m_2} \cdot F(s)_{|s=s_2} \tag{3.29m2a}$$

$$RB_{m_2 - 1} = \frac{d}{ds}[(s - s_2)^{m_2} \cdot F(s)]_{|s=s_2} \tag{3.29m2b}$$

$$\vdots$$

$$RB_1 = \frac{1}{(m_2 - 1)!} \frac{d^{m_2 - 1}}{ds^{m_2 - 1}}[(s - s_2)^{m_2} \cdot F(s)]_{|s=s_2} \tag{3.29m2z}$$

$$RR_{mk} = (s - s_k)^{m_k} \cdot F(s)_{|s=s_k} \tag{3.29mka}$$

$$RR_{m_k-1} = \frac{d}{ds}\left[(s - s_k)^{m_k} \cdot F(s)\right]_{|s=s_k} \tag{3.29mkb}$$

$$\vdots$$

$$RR_1 = \frac{1}{(m_k - 1)!} \frac{d^{m_k-1}}{ds^{m_k-1}}\left[(s - s_k)^{m_k} \cdot F(s)\right]_{|s=s_k}. \tag{3.29mkz}$$

Now, the inverse Laplace transform of (3.28) is created based on the linearity theorem expressed in Table 3.2. Simply, the proper formula of Table 3.1 is implemented (one at a time) to each summand to get the time domain sum.

To find the response to a step function, here denoted $u(t)$, one has first to multiply the transfer function by 1/s (as shown in the second row of Table 3.1) and then to implement the same procedure to the new function as would do with the original. In other words one is to find the inverse Laplace transform of $H(s)/s$.

3.4 Implementation Example

To analyze the properties of the transfer function and to draw the responses the user of the *RM* software may implement two programs *LP_analysis* and *TF_analysis*. The first is intended to be used for low-pass functions only while the second is covering all the rest. Since we want to see the time domain responses we will go with the former.

As an example a tenth order LSM lowpass filter with improved selectivity by 8 transmission zeros will be considered.

Below is part of the results produced in the textual report file. To save space some parts are edited into tables.

Start.E

Project name: LOW_pass_LSM_selective

Read in data

n (numerator)=8, m (denominator)=10

Without re-normalization

Zeros		Poles	
Real part	Imaginary part	Real part	Imaginary part
0.000000000e+000	1.125937675e+000	-4.969168539e-002	1.000223491e+000
0.000000000e+000	-1.125937675e+000	-4.969168539e-002	-1.000223491e+000
0.000000000e+000	1.222018681e+000	-1.842393894e-001	9.993769349e-001
0.000000000e+000	-1.222018681e+000	-1.842393894e-001	-9.993769349e-001
0.000000000e+000	1.503023406e+000	-4.379657354e-001	9.887732657e-001
0.000000000e+000	-1.503023406e+000	-4.379657354e-001	-9.887732657e-001
0.000000000e+000	2.371689504e+000	-9.030666269e-001	8.798249712e-001
0.000000000e+000	-2.371689504e+000	-9.030666269e-001	-8.798249712e-001

Residues (normalized) in the poles	
Real part	Imaginary part
1.627048079e-002	-6.571925584e-002
1.627048079e-002	6.571925584e-002
3.209986589e-001	7.613256117e-002
3.209986589e-001	-7.613256117e-002
-2.456262533e-001	1.125833931e+000
-2.456262533e-001	-1.125833931e+000
-3.060301376e+000	-8.545034869e-001
-3.060301376e+000	8.545034869e-001
2.968658489e+000	-4.864565649e+000
2.968658489e+000	4.864565649e+000

****

The cut-off frequency defined by 4.50000e+001 dB is equal to 1.11552e+005

****

Delay time (at 0.5)=6.52964e-006 rise time (0.1-0.9)=5.61798e-006

****

The number of oscillations at the beginning of the step response is=0

****

The number of oscillations at the end of the step response is=11

Extrema of the step response	
time (s)	Value
1.22140e-005	1.19159e+000
1.81380e-005	8.86115e-001
2.36398e-005	1.08061e+000
2.89668e-005	9.38123e-001
3.41852e-005	1.04980e+000
3.93295e-005	9.58826e-001
4.44238e-005	1.03455e+000
4.95168e-005	9.70778e-001
5.45247e-005	1.02484e+000
5.95540e-005	9.78840e-001
6.45664e-005	1.01806e+000

End.E

The same program creates a .csv file containing tables of time and frequency domain responses. Below is a set of figures (Fig. 3.5) obtained by the use of the MS Excel program.

3.5 Developer's Corner

We will stress two issues in this paragraph.

First, as already mentioned, there will be no transfer function synthesis and analysis as well as passive circuit synthesis program, without a routine for solving high-order polynomials. Care must be taken for accuracy since its results define the quality of all subsequent synthesis programs. It is as the tap water in a household. Nothing can be cleaner in a household than the tap water is.

Second, to visualize the analysis results one needs a program for creation drawings. It is our advice (what was done within the $\mathcal{RM}$ software) not to spend energy on writing a proprietary software for graphics. Creation of .csv files within the program for later analysis will allow making drawings on will, based on the MS Excel's drawing capabilities. It is worth mentioning that Excel allows for merging drawings created by separate analyses so allowing for better documentation and, possibly, comparisons of different solutions.

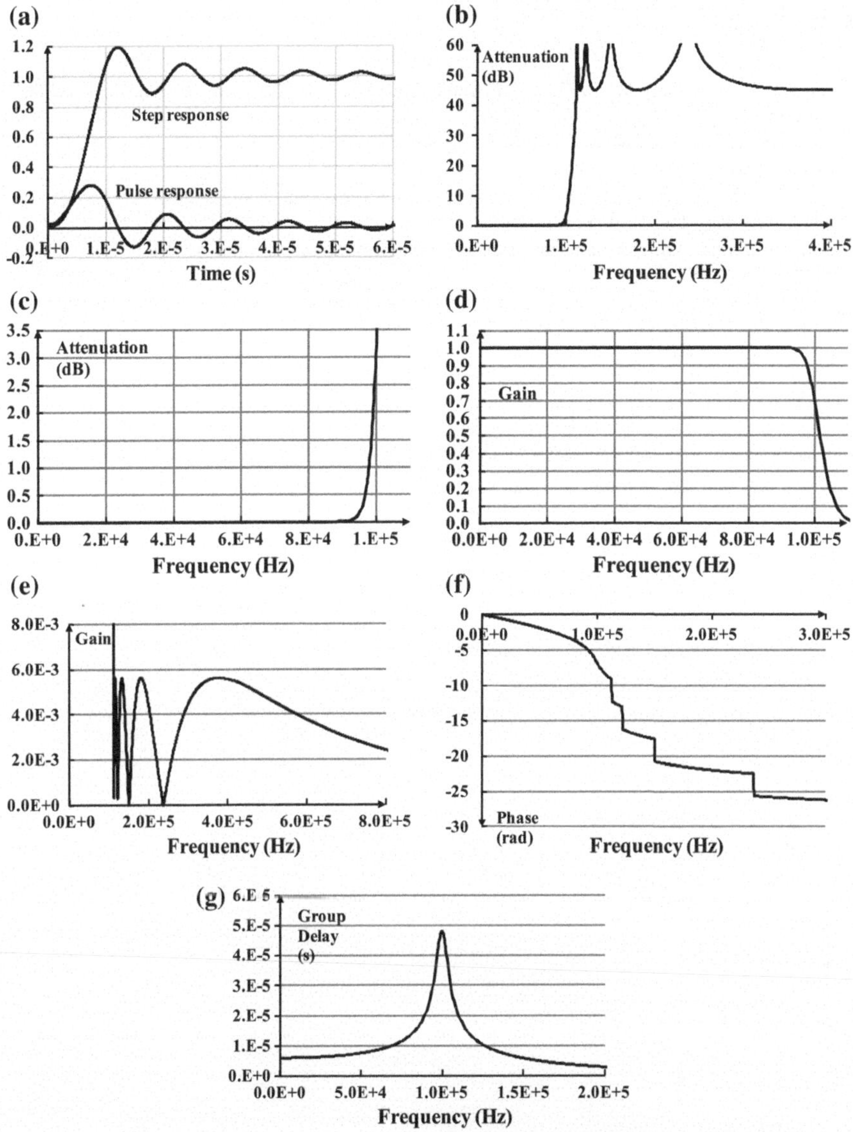

Fig. 3.5 Diagrams created by the MS Excel program for a low-pass filter. **a** Time domain responses, **b** overall attenuation characteristic, **c** passband attenuation, **d** passband amplitude characteristic, **e** stopband amplitude characteristic, **f** overall phase characteristic, and **g** overall group delay characteristic

References

1. Litovski V, Zwolinski M (1997) VLSI circuit simulation and optimization. Chapman and Hall, London
2. Linnér PLJ (1973) A note on the time response of multiple-pole transfer function. IEEE Trans Circuit Theory CT 20(5):617
3. Korn GA, Korn TM (1961) Mathematical handbook for scientists and engineers. McGraw-Hill Book Company Inc, New York
4. Roy SD (1968) On the transient response of all-pole low-pass filters. IEEE Trans Circuit Theory 15(4):485–488
5. Raut R, Swamy MNS (2010) Modern analog filters analysis and design, a practical approach. Wiley-VCH Verlag GmbH & Co, KGaA

Chapter 4
Frequency Transformations in the Analog Domain and Applications

To obtain analog transfer function based on low-pass prototypes one uses frequency transformations. Detailed theory is given for transformation of the zeros and poles of the prototype low-pass transfer function into a new one of a band-pass, high-pass, and band-stop type. Advice is given as to how an all-pass transfer function is obtained from a low-pass prototype. Examples are given with both tables and figures demonstrating the migration of the filter's characteristic frequencies. Nonlinearities introduced by the transformations are exemplified.

4.1 Introduction

This chapter is devoted to a special procedure for obtaining band-pass, band-stop (notch) and high-pass analog transfer function based on existing low-pass prototype [1, 2]. To that end the poles and zeros of the prototype function are used as input to a transformation formula or, better to say, equation, and the solution is a set of new poles and zeroes whose transfer function exhibits a transformed amplitude characteristic. The expressions or the formulas used in the transformation process are nonlinear so affecting the properties of the final result in the sense that partially distort both the amplitude and the phase characteristic.

There exists an additional analog transformation [3] performing linear frequency translation which, however, may be useful if implemented to existing circuit not to a function as such. It is of use in the design of so called polyphase filters which are out of the scope of this book.

As for the transformations relating the analog and the discrete (and digital) domain, discussion will be given in the corresponding chapters related to system synthesis.

V. Litovski, *Electronic Filters*, Lecture Notes in Electrical Engineering 596,
https://doi.org/10.1007/978-981-32-9852-1_4

In the next we will first describe the transformations and fully elaborate their implementation. Then, numerical examples will be given for poles and zeros transformations. Since, band-pass, band-stop, and high-pass filters will be synthesized many times during the elaboration of algorithms in this book, no frequency characteristics of the transformed filters will be discussed here. Nevertheless, to get insight into the distortions introduced by the low-pass to band-pass transformation (which is the most frequently used) we will depict the transformed characteristic for various values of the relative bandwidths of the resulting filters.

The following notation will be introduced:

s Prototype (low-pass) complex frequency.
z Mapped (transformed) complex frequency.
B_r Relative bandwidth.
BW Absolute bandwidth in radians/s.
ω_0 Central frequency in radians/s

4.2 Low-Pass to Band-Pass Transformation

The low-pass to band-pass transformation [1] is given by

$$\frac{s}{\omega_0} = \frac{1}{B_r}\left(\frac{z}{\omega_c} + \frac{\omega_c}{z}\right). \tag{4.1a}$$

It is emanated from

$$\frac{\omega}{\omega_0} = \frac{1}{B_r}\left(\frac{\Omega}{\omega_c} - \frac{\omega_c}{\Omega}\right). \tag{4.1b}$$

In the above expressions ω stands for the angular frequency in the prototype (low-pass) transfer function, Ω for the angular frequency in the mapped (band-pass) transfer function, ω_0 is the cut-off frequency of the low-pass transfer function while ω_c is the central frequency of the band-pass transfer function.

Note, geometrical symmetry is assumed which means that

$$\omega_c = \sqrt{\omega_l \omega_u}, \tag{4.2a}$$

$$BW = \omega_u - \omega_l \tag{4.2b}$$

and

$$B_r = BW/\omega_c \tag{4.2c}$$

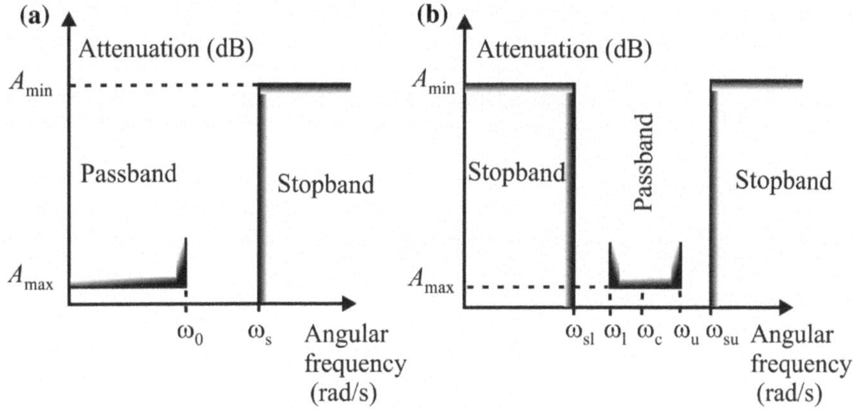

Fig. 4.1 The low-pass (**a**) and the band-pass (**b**) frequency mask

where ω_l and ω_u are the lower and upper cut-off frequencies of the band-pass filter, respectively.

Figure 4.1 illustrates the transformation.

Implementation of (4.1a) means that one is to substitute the right-hand side of the above expression in place of s/ω_0 whenever it appears in the low-pass transfer function. It is, again, obvious that the implementation of this and the rest of the transformations will be much easier if the transfer function numerator and denominator polynomials are expressed in a factored form.

In the next, to simplify, the normalized frequency in the prototype function will be used i.e. instead of s/ω_0, s will be substituted in (4.1a).

We will proceed with the so called Geffe algorithm [4].

To find the poles and zeros of the transformed transfer function solution of (4.1a) proceeds as follows:

$$s = \frac{1}{B_r \cdot \omega_c \cdot z}\left(z^2 + \omega_c^2\right) = > z^2 - B_r \cdot \omega_c \cdot s \cdot z + \omega_c^2 = 0 \tag{4.3a}$$

$$z = \frac{1}{2}\left\{B_r \cdot \omega_c \cdot s \pm \sqrt{(B_r \cdot \omega_c \cdot s)^2 - 4 \cdot \omega_c^2}\right\} \tag{4.3b}$$

$$z = \frac{B_r \cdot \omega_c \cdot s}{2}\left\{1 \pm \sqrt{1 - 4/(B_r \cdot s)^2}\right\} \tag{4.3c}$$

For the normalized frequency in the prototype plane: $s = \sigma + j\omega$ one gets

$$z = \frac{B_r \cdot \omega_c \cdot (\sigma + j\omega)}{2}\left\{1 \pm \sqrt{1 - 4/(B_r \cdot (\sigma + j\omega))^2}\right\} \tag{4.4a}$$

or

$$z = \left[\frac{B_r \cdot \omega_c \cdot \sigma}{2} + j\frac{B_r \cdot \omega_c \cdot \omega}{2}\right]\left\{1 \pm \sqrt{1 - \frac{4}{B_r^2} \cdot \frac{1}{\sigma^2 - \omega^2 + 2j \cdot \sigma \cdot \omega}}\right\}. \quad (4.4b)$$

To simplify we introduce

$$M = \frac{B_r \cdot \omega_c \cdot \sigma}{2}; \quad (4.5a)$$

$$N = \frac{B_r \cdot \omega_c \cdot \omega}{2}, \quad (4.5b)$$

so that the solution of (4.4b) is

$$z_{1/2} = [M + j \cdot N] \cdot \left\{1 \pm \sqrt{1 - \frac{4}{B_r^2} \cdot \frac{\sigma^2 - \omega^2 - 2j \cdot \sigma \cdot \omega}{(\sigma^2 - \omega^2)^2 + (2 \cdot \sigma \cdot \omega)^2}}\right\}. \quad (4.6)$$

To make the formula "programmable" we use

$$Q = \frac{4}{B_r^2} / \left[(\sigma^2 - \omega^2)^2 + (2 \cdot \sigma \cdot \omega)^2\right] \quad (4.7)$$

so that

$$z_{1/2} = [M + j \cdot N] \cdot \left\{1 \pm \sqrt{1 - Q \cdot (\sigma^2 - \omega^2 - 2j \cdot \sigma \cdot \omega)}\right\} \quad (4.8)$$

or

$$z_{1/2} = (M + \mathrm{j} \cdot N) \times \left\{ \begin{array}{c} 1 \pm \sqrt{4}\left[1 - Q \cdot (\sigma^2 - \omega^2)\right]^2 + \left[2Q\sigma \cdot \omega\right]^2 \\ \times\, \mathrm{e}^{\mathrm{j}\frac{1}{2}\mathrm{arctg}\frac{2Q \cdot \sigma \cdot \omega}{1 - Q \cdot (\sigma^2 - \omega^2)}} \end{array} \right\}. \quad (4.9)$$

To further simplify the programming we introduce

$$R = \sqrt{4}\left[1 - Q \cdot (\sigma^2 - \omega^2)\right]^2 + [2Q \cdot \sigma \cdot \omega]^2 \quad (4.10)$$

and

$$F = \frac{1}{2}\arctan\frac{2 \cdot Q \cdot \sigma \cdot \omega}{1 - Q \cdot (\sigma^2 - \omega^2)} \quad (4.11)$$

so that

$$z_{1/2} = [M + \mathrm{j} \cdot N] \cdot \left\{1 \pm R \cdot e^{\mathrm{j} \cdot F}\right\} \tag{4.12a}$$

$$z_{1/2} = [M + \mathrm{j} \cdot N] \cdot \{1 \pm R \cdot \cos(F) \pm \mathrm{j} \cdot R \cdot \sin(F)\} \tag{4.12b}$$

$$\begin{aligned} z_{1/2} = {} & M \cdot (1 \pm R \cdot \cos(F)) \pm N \cdot R \cdot \sin(F) \\ & + \mathrm{j} \cdot \{N \cdot (1 \pm R \cdot \cos(F)) \pm R \cdot M \cdot \sin(F)\} \end{aligned} \tag{4.12c}$$

Now we use the transformation

$$\cos(F) = \sqrt{(1 + \cos(2 \cdot F)/2} = > \cos(2 \cdot F) = \frac{1}{\sqrt{1 + \mathrm{tg}^2(2 \cdot F)}} \tag{4.13}$$

to introduce

$$\cos(2 \cdot F) = \frac{1}{\sqrt{1 + \left(\frac{2 \cdot Q \cdot \sigma \cdot \omega}{1 - Q \cdot (\sigma^2 - \omega^2)}\right)^2}} \tag{4.14a}$$

and

$$\sin(F) = \sqrt{(1 - \cos(2 \cdot F))/2}. \tag{4.14b}$$

With all that, the final solution is:

$$z_{1/2} = \Sigma_{1/2} + \mathrm{j} \cdot \Omega_{1/2} \tag{4.15a}$$

$$\Sigma_{1/2} - M \cdot (1 \pm R \cdot \cos(F)) \mp N \cdot R \cdot \sin(F) \tag{4.15b}$$

$$\Omega_{1/2} = N \cdot (1 \pm R \cdot \cos(F)) \mp R \cdot M \cdot \sin(F) \tag{4.15c}$$

4.3 Low-Pass to Band-Stop Transformation

The low-pass to band-stop transformation [1] is given by

$$\frac{s}{\omega_0} = \frac{B_r \cdot \omega_c \cdot z}{\left(z^2 + \omega_c^2\right)}. \tag{4.16a}$$

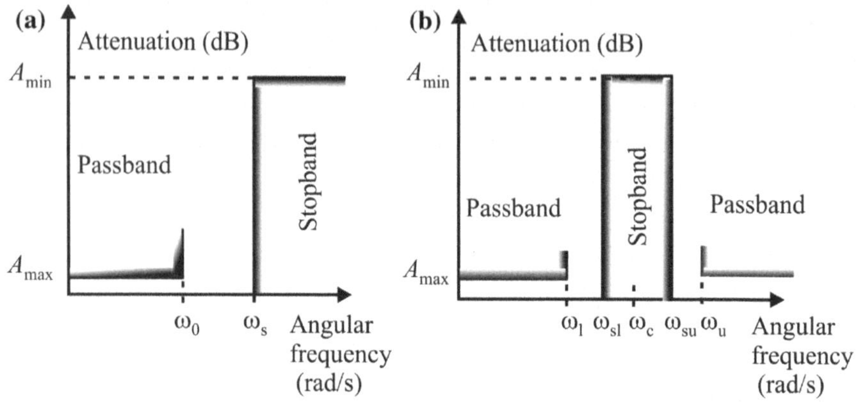

Fig. 4.2 The low-pass (**a**) and the band-stop (**b**) frequency mask

It is emanated from

$$\frac{\omega}{\omega_0} = \frac{B_r \cdot \omega_c \cdot \Omega}{\Omega^2 - \omega_c^2}. \tag{4.16b}$$

Figure 4.2 illustrates the transformation.

Implementation of (4.16a) means that one is to substitute the right-hand side of the above expression in place of s/ω_0 whenever it appears in the low-pass transfer function.

In the next, to simplify, the normalized frequency in the prototype function will be used i.e. instead of s/ω_0, s will be substituted in (4.16a).

To find the poles and zeros of the transformed transfer function solution of (4.16) proceeds as follows:

$$s \cdot \left(z^2 + \omega_c^2\right) = B_r \cdot \omega_c \cdot z \Rightarrow z^2 - \frac{1}{s} B_r \cdot \omega_c \cdot z + \omega_c^2 = 0. \tag{4.17}$$

The solution is

$$z_{1/2} = \frac{1}{2}\left\{\frac{1}{s}B_r \cdot \omega_c \pm \sqrt{\left[\left(\frac{1}{s}B_r \cdot \omega_c\right)^2 - 4 \cdot \omega_c^2\right]}\right\} \tag{4.18a}$$

or

$$z_{1/2} = \frac{1}{2} \cdot \frac{1}{s} B_r \cdot \omega_c \cdot \left\{1 \pm \sqrt{\left[1 - \frac{4 \cdot s^2}{B_r^2}\right]}\right\} \tag{4.18b}$$

For the normalized frequency in the prototype plane: $s = \sigma + j\omega$ one gets

$$z_{1/2} = \frac{B_r \cdot \omega_c \cdot (\sigma + j \cdot \omega)}{2 \cdot (\sigma^2 + \omega^2)} \cdot \left\{ 1 \pm \sqrt{\left[\frac{1}{B_r^2} \cdot \left[B_r^2 - 4 \cdot (\sigma^2 - \omega^2) - 8 \cdot j \cdot \sigma \cdot \omega \right] \right]} \right\} \tag{4.19}$$

with $M = \frac{B_r \cdot \omega_c \cdot \sigma}{2 \cdot (\sigma^2 + \omega^2)}$, $N = \frac{B_r \cdot \omega_c \cdot \omega}{2 \cdot (\sigma^2 + \omega^2)}$.

We introduce the following abbreviations

$$Q = B_r^2 - 4 \cdot \left(\sigma^2 - \omega^2\right) \tag{4.20a}$$

$$R = \cdot \frac{1}{B_r} \cdot \sqrt{4} Q^2 + (8 \cdot \sigma \cdot \omega)^2 \tag{4.20b}$$

$$F = \frac{1}{2} \cdot \mathrm{arctg} \frac{-8 \cdot \sigma \cdot \omega}{Q}. \tag{4.20c}$$

Now, having in mind that

$$\cos(F) = \sqrt{(1 + \cos(2 \cdot F)/2} \Rightarrow \cos(2 \cdot F) = \frac{1}{\sqrt{1 + \mathrm{tg}^2(2 \cdot F)}} \tag{4.21a}$$

we get

$$\cos(2 \cdot F) = \frac{1}{\sqrt{1 + \left(\frac{-8 \cdot \sigma \cdot \omega}{Q}\right)^2}} = \frac{Q}{B_r^2 \cdot R^2} \tag{4.21b}$$

enabling the calculation:

$$\sin(F) = \sqrt{(1 - \cos(2 \cdot F))/2}. \tag{4.21c}$$

After substitution in (4.19) we get

$$z_{1/2} = (M + j \cdot N) \cdot \{1 \pm R \cdot (\cos F + j \cdot \sin F))\} \tag{4.22a}$$

or

$$z_{1/2} = (M + j \cdot N) \cdot \{[1 \pm R \cdot \cos F] \pm j \cdot R \cdot \sin F\}, \tag{4.22b}$$

Finally, the solution is

$$z_{1/2} = \Sigma_{1/2} + \mathrm{j} \cdot \Omega_{1/2}, \tag{4.23a}$$

with

$$\Sigma_{1/2} = M \cdot [1 \pm R \cdot \cos F] \mp N \cdot R \cdot \sin F \tag{4.24a}$$

$$\Omega_{1/2} = N \cdot [1 \pm R \cdot \cos F] \pm M \cdot R \cdot \sin F \tag{4.24b}$$

4.4 Low-Pass to High-Pass Transformation

The low-pass to high-pass transform is

$$s/\omega_0 = \omega_c/z \tag{4.25a}$$

which is emanating from

$$\omega/\omega_0 = -\omega_c/\Omega. \tag{4.25b}$$

Figure 4.3 illustrates the transformation.

Implementation of (4.25a) means that one is to substitute the right-hand side of the above expression in place of s/ω_0 whenever it appears in the low-pass transfer function.

To find the poles and zeros of the transformed transfer function solution of (4.25a, 4.25b) proceeds as follows:

$$z = \Sigma + \mathrm{j}\Omega \tag{4.26a}$$

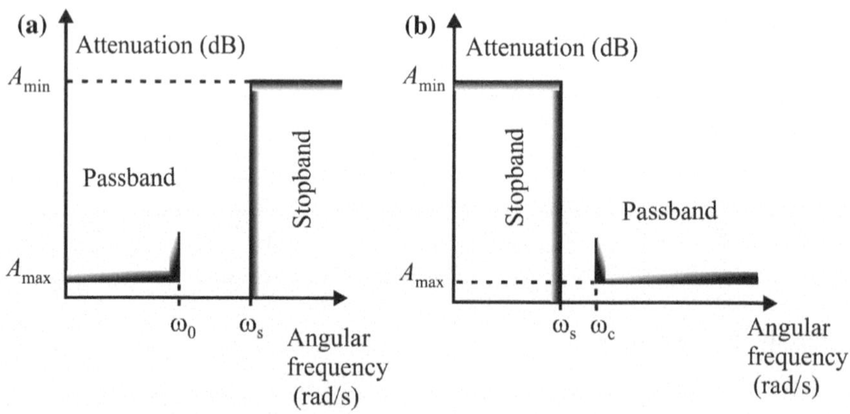

Fig. 4.3 The low-pass (**a**) and the high-pass (**b**) frequency mask

where, for the normalized frequency in the prototype plane: $s = \sigma + j\omega$, one gets

$$\Sigma = \sigma/\left(\sigma^2 + \omega^2\right) \tag{4.26b}$$

$$\Omega = -\omega/\left(\sigma^2 + \omega^2\right). \tag{4.26c}$$

4.5 Examples of Implementation

In this paragraph we will first give examples of the transformations with graphical presentation of the positions of the poles and zeros before and after transformation. One may get, in that way, a notion on the migration of the poles and zeros after transformation.

After that the influence of the band width to the distortion of the amplitude and the group delay characteristic will be investigated on a set of examples related to a lop-pass to band-pass transform.

Within the *RM* software for filter design this activity is performed by the *transformations* program.

4.5.1 Example No. 1. Low-Pass to Band-Pass Transformation

Here is an excerpt of the report produced by the *transformations* program. A 9th order polynomial LSM [5] low-pass will be used as a prototype. It was created by the *CMAC* program of the *RM* software. Figure 4.4 serves for visualization of the tables below.

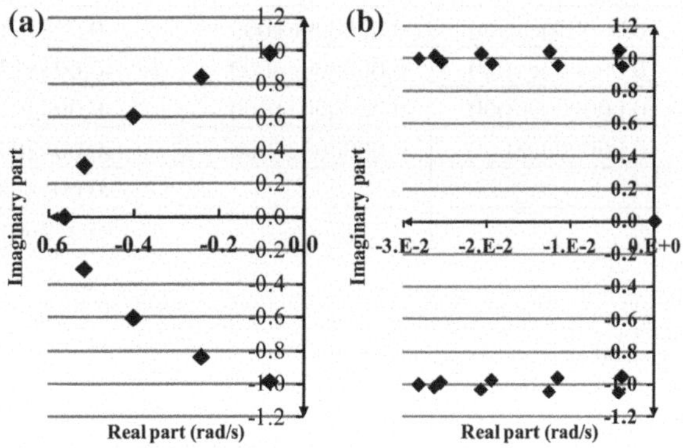

Fig. 4.4 Position of the prototype poles (**a**) and the transformed poles and zeros (**b**) for the LP-to-BP transformation depicted in the table above. All zeros of the BP are in the origin

Start.E

INPUT DATA
Project name: LSM 3dB to BP
n (order of the numerator)=0 m (order of the denominator)=9
Required passband type (ipb) is=1
Normalized central frequency will be w0=1.000000
Relative passband width will be br=0.100000

Poles	
Real part	Imaginary part
-0.40110200000	-0.60539300000
-0.40110200000	0.60539300000
-0.56303500000	0.00000000000
-0.08021700000	-0.98356600000
-0.08021700000	0.98356600000
-0.24265300000	0.83980600000
-0.24265300000	-0.83980600000
-0.51658800000	0.31146900000
-0.51658800000	-0.31146900000

AFTER TRANSFORMATION
The input LOW-PASS transfer function will be transformed into a BAND-PASS one
new order of the numerator is nbp=9
new order of the denominator is mbp=18
TRANSFORMED

Zeros	
Real part	Imaginary part
0.000000e+000	0.000000e+000
0.000000e+000	0.000000e+000
0.000000e+000	0.000000e+000
0.000000e+000	0.000000e+000
0.000000e+000	0.000000e+000
0.000000e+000	0.000000e+000
0.000000e+000	0.000000e+000
0.000000e+000	0.000000e+000
0.000000e+000	0.000000e+000

Poles	
Real part	Imaginary part
-0.01944819522	0.96998752344
-0.02066200478	-1.03052682344
-0.02066200478	1.03052682344
-0.01944819522	-0.96998752344
-0.02815175745	0.99960366094
-0.02815174255	-0.99960366094
-0.00381383973	0.95202220793
-0.00420786027	-1.05037880793
-0.00420786027	1.05037880793
-0.00381383973	-0.95202220793
-0.01264169241	1.04279809534

		-0.01162360759	-0.95881749534
		-0.01162360759	0.95881749534
		-0.01264169241	-1.04279809534
		-0.02623173827	1.01536119563
		-0.02542706173	-0.98421429563
		-0.02542706173	0.98421429563
		-0.02623173827	-1.01536119563

End.E

4.5.2 *Example No. 2. Low-Pass to High-Pass (LP-to-HP) Transformation*

Here is an excerpt of the report produced by the *transformations* program. As a low-pass prototype a 5th order modified elliptic [6] filter was used created by the *elliptic* program of the *RM* software. Figure 4.5 serves for visualization of the tables below.

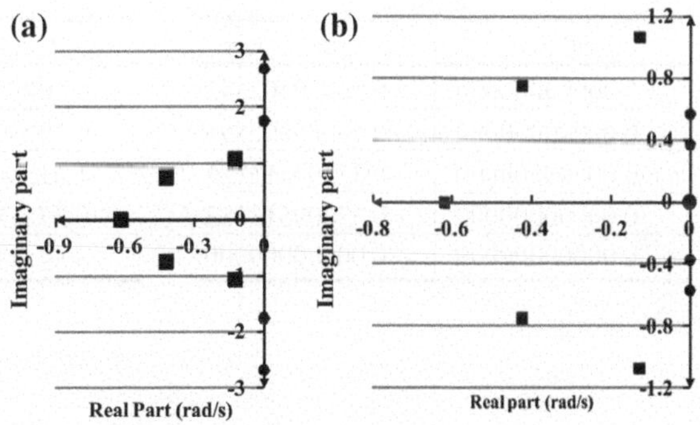

Fig. 4.5 Position of the prototype poles and zeros (**a**) and the transformed poles and zeros (**b**) for the LP-to-HP transformation depicted above. Poles are squares, zeros are circles

Start.E

INPUT DATA

Project name: LP_to_HP Modified elliptic
n (order of the numerator)=4m (order of the denominator)=5
Required passband type (ipb) is=2
Originally read

Zeros		Poles	
Real part	Imaginary part	Real part	Imaginary part
0.00000000000	1.75563329940	-0.12636700000	-1.06997800000
0.00000000000	-1.75563329940	-0.12636700000	1.06997800000
0.00000000000	2.68028155470	-0.61894820000	0.00000000000
0.00000000000	-2.68028155470	-0.42384720000	0.74807310000
		-0.42384720000	-0.74807310000

AFTER TRANSFORMATION

The input LOW-PASS transfer function will be transformed into a HIGH-PASS one
new order of the numerator is nhp=5
new order of the denominator is mhp=5

TRANSFORMED

Zeros		Poles	
Real part	Imaginary part	Real part	Imaginary part
0.00000000000	-0.56959502895	-0.10885997178	0.92174202829
0.00000000000	0.56959502895	-0.10885997178	-0.92174202829
0.00000000000	-0.37309513183	-1.61564408783	-0.00000000000
0.00000000000	0.37309513183	-0.57333997220	-1.01192177360
0.00000000000	0.00000000000	-0.57333997220	1.01192177360

End.E

4.5.3 Example No. 3. Low-Pass to Band-Stop (LP-to-BS) Transformation

Here is an excerpt of the report produced by the transformations program. As a low-pass prototype a 7th order Papoulis filter [7–10] with corrected group delay characteristic (by 6th order phase corrector) will be used. It was created by the *CMAC* and *corrector* programs of the *RM* software. Figure 4.6 serves for visualization of the tables below.

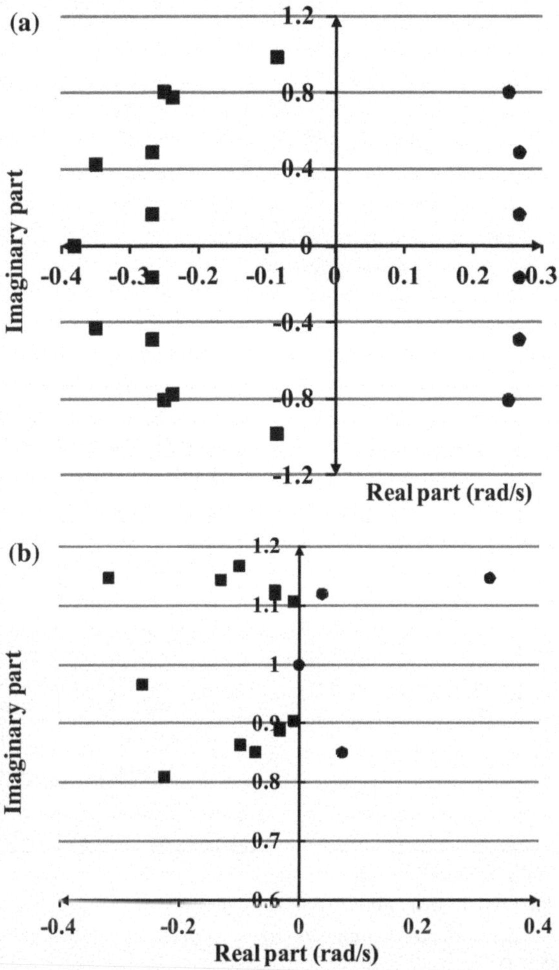

Fig. 4.6 Position of **a** the prototype poles and zeros and **b** the transformed poles and zeros (only the top half-plane shown) for the LP-to-BR transformation depicted above. Poles are squares, zeros are circles

Start.E

INPUT DATA

Project name: Corrector for Papoulis/6,7 5%
n (order of the numerator)=6 m (order of the denominator)=13
Required passband type (ipb) is=4
Normalized central frequency will be w0=1.000000
Relative passband width will be br=0.200000

Originally read

Zeros		Poles	
Real parts	Imaginary parts	Real parts	Imaginary parts
0.25077151064	0.80462225883	-0.38215996450	0.00000000000
0.26662801721	0.48974231189	-0.08622185217	0.98451675137
0.26695942271	0.16537172785	-0.08622185217	-0.98451675137
0.26695942271	-0.16537172785	-0.23747539792	0.77841703029
0.26662801721	-0.48974231189	-0.23747539792	-0.77841703029
0.25077151064	-0.80462225883	-0.34928406404	0.42905995537
		-0.34928406404	-0.42905995537
		-0.25077151064	0.80462225883
		-0.26662801721	0.48974231189
		-0.26695942271	0.16537172785
		-0.26695942271	-0.16537172785
		-0.26662801721	-0.48974231189
		-0.25077151064	-0.80462225883

AFTER TRANSFORMATION
new order of the numerator is nbr=26
new order of the denominator is mbr=26
TRANSFORMED

Zeros		Poles	
Real part	Imaginary part	Real part	Imaginary part
0.03132856008	-0.89250530885	-0.26167053898	0.96515725887
0.03928113582	1.11906267523	-0.26167051864	-0.96515725887
0.07236112949	-0.85127368463	-0.00971319384	1.10582861093
0.09913800545	1.16628327631	-0.00794242104	-0.90422950161
0.22430330194	-0.81056547987	-0.00794242104	0.90422950161
0.31711339733	1.14595358544	-0.00971319384	-1.10582861093

0.31711339733	-1.14595358544
0.22430330194	0.81056547987
0.09913800545	-1.16628327631
0.07236112949	0.85127368463
0.03928113582	-1.11906267523
0.03132856008	0.89250530885
0.00000000000	1.00000000000
0.00000000000	-1.00000000000
0.00000000000	1.00000000000
0.00000000000	-1.00000000000
0.00000000000	1.00000000000
0.00000000000	-1.00000000000
0.00000000000	1.00000000000
0.00000000000	-1.00000000000
0.00000000000	1.00000000000
0.00000000000	-1.00000000000
0.00000000000	1.00000000000
0.00000000000	-1.00000000000
0.00000000000	1.00000000000
0.00000000000	-1.00000000000

-0.04004239582	1.12378028350
-0.03166694693	-0.88872530889
-0.03166694693	0.88872530889
-0.04004239582	-1.12378028350
-0.13005146716	1.14360844508
-0.09817030223	-0.86326120834
-0.09817030223	0.86326120834
-0.13005146716	-1.14360844508
-0.03928113582	1.11906267523
-0.03132856008	-0.89250530885
-0.09913800545	1.16628327631
-0.07236112949	-0.85127368463
-0.31711339733	1.14595358544
-0.22430330194	-0.81056547987
-0.22430330194	0.81056547987
-0.31711339733	-1.14595358544
-0.07236112949	0.85127368463
-0.09913800545	-1.16628327631
-0.03132856008	0.89250530885
-0.03928113582	-1.11906267523

End.E

4.5.4 Low-Pass to Band-Pass Transformation Revisited

We will try here to demonstrate the effects of the nonlinearity of the LP-to-BP transform. As a prototype an 8th order LSM low-pass filter with increased selectivity (by six transmission zeros) will be used. It was created by the *CMAC* and the *Chang_1* programs of the *RM* software. These results were transformed into band-pass by the *transformation* program. Three values of the relative pass-band width were used. The goal was to have a narrow-band, a medium-band, and a broad-band solution. The results are depicted in Fig. 4.7. The overall attenuation characteristics are presented first in Fig. 4.7a. Not much of asymmetry may be recognized. That was the reason to look deeper into the attenuation characteristics. Figure 4.7b represents the pass-band attenuation properties.

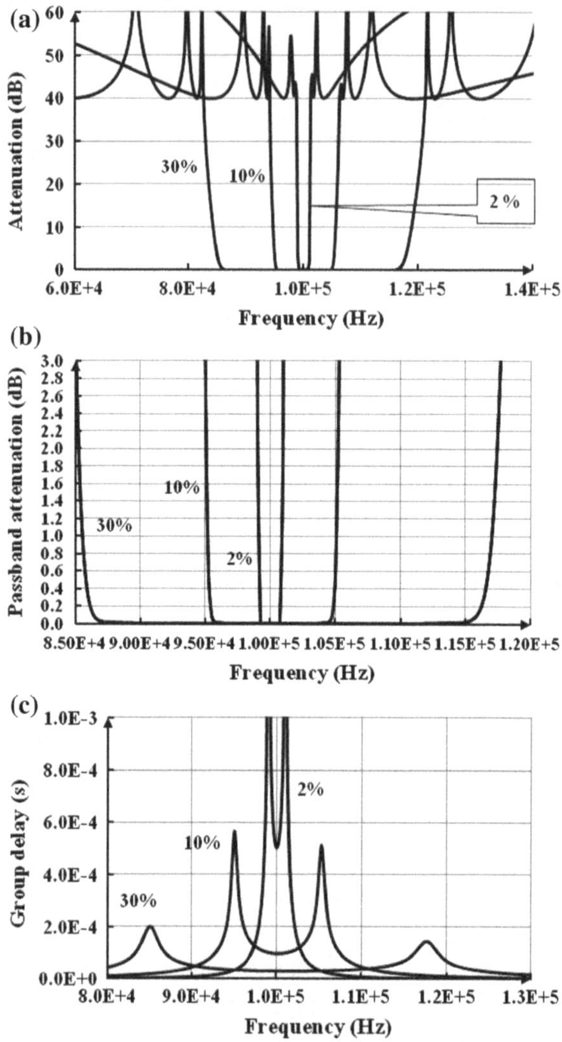

Fig. 4.7 **a** Attenuation characteristics of the transformed filters, **b** passband attenuation characteristics of the transformed filters, and **c** group delay characteristics of the transformed filters. The percentages shown are the planned relative bandwidths in percentage

Here is clearly noticeable that the asymmetry is increasing with the rise of the width of the relative bandwidth. Namely, for the case of relative bandwidth of 30%, while the lower edge of the pass-band is exactly at its place (on the 85 kHz mark), the upper edge is significantly below its place (which should be at 115 kHz). The difference of the required and the obtained frequency of the upper edge of the passband is about 2.5 kHz which is significant as compared to the 30 kHz of required band-width.

On the other end, when the relative bandwidth is small, as is 2%, the asymmetry is hardly noticeable on Fig. 4.7b.

The phenomenon is also noticeable when the group delay characteristic is considered. Here, both the locations and the relative values of the peaks of group delay are affected. Broader band-width leads to larger asymmetry of the peaks and their locations.

The reason for these distortions is in the influence of the lower half of the transfer function. Namely, both the amplitude characteristic and the group delay are even functions and, in the case of band-pass filter, may be observed as sum of two: the function for negative and the function for positive frequencies. Their parts on the opposite sides on the frequency axis influence each other to produce distortion. The nearer the functions (the broader the pass-band) the larger the mutual influence becomes.

4.6 Developer's Corner

The reader probably noticed that all transformations were performed with normalized poles and zeros, and properly normalized central frequency of the final filter. That is recommendable since further re-normalization (e.g. relocation on the frequency axis for band-pass and band-stop filters) is becoming trivial. In addition, one may expect better numerical results since so many subtractions are present in (4.15a, 4.15b, 4.15c) and (4.24a, 4.24b) the result of which is thereafter multiplied by large numbers (if non-normalized quantities were to be used).

We will use the opportunity here to demonstrate how a band-stop filter is eliminating a spectral component which was first mentioned in relation to Fig. 2.5b. For that purpose we have chosen a low-pass filter (Fourth order Papoulis, to be regularly introduce later on.) and transformed it into a band-stop filter with relative width of the stopband of 50% of the central frequency of the final solution. That is 25% above the central frequency and 25% below. Its gain as a function of the normalized angular frequency is depicted in Fig. 4.8a. Since the LP-to-BS transformation is nonlinear as is the LP-to-BP transformation, the amplitude characteristic is not arithmetically symmetrical, however.

To demonstrate how a spectral component is eliminated from the signal we used as an excitation the following

$$v_{\text{in}}(t) = \sin(t) + \sin(1.2 \cdot t). \tag{4.27}$$

which is depicted in Fig. 4.8b as input signal. It contains two components. The frequency of the first one is 1 rad/s which falls into the middle of the stopband while the second is shifted towards the passband by 0.2 rad/s. Since the stopband width requested by the transformation was $2 \cdot 0.25$ rad/s the second component belongs to the upper edge of the stopband. In other words, since the transformation does not produce abrupt change of the gain, the second component belongs to the transition region of the filter's amplitude characteristic.

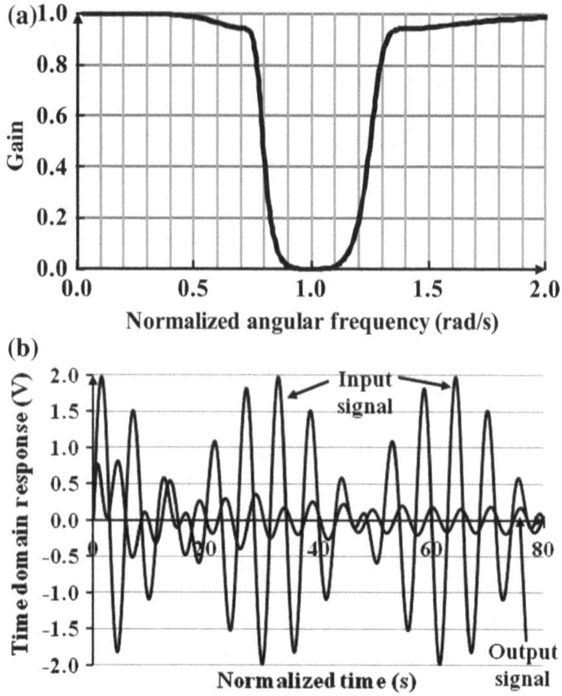

Fig. 4.8 **a** Amplitude (gain) characteristic of a band-stop filter and **b** time domain response demonstrating filtering. Order of the filter $n = 8$

This signal was used as an input to calculate the transient response of the filter obtained by convolution and inverse Laplace transform. The results are depicted in Fig. 4.8b. As can be seen, after a vigorous transient, the 1 rad/s component is fully eliminated and the output signal is settled to a monochromatic one (single frequency of 1.2 rad/s) with an amplitude which is approximately 20% of the amplitude of the corresponding component of the input signal (which may be verified from Fig. 4.8a).

As a corollary of these analyses one may draw conclusions as to how low selectivity (broad transition region) may influence the final spectrum. Here, a component which was supposed to belong to the stopband is reduced for only $20 \cdot \log(1/0.2) \approx 14$ dB.

References

1. Orchard H, Temes GC (1968) Filter design using transformed variables. IEEE Trans Circuit Theory 15(4):385–408
2. Temes GC, Mitra SK (eds) (1973) Modern filter theory and design. Wiley, New York
3. Andreani P, Mattisson S, Essink B (2000) A CMOS gm-C polyphase filter with high image band rejection. In: Proceedings of ESSCIRC, pp 244–247
4. Geffe P (1974) Designers guide to active bandpass filters part III. EDN 19(7):46–52

5. Raković BD, Litovski VB (1973) Least-squares monotonic low-pass filters with sharp cutoff. Electron Lett 9(4):75–76
6. Zolotarev EI (1877) Application of elliptic functions to problems about functions with least and greatest deviation from zero. Zapi Imp Akad Nauk St Petersburg 30(5) (in Russian). www.math.technion.ac.il/hat/fpapers/zol1.pdf
7. Papoulis A (1958) Optimum filters with monotonic response. Proc IRE 46:606–609
8. Papoulis A (1959) On monotonic response filters. Proc IRE 47:332–333
9. Fukada M (1959) Optimum filters of even orders with monotonic response. IRE Trans Circuit Theory 6(3):277–281
10. Topisirović D, Litovski V, Andrejevic Stošović M (2015) Unified theory and state-variable implementation of critical-monotonic all-pole filters. Int J Circuit Theory Appl 43(4):502–515

Chapter 5
Introduction to the $\mathcal{RM}$ Software for Filter Design

A brief description of the $\mathcal{RM}$ software for filter design will be given. We will start with the intentions we had at the beginning of its development and then describe how it is conceived to work. In fact it is integrated software where the (user friendly) interactive shell is created as a set of .html files while the programs for transfer function and system synthesis are written in C. All programs output a .csv file enabling the use of the MS Excel's capabilities to create drawings. In addition, where appropriate, SPICE net-list is created serving for both verification of the system synthesis and further design activities such as PCB or IC design. An .html file is always created containing the design report. Where appropriate it contains the circuit's schematic. The $\mathcal{RM}$ software is creating polynomial and rational transfer functions obeying requirements in the frequency domain regarding amplitude, group delay or both. Low-pas, band-pass, high-pass, all-pass and band-stop functions may be synthesized. As for physical synthesis one may get doubly terminated passive LC; active RC; active SC; active GM-C, and digital recursive (IIR) filters. A special feature of the $\mathcal{RM}$ software is the availability of parallel circuit solutions.

5.1 Introduction

This chapter is intended to give to the reader an overview of the $\mathcal{RM}$ software for filter design. That will include mainly its capabilities both in hardware design and in filter function synthesis. The user and designers interface will be discussed shortly, too.

As will be seen a vast variety of analogue filter functions may be synthesized by the $\mathcal{RM}$ software. One may say practically most of the existing solutions of the filter function approximation are offered.

The physical realization i.e. the implementation technology of the analogue filters is limited to passive LC, active RC, active GM-C, and switched capacitor solutions.

V. Litovski, *Electronic Filters*, Lecture Notes in Electrical Engineering 596,
https://doi.org/10.1007/978-981-32-9852-1_5

As for the digital filter design, only IIR filters obtained from their analogue prototypes represented in a form of a parallel connection of biquad cells is offered.

The synthesis procedures and the results are well documented and where appropriate SPICE files are produced for verification of the filter design and further embedding the filter circuit into a larger system.

To help the users and the designers, tutorials are available explaining the underlying procedures implemented in the $\mathcal{RM}$ software. That includes a list of references some of which are addressed properly for downloading from the internet. This book may be considered as kind of documentation for this software.

The $\mathcal{RM}$ software described here is a result of scholar and programming efforts of a large group of engineers and is verified through several decades of in-house implementation.

Part of the $\mathcal{RM}$ software related to the transfer function synthesis and system synthesis was written in C while for the interactive part HTML and Javascript were used. All together, the software occupies more than 40 thousand programming lines.

This chapter has a complement in a form of 34 video presentations and animations [1, 2].

5.2 What Is the Intention?

It was our intention to provide a design tool and consequently opportunities for design which encompasses very wide variety of approximants some of which, in usual educational and commercial design approaches, are overlooked despite the fact that they provide excellent solutions for some specific filter design problems. In that way we are offering more complete and excellent design opportunities.

Namely, a limited offer is generally made available to the users with Butterwort, Chebyshev, Inverse Chebyshev, Bessel and elliptic (Cauer) solutions being mainly recommended. All-pass correctors are available too.

The space of available filter functions which are to meet the system designer's requirements, however, is much broader and it is really a pity not to use these opportunities.

The $\mathcal{RM}$ software is up to encompass as broad the space of approximants as possible including monotonic amplitude characteristics in the passband, variable number and types of transmission zeroes in the passband and stopband, and equi-ripple approximation of the group delay. In that way minimum and non-minimum phase transfer function may be created and synthesis of proper circuits to be performed.

As for the physical realization, which is referred throughout this book as system design, an attempt was made to encompass as many types as possible. Only very specific and highly specialized technologies are omitted. From the choice of circuit's topologies point of view, the author took the freedom to decide which of the almost unlimited set of variants, realizing specific second order function, will be chosen to be implemented in the $\mathcal{RM}$ software. In that way the reader is freed of doubts as to what to do when a choice comes forward and, in the same time, the author puts himself in a position to take the responsibility if the decision (the choice he made) is not the most appropriate.

To repeat, the book is limited to parallel implementation of IIR digital filters to keep it within tractable limits.

Finally, in most cases, a SPICE net-list is created automatically which allows for the connection of the $\mathcal{RM}$ software with PCB and IC design systems.

5.3 How Is It Conceived to Work?

A user friendly interface was developed to guide the user through the variants and to offer choices and advice during the upload of the customer's order.

To start with, a special graphics is offered to the user at the beginning of the design process which gives advice as to how to proceed through the maze of actions and variants. It is depicted in Fig. 5.1.

In the next some aspect of the functionality of the $\mathcal{RM}$ software will be highlighted.

5.3.1 Designer's Working Environment

The designer's graphical user interface is intended to guide the designer through the process including:

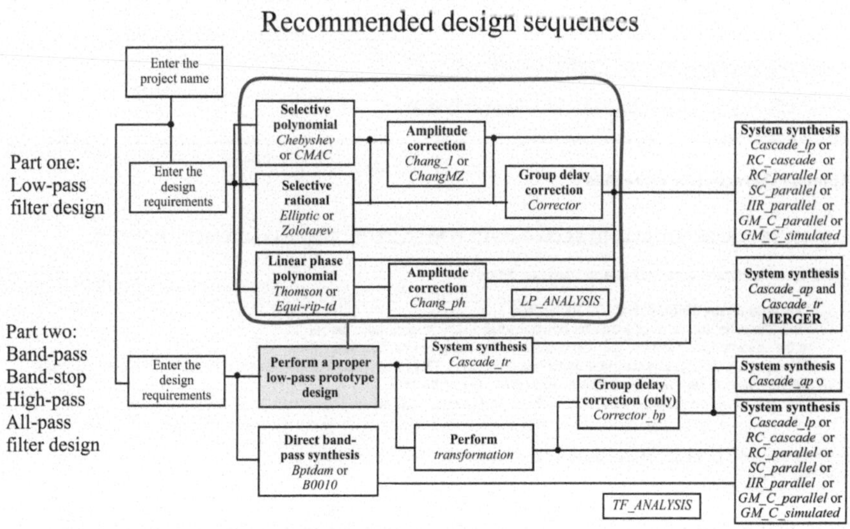

Fig. 5.1 Recommended design sequences

- Design specification transformation into design data,
- Transfer function synthesis and verification with possibilities to choose among alternatives,
- Transfer function analysis including response parameter evaluation in frequency and time domain and response graphical presentation,
- Circuit and system synthesis and verification including SPICE scripts creation when applicable.

The graphical interface is intended to inform the user about the possibilities available and about the location of the results, to suggest literature, to offer some tutorials, to allow for interactive upload of the design requirements, and, most importantly, to activate the programs in the course of the design process.

The interaction during the design process is controlled by an .html program which guides the designer through the maze depicted in Fig. 5.1. Figure 5.2 depicts a screenshot of a phase of the design process in which the designer is to choose among different actions related to amplitude or phase correction. By the way, amplitude correction is a unique capability of the $\mathcal{RM}$ software. As can be seen, not only actions are enabled but advice is given to the designer as to what are the capabilities of the program and its limitations, of course. After action performed, in a similar way, the designer is guided to the next one until the end of the design process.

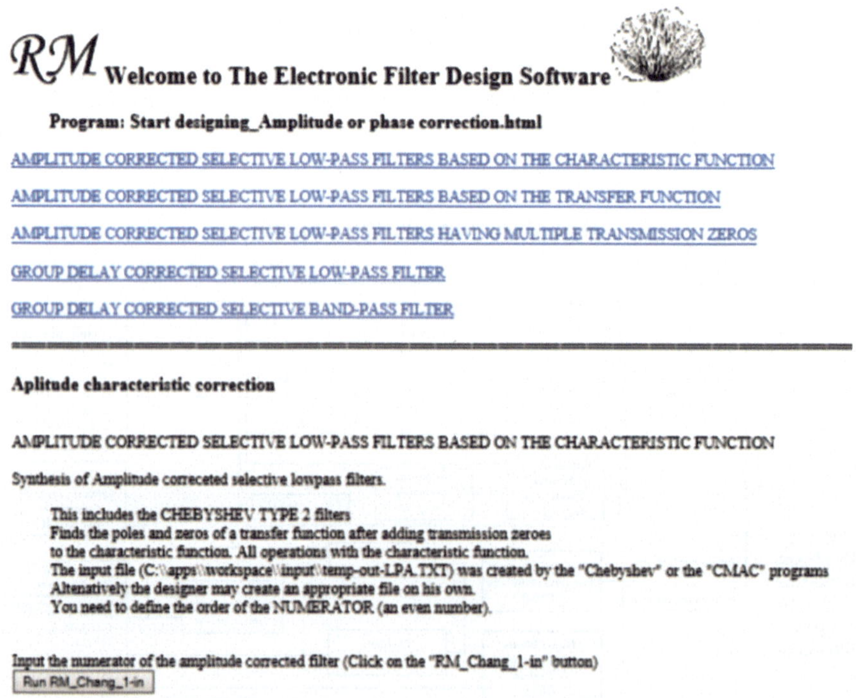

Fig. 5.2 Part of the screen seen by the designer in the amplitude or phase correction part of the design process

5.3.2 Graphical Infrastructure

The $\mathcal{RM}$ software uses the graphic capabilities of Microsoft Excel for presentation of the filter's performances in the frequency and the time domain by creating a comma separated .csv file. These will be used in the filter transfer function approximation phase. In fact, every action of the transfer function approximation process is followed with a creation of an Excel table containing the filter response. The user is expected to create a drawing following the "insert linear or area charts" icon in the Microsoft Excel.

The design report being an .html file contains full information about the design process, the properties of the solution in the frequency and time domain in the pre-hardware phase, table containing the filter responses, and, where appropriate, the circuit schematic obtained in the hardware phase.

Along with the .html report a .txt report is created containing all information given in the .html except for the drawings of the circuit schematics.

As for the performances of the filter after the physical synthesis part, for the analogue filters, SPICE files are created and delivered as part of the report. Running these files produces graphical representation of the filter response obtained by circuit simulation.

In the case of the digital IIR filters, frequency domain simulation results are reported after synthesis in a way similar to the one used in the approximation phase. The difference will be related to the number of digits used for evaluation of the filter's characteristic.

5.4 System Synthesis

In the next we will shortly describe the main issues related to the aspect of physical design without intention to make a full picture.

5.4.1 Passive LC Cascade Realization

For realization of doubly terminated passive lossless LC filters the cascade procedure is implemented in the $\mathcal{RM}$ software. Transmission zeros located on the imaginary axis (finite, infinite or zero) and complex transmission zeros are accepted. The program implements a proper order of extraction procedure in order to keep the element values spread as small as possible.

Low-pass and all-pass filters are synthesized directly while high-pas, band-pas and band-stop networks are obtained via circuit element transformation.

The program extracts the value of the load resistor while the source resistor is considered equal to be the normalization value.

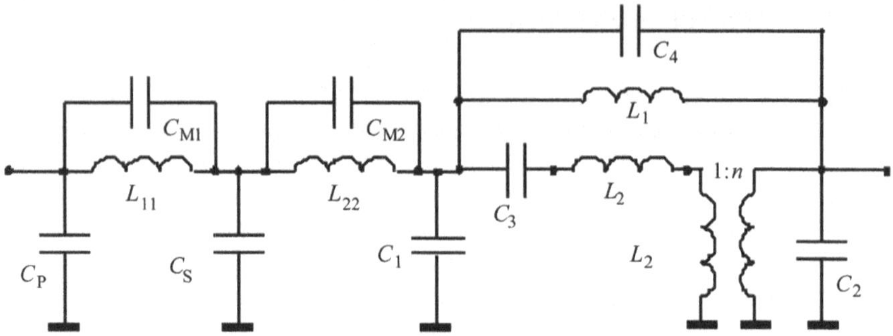

Fig. 5.3 Passive LC realization of a transfer function having transmission zeroes at infinity, at the imaginary axis, and in the right-hand half of the complex plane

Deliverables are:

- Transfer function (polynomial coefficents and pole/zeroe locations)
- Schematic with element values in a form of an .html file
- SPICE code for frequency domain simulation
- Textual report covering the complete synthesis process.

Figure 5.3 depicts one (post processed) schematic obtained by the 𝓡𝓜 software wile Fig. 5.4 depict an excerpt from the SPICE file created automatically.

5.4.2 Cascade Realization of Active RC Circuits

The cascade realization does not impose any limitations as to where the transmission zeros may be located. However, proper pole-zero pairing and order of extraction algorithms must be implemented in order to produce low noise and high-dynamics linear system.

A large variety of cells is used in order to accommodate to the specifics of the transfer functions to be converted into circuits. Figure 5.5 depicts one example of a symmetrical notch cell as implemented by the 𝓡𝓜 software.

Figure 5.6 represents part of one of the reports (.txt file) which is produced by the 𝓡𝓜 software. It is intended to be used for SPICE simulation of the circuit just synthesized. AC analysis is presumed while time domain (.tr) simulation may be enabled by the user after adding a short list of commands (after disabling the .ac analysis).

Deliverables are:

- Transfer function (polynomial coefficents and pole/zeroe locations)
- Schematic and element values in a form of a .html file
- Spice code (net-list) for frequency domain simulation.
- Textual report covering the complete synthesis process.

```
PASSIVE LOW-PASS LC FILTERS.
*PROJECT NAME:          29_lsm 45

*Welcome to the RM software for filter design

v       n0      0       ac      2       sin
Rgen    n0      1       1.0
*
* ZERO AT INFINITY
* parallel capacitance
C011    1       0       1.0408817158e+000
*
* Zero at the imaginary axis, w=1.6332740e+000
*Variant with no transformer-parallel LC in the series branch
CP2     1       0       -2.3667673580e-001
L2      1       2       1.3666813725e+000
CM2     1       2       2.7429293884e-001
CS2     2       0       1.7258189868e+000
*
* ZERO AT INFINITY
* serial inductance
L013    2       3       2.0651116558e+000
RP3     3       0       9.5081310049e-001
CP3     3       0       4.6823532318e-001
* Simulation settings--------------------------
.ac     dec     700     0.01    .6
.print ac v(3)
.end
```

Fig. 5.4 A SPICE file generated by the $\mathcal{RM}$ software

5.4.3 Parallel Realization of Active RC Circuits

Parallel realization of active RC filters is recommended/preferred by the $\mathcal{RM}$ software. The reason for that is the unification of the structure (always the same -by structure- second order cell is used), the lack of uncertainty, and avoiding all sorts of problems related to the signal dynamics and noise which are to be taken into account when cascade RC filters are designed.

The structure used for parallel RC filter synthesis by the $\mathcal{RM}$ software is depicted in Fig. 5.7. The node enumeration depicted is used for SPICE simulation later on. The second order cell depicted in Fig. 5.8 is the one used by the $\mathcal{RM}$ software.

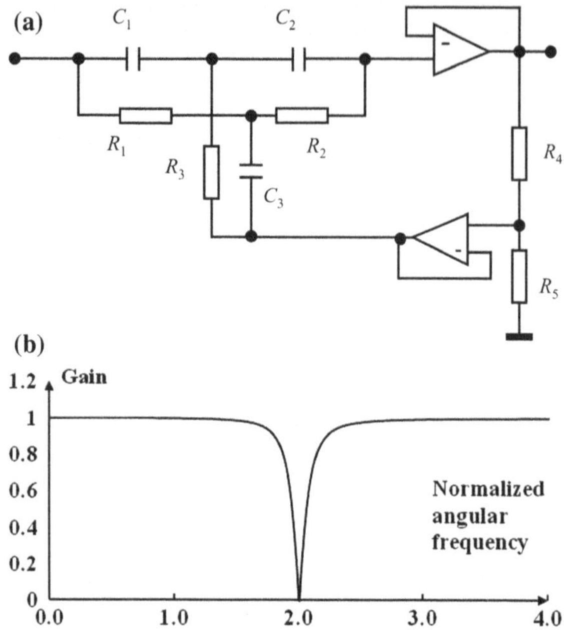

Fig. 5.5 A notch cell with symmetrical amplitude characteristic used in the $\mathcal{RM}$ software. The active RC realization **a** and the frequency response **b**

```
* A second order cell-------------------
r016   ninta6   nintb6   4.040495
r026   60       nintd6   1.177066
E126   nintb6   0        0         ninta6   99999999.990000
r036   ninta6   ninte6   1.177066
r046   50       ninta6   4.040495
r056   50       nintd6   2.491503
r066   50       nintc6   1.000000
r076   nintb6   nintc6   1.000000
r086   60       nintc6   1.000000
C016   nintb6   ninta6   1.000000
C026   nintd6   ninte6   1.000000
E106   ninte6   0        0         nintd6   99999999.990000
E116   60       0        0         nintc6   99999999.990000
* A Second order inverting low-pass high Q cell-----
r17    ninta7   60       14.318099
r27    ninta7   nintb7   0.166287
r37    nintc7   70       1.000000
r47    nintc7   0        1.408896
C017   ninta7   70       1.000000
C027   nintb7   0        0.718115
E7     70       0        nintb7    nintc7   99999999.990000
* A Second order inverting high Q cell high-pass-----
```

Fig. 5.6 Part of the SPICE description (net-list) produced by the $\mathcal{RM}$ software's cascade synthesis of active RC filters routine

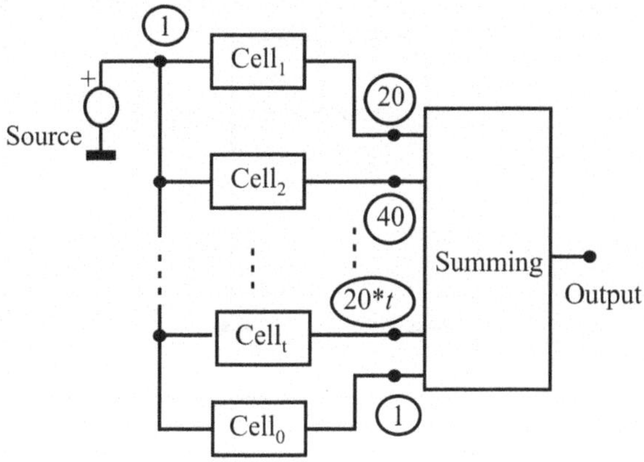

Fig. 5.7 Structure of the parallel RC active hardware solution

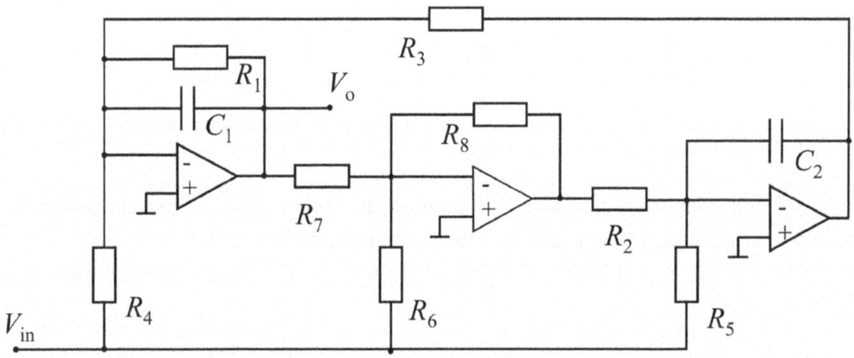

Fig. 5.8 The Tow Thomas biquad cell and node notation in the $\mathcal{RM}$ software

Deliverables are:

- Transfer function (polynomial coefficients and pole/zero locations in the s-domains
- Schematic with element values in a form of a .html file
- Spice code (net-list) for frequency domain simulation
- Textual report covering the complete synthesis process.

5.4.4 Gm-C Filter Synthesis Based on LC Prototypes

The Gm-C version of filter synthesis by structure is the same as the cascaded LC as depicted in Fig. 5.3. The difference is in the cells used. Here equivalence is used which allows for substitution of the inductor by a transconductor-based cell. In that

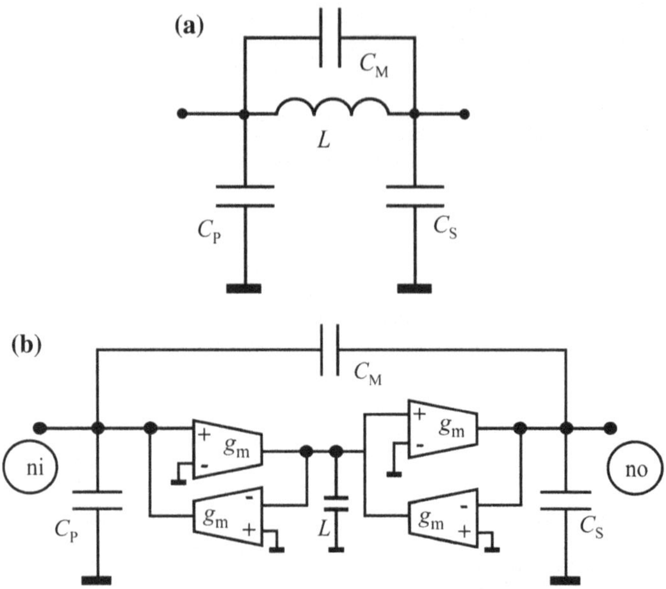

Fig. 5.9 **a** Passive LC cell realizing a transmission zero at the ω-axis and **b** its Gm-C equivalent

way a circuit containing transconductors and capacitors only is obtained. Figure 5.9 depicts one original LC cell and its GM-C counterpart while Fig. 5.10 depicts part of the textual report containing full information about the design process including the schematic.

EXTRACTION OF THE CELLS

---

k=1 (th) ZERO AT INFINITY: parallel capacitance C=7.0806597e-010
Node_UP=1 Node_Down=0

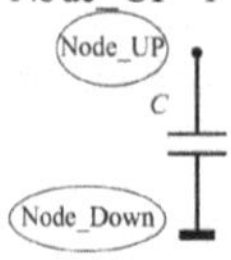

---

k=2 (th) Zero at infinity: SIMULATED series inductance L=1.9839568e-013
Node_L=1 Node_R=2

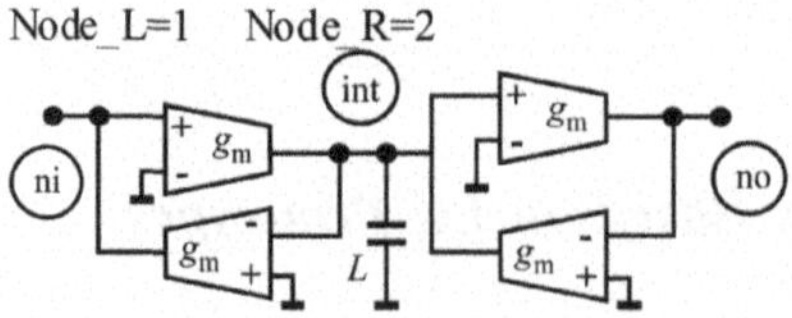

---

k=3 (th) ZERO AT INFINITY: parallel capacitance C=2.8669006e-009
Node UP=2 Node Down=0

Fig. 5.10 Excerpt from the .html report delivered by the 𝓡𝓜 software for the GM-C filter synthesis

Deliverables are:

- Transfer function (polynomial coefficients and pole/zero locations in the s-domains
- Schematic with element values in a form of an .html file
- Spice code (net-list) for frequency domain simulation
- Textual report covering the complete synthesis process.

5.4.5 Parallel Realization of Switched Capacitor Version

The issues related to the pole-zero pairing and order of extraction of cells in cascaded SC filter design are even more pronounced than in the usual RC design. That is why the *RM* software offers only parallel realization of SC filters. Consequently the structure of the hardware solution of the switched capacitor filter synthesis is the same as in the case of parallel RC as depicted in Fig. 5.7.

Figure 5.11 represents part of the report (.html file) which is produced by the *RM* software. Here, for every cell and for the whole, capacitance values and schematic is given. Inside that report one may recognize an emanation of a

--------------------End of the 1-th cell description ---------------------------

The input biquad function parameters after z-transform: (T(z)=(a2*z^2+a1*z+a0)/(b2*z^2+b1*z+1))

beta=7.630188e-001 b0=6.621782e-001 b1=-1.310584e+000 b2=1.000000e+000 a0=-5.499805e-005 a1=1.707628e-001 a2=1.708178e-001

Carusone-Johns-Martin circuit- Second order- high Q--- 2-------

Capacitances: c0=c1=c2=1.000000e-011, k1c1=5.759730e-012 , k2c1=2.881720e-012, k3c2=1.708178e-012 , k4c1=5.929541e-012 , k5c2=5.929541e-012 , k6c1=5.697268e-012

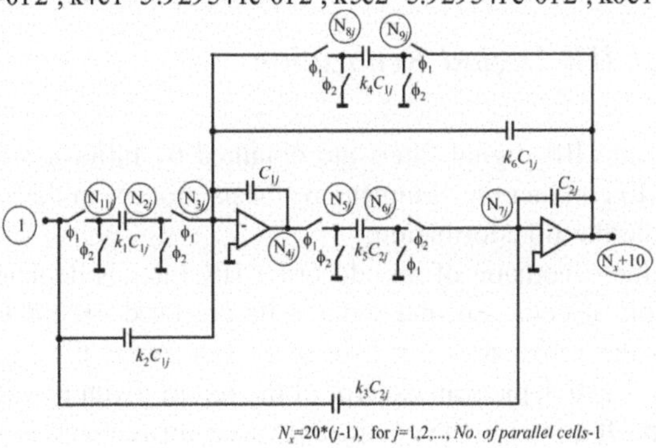

--------------------End of the 2-th cell description ---------------------------

Fig. 5.11 Part of the report produced by the *RM* software's SC filter synthesis routine

E111	10	0	0	7	9.9999999e+005	
* Carusone-Johns-Martin circuit- Second order- high Q----2--------						
C12	23	24	1.0000000e-011			
C22	27	30	1.0000000e-011			
C1K12	31	22	9.3695420e-017			
⋮						
C1K62	23	30	1.2493745e-012			
SL12	1	31	pulse1	0	S1	ON
SL22	31	0	pulse2	0	S1	off
⋮						
SR62	29	0	pulse2	0	S1	off
E102	24	0	0	23	9.9999999e+005	
E112	30	0	0	27	9.9999999e+005	
* Carusone-Johns-Martin circuit- Second order- high Q----3--------						
C13	43	44	1.0000000e-011			

Fig. 5.12 Part of the SPICE descrioption produced by the $\mathcal{RM}$ software's SC filter synthesis routine

universal second order switched capacitor biquad cell which is used by the $\mathcal{RM}$ software.

Figure 5.12 depicts part of the SPICE net-list describing the time domain simulation of the synthetized filter with sinusoidal excitation.

Deliverables are:

- Transfer function (polynomial coefficients and pole/zeroe locations in both s- and z-domains)
- Schematic with capacitance values in a form of an .html file
- Spice code (net-list) for time domain simulation (Non-ideal switches are implemented which may affect the simulation run and partly the results).

5.4.6 *Parallel IIR Digital Realization*

In the $\mathcal{RM}$ software IIR digital filters are obtained by bilinear s-to-z transformation. So, one has to conduct the complete synthesis process in the s-domain and, at the end, to activate the transformation.

In Fig. 5.13a the schematic of an odd order IIR filter is depicted realized as a parallel connection of cells. Second order cells are used except for a single one which completes the schematic in a case of an odd order filter as is the case in Fig. 5.13. Figure 5.13b depicts an excerpt of the report given by the proper system synthesis program. It contains the coefficients used for realization the operators in Fig. 5.13a.

(a)

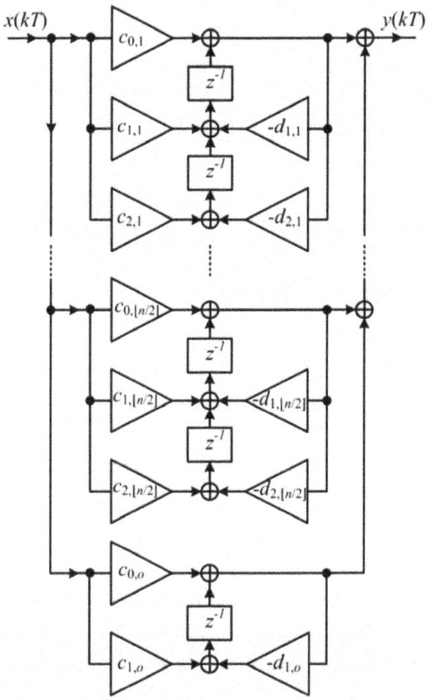

(b)

```
Ordered vector of analog transrer functions to be parallelized
      T(s)=(g*s+q)/(s**2+a*s+b)
  g            q                a             b
    -2.98349 0.0003670645          1.802        1.00007
    -2.06544     -3.312121         1.247      0.9999635
   0.7369042 -0.0001504324          0.445      0.9999363
    4.312025             1             0              0
------------------------------------------------------------------------
Transformation in the z-domain


                          BILINEAR TRANSFORM
T(z)=(gamma*z**2+epsilon*z+delta)/(ksi*z**2+alfa*z+beta)

NORMALIZED EXACT COEFFICIENTS
cell no.         beta        alfa         ksi        delta       epsilon         gamma
 1          0.3594004          -1    0.7161324    0.2953195 1.386173e-005    -0.2953056
 2          0.4143318          -1    0.6611926    0.1419023    -0.1250776    -0.2669799
 3          0.4937143          -1     0.581808  -0.07294292 -5.680863e-006    0.07293724
 4         -0.6795993           1            0    0.6907879     0.6907879             0

NORMALIZED 16 bit COEFFICIENTS
cell no.         beta        alfa                  ksi     delta          epsilon          gamma
1            0.359375          -1    0.7161255    0.2953186             0    -0.2952881
2           0.4143066          -1    0.6611633    0.1418762     -0.125061    -0.2669678
3           0.4937134          -1    0.5817871  -0.07293701            -0    0.07293701
4          -0.6795959           1            0    0.6907654     0.6907654             0

NORMALIZED 32 bit COEFFICIENTS
cell no.         beta        alfa                  ksi     delta          epsilon          gamma
 1          0.3594004          -1    0.7161324    0.2953195 1.386134e-005    -0.2953056
 2          0.4143318          -1    0.6611926    0.1419023    -0.1250776    -0.2669799
 3          0.4937143          -1     0.581808  -0.07294292 -5.680602e-006    0.07293724
 4         -0.6795993           1            0    0.6907879     0.6907879             0

========================================================================
FREQUENCY DOMAIN CHARACTERISTIC OF THE FILTER
 omegastart=1.9076020e+001 (1/s)       omegaend=7.6304081e+004 (1/s)   increment=7.6304081e+001 (1/s)


========================================================================
Omega (rad/s)   Amp_log (dB)=   Phase (degree)  Gr. delay (s)   Amplitude
========================================================================
3.0360429e+000  5.9861784e-008  -4.4939462e-003 8.414802e+002   1.000000e+000
```

Fig. 5.13 Parallel realization of a nth order IIR filter for n odd. a Schematic and b part of the report

Deliverables are:

- Transfer function (polynomial coefficients with 16- and 32- bit approximation and pole/zeroe locations)
- Schematic (as the one depicted in Fig. 5.13a)
- Frequency characteristic in a form of a table.

5.5 Possible Transfer Functions

In this paragraph a complete review of the types of transfer functions which may be synthesized by the $\mathcal{RM}$ software will be given.

5.5.1 Low-Pass Filters

We will separate the category of low-pass filters into polynomial and rational. The transfer functions having all their transmission zeros at infinity are referred to as all-pole, too. Among the polynomial filters we will distinguish the selective and the linear phase filters.

5.5.1.1 Selective Polynomial Filters

The $\mathcal{RM}$ software is capable to synthesize the most important classes of selective polynomial filters. These are divided into two categories: monotonic and Chebyshev.

When monotonic amplitude characteristic is considered one in fact has in mind the critical monotonic amplitude of selective polynomial filters. The term critical is used to denote that the amplitude characteristic of this kind of filters has maximum number of inflection points in the passband in which the derivative touches zero while not changing its sign. We will refer to this category as the CMAC (Critical Monotonic Amplitude Characteristic) filters. Four major sub-classes of CMAC filters are available within the $\mathcal{RM}$ software.

- LSM. The LSM filters which stand for "Least Squares Monotonic" are developed under the condition of minimum reflected power in the passband.
- Halpern. The Halpern or, for short, H-filters exhibit monotonic amplitude characteristic with a maximum asymptotic selectivity.
- Papoulis. This class is often referred to as the L-filters or also as optimal filters. Its main property is to have maximum slope of the amplitude characteristic at the edge of the passband.

– Butterworth (B). This is the oldest and the best know class of filters exhibiting monotonic amplitude characteristic. Its main property is that it is maximally flat at the origin meaning that all the inflection points are located there.

The most popular non-monotonic amplitude characteristic is the one exhibiting equi-ripple amplitude characteristic in the passband. Since it is obtained by use of Chebyshev polynomials it was named Chebyshev (C). For a given variation of the amplitude in the passband this characteristic exhibits the largest selectivity.

It is important to mention that for even order Chebyshev filters the amplitude characteristic does not start with unity gain. In fact, the gain at the origin is equal to the gain at the cut-off frequency. To avoid this, the $\mathcal{RM}$ software provides with a special routine which allows for the gain to start at the unity, for the price of slight reduction of the selectivity since one order of freedom is sacrificed so that the amplitude characteristic has an additional inflection point at the origin.

The passband amplitude characteristics of the selective 10th order polynomial filters is depicted in Fig. 5.14. In the case of the Chebyshev filter 0.5 dB passband attenuation was used with final renormalization of the cut-off frequency so that it produces 3 dB allowing comparison with the monotonic ones.

For illustration purposes in Fig. 5.15 represents pulse responses of the CMAC and Chebyshev filters which are produced by the transfer-function-analysis program of the $\mathcal{RM}$ software. Along with tables convenient to study and draw the characteristics of the filter this program is extracting properties of the responses such as rise time, overshoot, frequency of the lower edge of the stopband and similar.

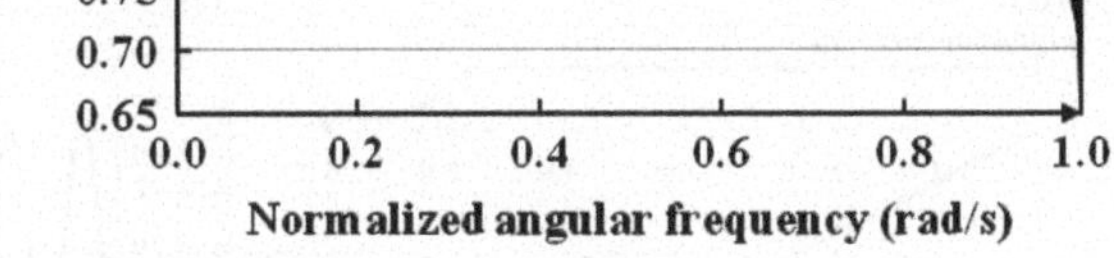

Fig. 5.14 Attenuation characteristics of the CMAC and Chebyshev filter for $n = 10$

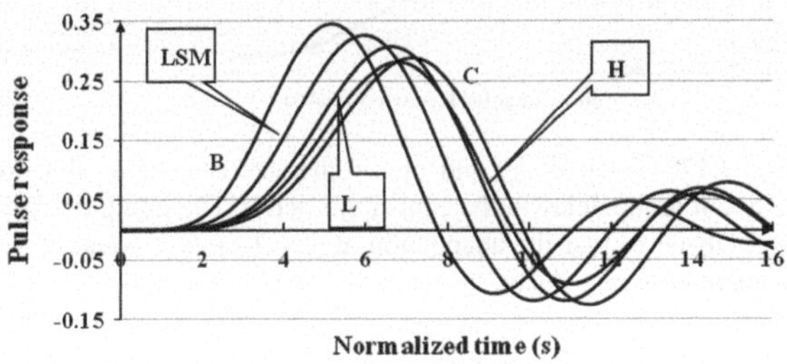

Fig. 5.15 Pulse responses of the 7th order monotonic (CMACs) and Chebyshev filters

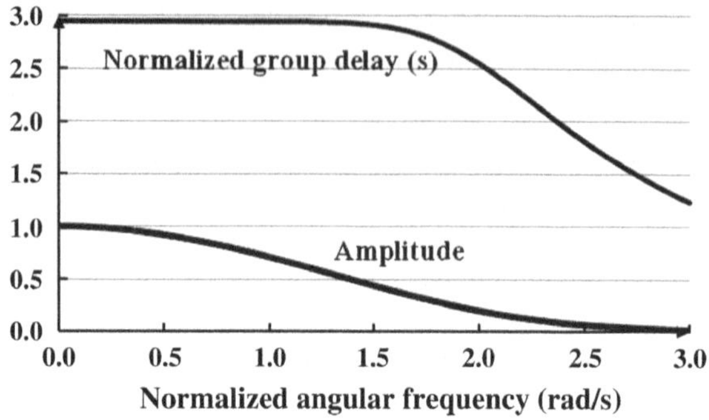

Fig. 5.16 A plot of the gain and group delay for a 7th order low-pass Bessel filter having 3 dB attenuation at $\omega = 1$

5.5.1.2 Polynomial Linear Phase Monotonic Pass-Band-Amplitude Filters

Two categories of polynomial filters approximating linear phase are available by the $\mathcal{RM}$ software: maximally flat and equi-ripple approximants.

Using Bessel polynomials one may produce a transfer function exhibiting maximally flat approximation of the group delay at the origin. Following the authors name these are frequently referred to as the Thomson filters. The result for $n = 7$ is depicted in Fig. 5.16.

Better approximation of constant group delay may be achieved if the equi-ripple criterion of approximation is implemented. For a given delay error and order of the filter, broader approximation interval is obtained. The group delay and phase characteristics of such a functions obtained by the $\mathcal{RM}$ software are depicted in Fig. 5.17a. Figure 5.17b depicts enlarged part of the group delay characteristic.

A special feature of the constant group delay approximant being of the equi-ripple type is enabled by the $\mathcal{RM}$ software. Namely, for even order filters, the

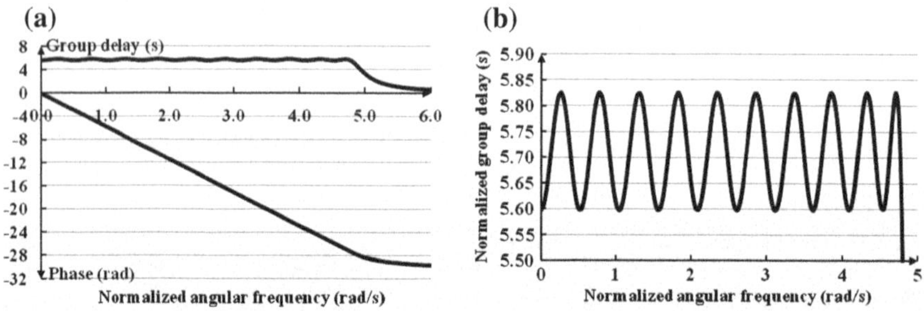

Fig. 5.17 Group delay and phase characteristic of a 20th order low-pass filter approximating constant group delay in the equi-ripple sense. The group delay error is 2%. The transfer function is normalized so that the attenuation at $\omega = 1$ rad/s is equal to 3 dB. **a** Group delay and phase and **b** enlarged group delay

designer is allowed to choose whether the error in the origin will be negative (proper to even order filters) or positive (which resembles to odd order approximants).

5.5.1.3 Amplitude Correction of Selective Polynomial Filters

To produce this kind of amplitude characteristic the $\mathcal{RM}$ software starts with the characteristic function of the polynomial filter to create new characteristic function of a filter with finite number of transmission zeroes.

Due to the influence of the extension of the characteristic function the original shape of the characteristic function in the passband is changed in that

- For critical monotonic response no more inflection points may be found in the passband while the general shape of the original polynomial characteristic function is mainly preserved.
- For equi-ripple response the equi-ripple amplitude in the passband is not preserved.

Figure 5.18 illustrtes the attenuation characteristics obtained by this method which is unique to the $\mathcal{RM}$ software. The sufix "Z" is used to denote the presence of transmission zeros.

It is worth mentioning that the so called Chebyshev II filters or Inverse Chebyshev filters are a special case of Butterworth filters extended with maximum number of transmission zeros. The $\mathcal{RM}$ software allows for maximally flat amplitude characteristic with number of zeros smaller than maximum.

Multiple zeros at real frequencies produce a wide interval in the stopband where the attenuation takes extremely high values. Such functions may be of interest if a component of the signal's spectrum is to be suppressed. It would belong to the stopband but its location would vary in time so that very high attenuation is needed in broader interval than usual.

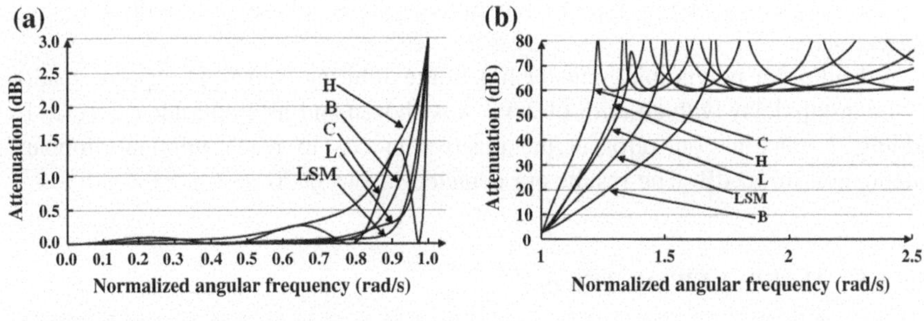

Fig. 5.18 **a** Passband and **b** stopband attenuation of the amplitude corrected selective 7th order filters. Three transmission zeros at real frequencies were used. The stopband attenuation is 60 dB. H = Halpern-Z, C = Chebyshev-Z (originally 3 dB of ripple), B = Butterworth-Z, L = Papoulis-Z, and LSM = LSM-Z

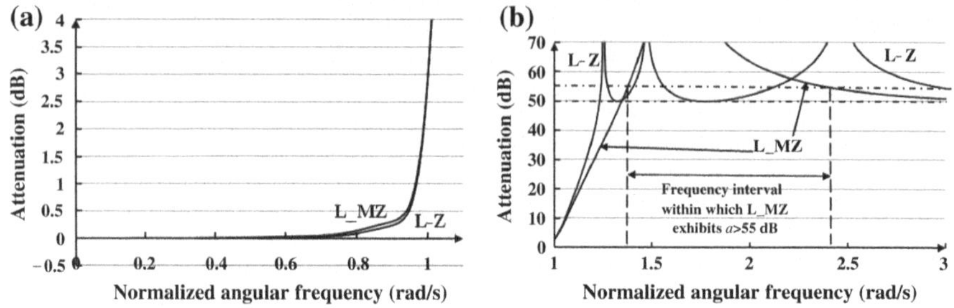

Fig. 5.19 Attenuation characteristic of 7th order Papoulis-Z and Papoulis-MZ filters extended with three transmission zeros at the ω-axis. $a_{\min}$ = 50 dB. **a** Passband and **b** stopband attenuation

Results of the implementation of this kind of approximation are presented below. The newly created transfer functions are labeled –MZ (coming from multiple zeros).

A 7th order filter with a triple transmission zero, which uses the Papoulis's characteristic function as a base, exhibiting 50 dB of minimum stopband attenuation, is presented in Fig. 5.19 and denoted as L-MZ. To demonstrate the difference the Papoulis-Z (in Fig. 5.19 denoted as L-Z) counterpart is depicted in the same figure. One may observe that there is a negligible difference in the passband attenuation characteristics of both filters. In the stopband, however, the situation is much more different.

5.5.1.4 Amplitude Correction of Linear Phase Polynomial Filters

To increase the selectivity of linear phase filters while preserving their phase characteristic one may use the transfer function correction. In this approach the poles are kept and transmission zeros are added. An algorithm was developed to find the positions of the transmission zeroes while obeying the stop-band attenuation requirements.

One may consider this type of functions as linear phase filter with amplitude correction.

A 14th order polynomial transfer function exhibiting equi-ripple approximation of the group delay with an error of 2.5% was chosen and its amplitude corrected by adding 7 transmission zeros at the positive ω-axis to reach minimal stopband attenuation of 50 dB. The results are depicted in Fig. 5.20.

5.5.1.5 Modified Elliptic Filters

The greatest selectivity i.e. the smallest width of the transition band (frequency interval between the pass-band and the stop-band), for a given complexity of the transfer function, is obtained by use of elliptic (or Zolotarev) filters. To construct a

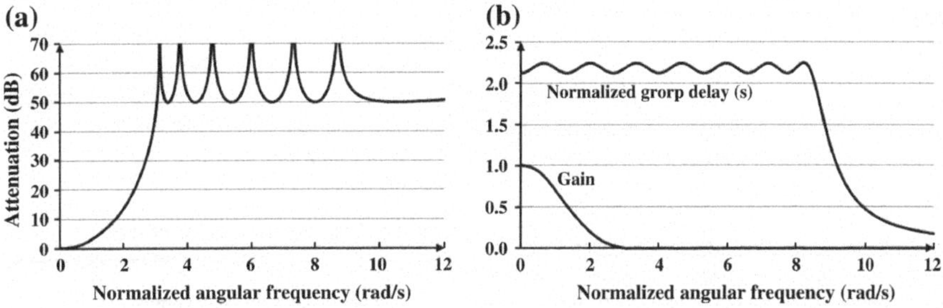

Fig. 5.20 Amplitude corrected 14th order linear phase filter approximating group delay. **a** Attenuation and **b** Gain and group delay

transfer function of this kind one needs to put all transmission zeros (in the stop-band) and all attenuation zeros (in the passband) on the ω-axis. This class exhibits equiripple approximation in both the pass-band and the stop-band. It is known as the most selective of all.

Since, however, the synthesis of this kind of functions is related to the use of elliptic integrals; since in the original solution the maximum passband and minimum stopband attenuations are inter-related; and since the maximum assymptotic attenuation is 12 dB only (6 dB for odd functions), the $\mathcal{RM}$ software creates these functions through an iterative process allowing for independent values of the maximum passband and minimum stopband attenuations and in the same time for the number of transmission zeros to be less than maximum (so offering larger assymtotic attenuation if necessary).

Example of modified elliptic filter synthesis is depicted in Fig. 5.21. It is a 7th order filter with 3 transmission zeros on the positive half of the ω-axis. The minimum stopband attenuation was set to 60 dB while the maximum passband attenuation was imposed to be 0.1 dB.

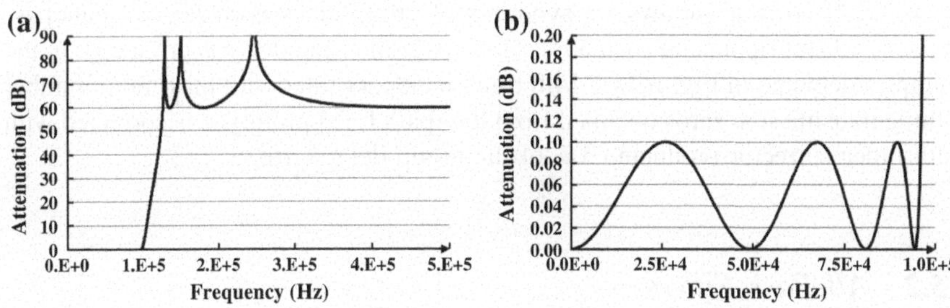

Fig. 5.21 Attenuation of a 7th order modified ellipric filter exhibiting minimum stopband attenuation $a_{min} = 60$ dB and having maximum number of transmission zeros. Passband attenuation $a_{max} = 0.1$ dB. The transfer function is renormalized to exhibit 3 dB at $\omega = 1$. **a** Overall attenuation and **b** pasband attenuation

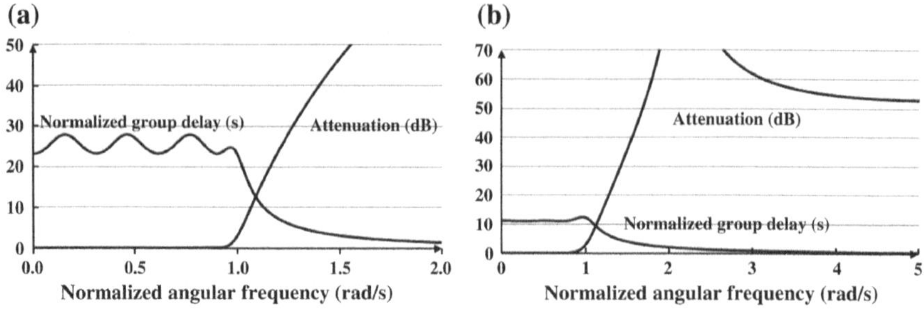

Fig. 5.22 Group delay and attenuation of **a** of 9th order polynomial LSM filter corrected with 6th order corrector introducing approximation error of 10% and **b** of a 9th order maximally flat amplitude approximant with 4 multiple transmission zeros having stopband attenuation of 50 dB and group delay error of 1% with a 4th order corrector

5.5.1.6 Selective Low-Pass Filters with a Corrected Group Delay Characteristic

The subject of correcting a group delay characteristic of a selective filter is one of the most intriguing among the transfer function approximation tasks. Here one is to compensate the passband variations of the group delay to the maximal extent for a given approximation freedom which is expressed by the order of the corrector.

Usually one is after equi-ripple approximation of the overall group delay which is obtained as a sum of the group delay of the selective filter and the one of the corrector. The relative approximation error of the overall group delay, $\Delta\tau_r$ (%), is an input parameters here while at the end of the approximation, the interval in which the group delay is corrected is obtained as a result.

An iterative approximation procedure is implemented within the $\mathcal{RM}$ software for these purposes. Two results are depicted in Fig. 5.22. The first one is related to correction of the group delay of a polynomial selective filter while the second represents a corrected maximally flat amplitude approximant with multiple zeros.

The $\mathcal{RM}$ software allows for synthesis of group delay correctors of bandpass filters, too. Equi-ripple approximation of the overall group delay is produced. In the examples depicted in Fig. 5.23 a 14th order bandpass filter with maximum number of finite transmission zeros obtained from low-pass LSM prototype is corrected with a 4th order corrector producing 1% of the group delay error.

5.5.2 All-Pass Filters

An all-pass filter is a function whose amplitude characteristic is equal to unity for all frequencies while its phase may vary. If constant group delay is produced these are frequently referred to as delay lines. The above mentioned correctors which were

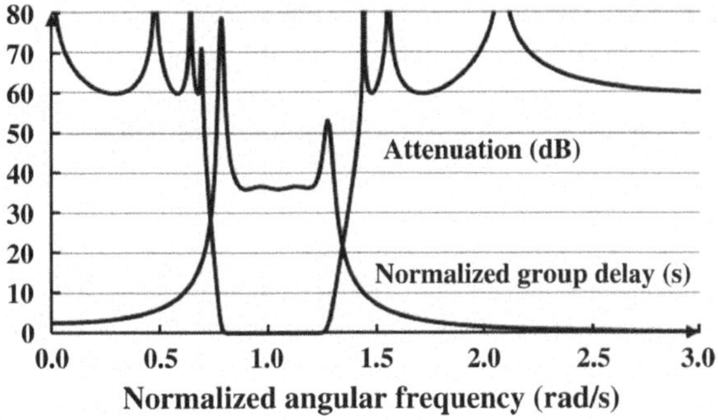

Fig. 5.23 Band-pass LSM filter of 14th order with corrected group delay

used for improving the phase characteristic of selective filters were in fact all-pass networks.

As could be recognized there are two categories of all-pass filters. The first one is correcting the group delay of low-pass selective filters. We will refer to this kind as the low-pass all-pass. These are synthesized simply by extending the transfer function of a low-pass linear phase filter by an equal number of transmission zeros as the number of poles is. The new zeros are located at positions in the right-half of the complex frequency plane mirrored to the poles.

The final group delay characteristic is simply the one of the low-pass filter with doubled value of the group delay and no attenuation.

If the group delay requirements are dislocated around a given frequency, the so called band-pass all-pass filters arise. The $\mathcal{RM}$ software produces band-pass all-pass filters approximating constant and linearly changing (rising and falling) group delay. Figure 5.24 depicts one solution of a 10th order filter exhibiting parabolic phase (linearly decreasing group delay) characteristic around the central frequency of 100 kHz. Zero valued, positive and negative slopes of the group delay may be produced as will be demonstrated later on in the book.

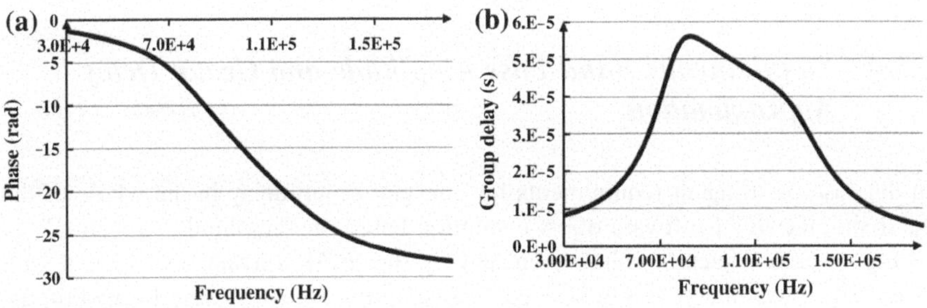

Fig. 5.24 A 10th order band-pass all-pass approximant of linearly decreasing group delay. **a** Phase and **b** group delay characteristics

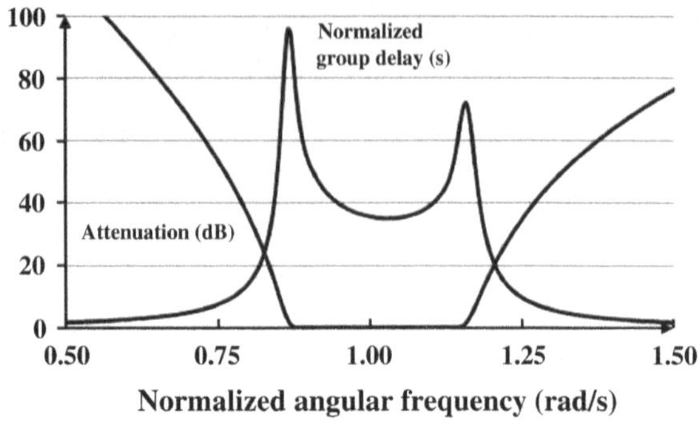

Fig. 5.25 14th order band-pass filter obtained by transformation from a 7th order LSM prototype. The requested relative bandwidth was 30%

5.5.3 *Band-Pass Filters*

The design of this kind of filters based on the existing low-pass prototype is performed by a so called low-pass-to-band-pass transformation. By its nature, this transform produces a geometrically symmetrical (around the central frequency) amplitude characteristic. There are two situations to be distinguished when designing band-pass filter s: (a) narrow-band filters and (b) broad-band filters. Due to the influence of the bottom (negative frequency) half of the characteristic, the resulting function in the case of broad-band solution is more (arithmetically) asymmetric then the narrow-band case.

Figure 5.25 illustrates the performances of a 14th order band-pass filter obtained by transformation of a low-pass 7th order LSM prototype. The requested relative bandwidth was 30%. As a result of the transformation the passband of the filter is slightly distorted. Namely the lower cut-off frequency of the passband is 0.861 rad/s (instead of 0.85 rad/s) while the upper is 1.161 rad/s (instead of 1.15 rad/s). Of course, the band-width is preserved to be 0.299941 rad/s. In addition, one may observe that the group delay is also affected by the transformation and the asymmetry is strong.

5.5.4 *Simultaneous Band-Pass Amplitude and Group Delay Approximation*

In the case of stringent requirements for constant group delay in the whole of the passband, the $\mathcal{RM}$ software offers a solution based on "amplitude correction".

Figure 5.26 depicts a solution created by the $\mathcal{RM}$ software in which a 14th order band-pass filter was synthesized. Transmission zeros (8) were located on the ω-axis with two additional in the origin. The stopband attenuation was 40 dB and the requested passband width was 20% of the central frequency.

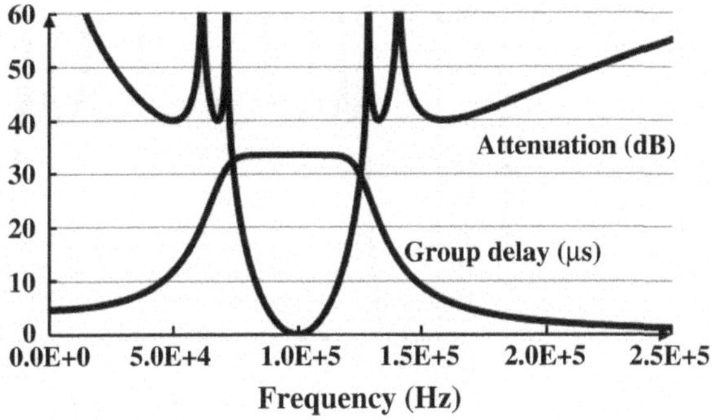

Fig. 5.26 Results of the simultaneous band-pass group delay and amplitude approximation

5.5.5 High-Pass Filters Obtained by LP-HP Transformation

High-pass filters are usually obtained from a proper low-pass prototype using the low-pass-to-high-pass transformation. Figure 5.27 depicts a 15th order modified elliptic high-pass filter obtained in that way.

In this example the maximum passband attenuation was chosen to be 0.1 dB while the minimum stopband attenuation is 80 dB.

5.5.6 Band-Stop Filters Obtained by LP-BS Transformation

Band-stop filters are also obtained from a proper low-pass prototype using the low-pass-to-band-stop transformation. Figure 5.28 depicts the attenuation characteristic of a band-stop filter obtained in that way. It is a 22nd order filter produced from an 11th order prototype of amplitude corrected Papoulis (L) filter.

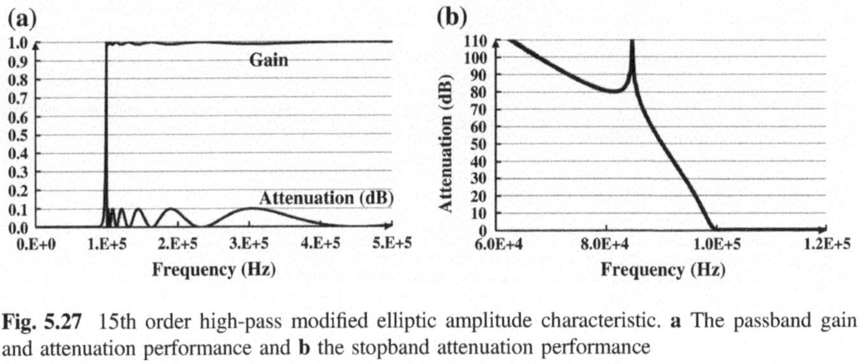

Fig. 5.27 15th order high-pass modified elliptic amplitude characteristic. **a** The passband gain and attenuation performance and **b** the stopband attenuation performance

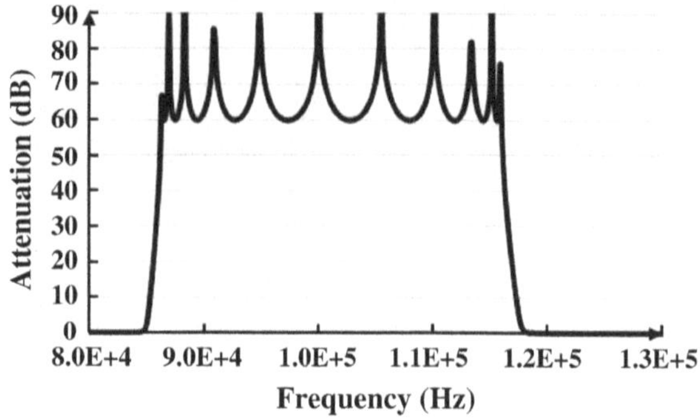

Fig. 5.28 A 22nd order band-stop filter obtained by transformation of L_Z prototype

5.6 Developer's Corner

Use of the support offered by [1] may be very helpful in understanding what the capabilities of the $\mathcal{RM}$ software are and, more importantly, how it may be used successfully. There are 17 video presentations covering this chapter completely with much larger number of examples. In addition, there are 17 additional animations which may serve as a user's guide. Namely, design procedures are exemplified in full detail allowing the designer to follow the process in parallel while running his/hers design.

Since we learned that not all countries have access to YouTube, copies of all 34 videos were uploaded at [2].

References

1. https://www.youtube.com/channel/UCF_Ipw_YD2gwrRpJDUJJULw/playlists?disable_polymer = 1
2. http://skysupervisor.com/projects/RM%20software/RM%20software.htm

Part II
Transfer Function Synthesis

Chapter 6
Low-Pass Selective Filters with Critical Monotonic Amplitude Characteristic (CMAC) in the Passband

Low-pass filters with critically monotone amplitude characteristic (CMAC) in the passband are subject of this chapter. Their unique property is that the first derivative of the amplitude characteristic touches zero (without changing its sign) in maximum number of points at the ω-axis. Unified theory of synthesizing their transfer functions will be given so creating the Butterworth, Papoulis (Legendre), Halpern, and LSM filters. Tables will be given (for the first time) containing data on the pole positions of these filters up to the 10th order. Properties of the attenuation characteristics of all CMAC functions will be studied and comparisons will be given based on several criteria including the passband and stopband behaviour. Comparisons will be given related to the group delay characteristics and time domain performance of all CMAC filters. Design example will be given. An alternative method for creating the CMAC characteristics will be cited.

6.1 Introduction

Critical monotonic amplitude characteristic (CMAC) has the property that the amplitude characteristic in the pass-band has monotonic character with maximal number of inflection points with different abscissas. In that way, the first derivative of amplitude characteristic is equal to zero for maximum number of times, without changing its sign, meaning that its value is limited, and the sensitivity of the amplitude characteristic to changes of circuit parameters is reduced.

Since, besides their excellent properties, the CMAC filters are mostly ignored by the designers (except for the Butterworth filters) in this document more detailed information will be given about their properties in the frequency and time domain. Namely, it is our general intention to encourage the reader to write his/hers own

V. Litovski, *Electronic Filters*, Lecture Notes in Electrical Engineering 596,
https://doi.org/10.1007/978-981-32-9852-1_6

software for filter design for which we think we are making available full information. However, to reinforce the knowledge related to CMAC filters we will give here much more information than theory and algorithm. In fact small catalog will be given for every sub-class of CMAC.

6.2 Characteristic Function of the CMAC Filters

In this section a short overview of the theory leading to the class of CMAC filters will be given. It is based on the newest developments reported in [1], where the subject was considered in full detail.

The normalized squared modulus of the amplitude characteristic of a low-pass filter may be written as

$$|H(j\omega)|^2 = \frac{1}{1+\varepsilon^2 K(\omega^2)} \tag{6.1}$$

where ε^2 is a constant used to control its value at the edge of the passband (cut-off frequency ω_0) which for now will be considered equal to unity (i.e. $\omega_0 = 1$ rad/s). $K(\omega^2)$ is the characteristic function which is here restricted so as $K(1) = 1$. In the case of selective polynomial filters it is frequently represented as a squared polynomial of ω^2.

Halpern [2] proposed to write the characteristic function in the following form:

$$K(\omega^2) = L_n(\omega^2) = \int_0^{\omega} E_{n-1}(x^2) \cdot dx. \tag{6.2}$$

Since here we are looking for a critical monotonic amplitude characteristic, $E_{n-1}(x^2)$ is to be chosen so that it enables critical monotonicity. The first derivative of a critical-monotonic function never changes its sign while having maximal number of zeros. The first condition for that is $E_{n-1}(x^2)$ to be a full square, i.e. to be expressed as a square of another polynomial: $E_{n-1}(x^2) = V_{n-1}^2(x)$, where $V_{n-1}(x)$ is to be odd or even polynomial. The second condition was that all the zeros of $V_{n-1}(x)$ have to be real and to be located in the interval $\{0, 1\}$. To that end, $V_{n-1}(x)$ was expressed as a sum of orthogonal polynomials with the interval of orthogonality defined by the normalized pass-band of the filter, i.e. $\omega \in \{0, 1\}$.

The following expression for $V_{n-1}(x)$ was proposed in [2]:

$$V_{n-1} = \sum C_i \cdot U_i(x) \tag{6.3}$$

where C_i are properly chosen constants, and $U_i(x)$ are Jacobi polynomials satisfying the following relation

$$\int_0^1 x \cdot U_j \cdot U_k = \begin{cases} 0 & \text{for}\, j \neq k \\ 1 & \text{for}\, j = k \end{cases} \tag{6.4}$$

while j and k are both even or both odd natural numbers. In that way, we get

$$L_n\left(\omega^2\right) = \int_0^{\omega} x \left\{ \sum_{i=0}^{(n-2)/2} C_{2i+1} \cdot U_{2i+1}(x) \right\}^2 \cdot dx \quad \text{for}\, n\text{—even} \tag{6.5}$$

$$L_n\left(\omega^2\right) = \int_0^{\omega} x \left\{ \sum_{i=0}^{(n-1)/2} C_{2i} \cdot U_{2i}(x) \right\}^2 \cdot dx \quad \text{for}\, n\text{—odd,} \tag{6.6}$$

where n is the order of the filter.

The first 10 Jacobi polynomials are given in Table 6.1. In general one may use the following expressions

$$U_{2i+1}(x) = 2\sqrt{i+1} \sum_{m=0}^{i} (-1)^m \frac{(2i+1-m)!}{m! \cdot (i+1-m)! \cdot (i-m)!} x^{2(i-m)+1} \tag{6.7}$$

$$U_{2i}(x) = \sqrt{4i+2} \sum_{m=0}^{i} (-1)^{i-m} \frac{(i+m)!}{(m!)^2 \cdot (i-m)!} x^{2m} \tag{6.8}$$

Table 6.1 The firsts 10 Jacobi polynomials

k	$U_k(x)$
0	$\sqrt{2}$
1	$\sqrt{4} \cdot x$
2	$\sqrt{6} \cdot (2x^2 - 1)$
3	$\sqrt{8} \cdot (3x^3 - 2x)$
4	$\sqrt{10} \cdot (6x^4 - 6x^2 + 1)$
5	$\sqrt{12} \cdot (10x^5 - 12x^3 + 3x)$
6	$\sqrt{14} \cdot (20x^6 - 30x^4 + 12x^2 - 1)$
7	$\sqrt{16} \cdot (35x^7 - 60x^5 + 30x^3 - 4x)$
8	$\sqrt{18} \cdot (70x^8 - 140x^6 + 90x^4 - 20x^2 + 1)$
9	$\sqrt{20} \cdot (126x^9 - 280x^7 + 210x^5 - 60x^3 + 5x)$

Finally, the constants C_i (elements of the **C** vector) have to be chosen so that the normalization criterion $L_n(1) = 1$ is satisfied, i.e.:

$$\sum_k C_k^2 = 1. \tag{6.9}$$

One should mention that in (6.3) in the place of U_i, one may use any other class of orthogonal polynomials having simple zeros in the interval $\{0, 1\}$.

To get the values of **C** one is to implement some design criterion. Here we mention the four main criteria implemented in the past and the corresponding classes of filters that arise. As stated in [1] they may be generated by a unified procedure. A common name was given to all of them as Critical Monotonic Amplitude Characteristic (CMAC) filters. The criteria and the resulting nomenclature are listed below.

1. Maximally flat in the origin. This means all derivatives of $L_n(\omega^2)$ at the origin are to be zero. The class of filters thus obtained is called Butterworth's after the author [3]. These will be here referred to as *B-filters*.
2. Maximum slope of the characteristic function at the edge of the pass-band. The class of filters so obtained is called *L-filters* and was introduced by Papoulis [4, 5]. The name L comes from the fact that in the original derivation Legendre polynomials were used. In some references [6] it is stated as "optimal filters" which is arbitrary.
3. Maximum asymptotic attenuation. This means the higher order coefficient in $L_n(\omega^2)$ has to be maximal. This class of filters was introduced by Halpern [2]. These will be here referred to as *H-filters*.
4. Least-squares-monotonic. In this case the returned power in the pass-band was minimized under the critical monotonicity criterion. This class was introduced by Raković and Litovski [7] and named *LSM-filters*.

6.3 Unified Theory of Critical Monotonic Passband Filters

In this paragraph procedures for evaluation the **C** vector will be given in some details.

6.3.1 Butterworth Filters

Butterworth [3] introduced the simplest form of the characteristic function of polynomial filters:

$$L_n\left(\omega^2\right) = \omega^{2n}. \tag{6.10}$$

The squared modulus of the transfer function of the B-filters has (2*n*–1) derivatives at the origin equal to zero so it is referred to as the maximally flat solution of the approximation problem of ideal low-pass amplitude characteristic.

In the next we will demonstrate how (6.10) may be expressed in the form of the Halpern's polynomial given by (6.2). In fact, we need to find the **C** vector in (6.5) and (6.6) so that the following is satisfied.

$$\int_0^\omega x \left\{ \sum_{i=0}^{(n-2)/2} C_{2i+1} \cdot U_{2i+1}(x) \right\}^2 = \omega^{2n} \quad \text{for } n\text{—even} \tag{6.11}$$

and

$$\int_0^\omega x \left\{ \sum_{i=0}^{(n-1)/2} C_{2i} \cdot U_{2i}(x) \right\}^2 = \omega^{2n} \quad \text{for } n\text{—odd.} \tag{6.12}$$

After differentiating these two expressions with respect to ω one gets:

$$\omega \cdot \left\{ \sum_{i=0}^{(n-2)/2} C_{2i+1} \cdot U_{2i+1}(\omega) \right\}^2 = 2n \cdot \omega^{2n-1} \tag{6.13}$$

or

$$\left\{ \sum_{i=0}^{(n-2)/2} C_{2i+1} \cdot U_{2i+1}(\omega) \right\} = (2n)^{1/2} \cdot \omega^{n-1}. \tag{6.14}$$

Now we rewrite (6.7), the $U_{2i+1}(x)$ polynomial, as:

$$U_{2i+1}(\omega) = \sum_{m=0}^{i} a_{2i+1,m} \cdot \omega^{2m+1} \tag{6.15}$$

where the $a_{2i+1,m}$ is the coefficient of the polynomial $U_{2i+1}(\omega)$ by ω^{2m+1}. After substituting and rearranging the left hand side of (6.14) and comparing the coefficients by the same exponents of ω from the left and the right hand side, we get the following system of linear equations for the coefficients C_{2i+1}:

$$\sum_{i=m}^{\frac{n-2}{2}} a_{2i+1,m} C_{2i+1} = 0 \qquad \text{for} \quad m = 0, 1, \ldots, \tfrac{n-4}{2} \tag{6.16a}$$
$$a_{n-1,(n-2)/2} C_{n-1} = \sqrt{2n}$$

for n-even, and by a similar procedure,

$$\sum_{i=m}^{\frac{n-1}{2}} a_{2i,m} C_{2i} = 0 \qquad \text{for} \quad m = 0, 1, \ldots, \tfrac{n-3}{2} \tag{6.16b}$$
$$a_{n-1,(n-1)/2} C_{n-1} = \sqrt{2n}$$

for n-odd.

By solving (6.16a), (6.16b) and (6.17) we get the **C** constants allowing expressing the Butterworth characteristic function in the form proposed by Halpern. Table 6.2 contains the solutions for polynomials up to 10th order.

After substitution of the **C** vector in (6.5) and (6.6) the characteristic function may be obtained. Part of the solutions of (the Feldkeller equation)

$$1 + L_n(\omega^2)_{|\omega^2 = -s^2} = 0 \tag{6.17}$$

belonging to the left half of the complex frequency plane, are the poles of the transfer function of the B-filters. Note, in (6.17) $\varepsilon^2 = 1$ was used which means we will get attenuation of 3 dB at the cut-off. The poles so obtained are listed in Table 6.3 for polynomials up to the 10th order. Of course, to produce these solutions one has to have a potent (high order) polynomial solver.

To get the coefficients of the denominator of the transfer function based on the data of Table 6.3 one has to develop a proper routine. Results of such calculations are given in Table 6.4.

Table 6.2 The **C** vector for the case of a maximally flat amplitude characteristic

n	$C_0(C_1)$	$C_2(C_3)$	$C_4(C_5)$	$C_6(C_7)$	$C_8(C_9)$
3	0.866025	0.5			
4	0.942809	0.333333			
5	0.745355	0.645497	0.166666		
6	0.866025	0.489897	0.1		
7	0.661437	0.687386	0.295803	0.05	
8	0.8	0.565685	0.197948	0.028571	
9	0.6	0.692820	0.383325	0.113389	0.014285
10	0.745355	0.602338	0.276641	0.070986	0.007936

Table 6.3 Poles of the transfer function of the Butterworth filters (Top are the real and bottom the imaginary parts)

$n = 3$	–0.5	–1			
	±0.866025	0			
$n = 4$	–0.382683	–0.92388			
	±0.92388	±0.382683			
$n = 5$	–0.309017	–0.809017	-1		
	±0.951057	±0.587785	0		
$n = 6$	–0.258819	–0.707107	–0.965926		
	±0.965926	±0.707107	±0.258819		
$n = 7$	–0.222521	–0.62349	–0.900969	–1	
	±0.974928	±0.781831	±0.433884	0	
$n = 8$	–0.19509	–0.55557	–0.83147	–0.980785	
	±0.980785	±0.83147	±0.55557	±0.19509	
$n = 9$	–0.173648	–0.5	–0.766044	–0.939693	–1
	±0.984808	±0.866025	±0.642788	±0.34202	0
$n = 10$	–0.156434	±–0.45399	–0.707107	–0.891007	–0.987688
	±0.987688	±0.891007	±0.707107	±0.45399	±0.156434

Table 6.4 Coefficient of the denominator of the transfer function of the Butterwort filters

	s^5	s^4	s^3	s^2	s^1	s^0
$n = 2$				1.	1.414214	1
$n = 3$			1.	2.000000	2.000000	1
$n = 4$		1.	2.613126	3.414214	2.613126	1
$n = 5$	1.	3.236068	5.236068	5.236068	3.236068	1
$n = 6$	3.863703	7.464102	9.141620	7.464102	3.863703	1
$n = 7$	10.09783	14.59179	14.59179	10.09783	4.493959	1
$n = 8$	21.84615	25.68835	21.84615	13.13707	5.125831	1
$n = 9$	41.98638	41.98638	31.16343	16.58171	5.758770	1
$n = 10$	74.23342	64.88239	42.80206	20.43172	6.392453	1
	s^{10}	s^9	s^8	s^7	s^6	
$n = 6$					1.	
$n = 7$				1.	4.493959	
$n = 8$			1.	5.125831	13.13707	
$n = 9$		1.	5.758770	16.58171	31.16343	
$n = 10$	1	6.392453	20.43172	42.80206	64.88239	

We will mention here that the so obtained transfer function has only finite poles. All zeros are at infinity. In other words the transfer function converges to $(1/a_n) \cdot s^{-n}$ as $s \to \infty$, where a_n is the coefficient by the highest degree of s. The same type of transfer function is frequently referred to as an all-pole function.

6.3.2 Legendre Filters

Papoulis [4, 5] introduced a new class of characteristic functions by maximizing the slope of the gain at the cut-off so indirectly trying to get maximum selectivity under the condition of critical monotonicity. The class of filters so obtained was named L-filters since Papoulis used Legendre polynomials. Note that Papoulis developed this class of filters in two separate papers [4, 5] devoted to odd and even order transfer functions. The same results for even order were published by [6].

He used a slightly different form for $L_n(\omega^2)$:

$$L_n\left(\omega^2\right) = \int_{-1}^{2\cdot\omega-1} v^2(x)\cdot dx \tag{6.18a}$$

or

$$L_n\left(\omega^2\right) = \int_{-1}^{2\cdot\omega^2-1} B(x+1)\cdot\left(\frac{dP_{k+1}(x)}{dx}\right)^2 dx, \tag{6.18b}$$

where the polynomial $v(x)$ was represented in a form of sum of Legendre polynomials of the first kind:

$$v(x) = a_0 + a_1 P_1(x) + \cdots + a_k P_k(x). \tag{6.19}$$

The constants a_k are chosen so that:

$$a_0 = a_1/3 = a_2/5 = \cdots = \frac{a_k}{2k+1} = \frac{1}{\sqrt{2}(k+1)}, \tag{6.20}$$

while $B(x+1)$ is created to satisfy $L_n(1) = 1$. The index k is found as $k = (n-1)/2$ for n-odd and $k = n/2 - 1$ for n-even.

Halpern [2] demonstrated, however, that by proper choice of the **C** constants in (6.5) and (6.6) one can produce the characteristic function of the L-filters. According to Halpern it is necessary to have:

$$C_i = U_i(1)\sqrt{\sum\nolimits_k U_k^2(1)}. \tag{6.21}$$

The values of the C constants obtained in that way are given in Table 6.5 for polynomials up to 10th order. It is followed by Table 6.6 containing the poles and Table 6.7 containing the coefficients of the transfer function of the Legendre filters.

In the next we will perform the developments that lead Halpern to the result expressed in (6.21). Mathematically we have the problem of maximizing the derivative at the cut-off under the constraint (6.9).

Table 6.5 The **C** vector for the L-filters

n	$C_0(C_1)$	$C_2(C_3)$	$C_4(C_5)$	$C_6(C_7)$	$C_8(C_9)$
3	0.5	0.866025			
4	0.577350	0.816496			
5	0.333333	0.577350	0.745355		
6	0.408248	0.577350	0.707106		
7	0.25	0.433012	0.559016	0.661437	
8	0.316227	0.447213	0.547722	0.632455	
9	0.2	0.346410	0.447213	0.529150	0.6
10	0.258198	0.365148	0.447213	0.516397	0.577350

Table 6.6 Poles of the transfer function of the Legendre filters (Top are the real and bottom the imaginary parts)

$n = 3$	–0.345186	–0.620332			
	±0.900866	0.0			
$n = 4$	–0.231689	–0.549744			
	±0.945511	±0.358572			
$n = 5$	–0.153587	–0.388140	–0.468090		
	±0.968146	±0.588632	0.0		
$n = 6$	–0.115193	–0.308961	–0.438902		
	±0.977922	±0.698167	±0.239981		
$n = 7$	–0.086209	–0.237440	–0.349232	–0.382103	
	±0.984370	±0.778301	±0.428996	0.0	
$n = 8$	–0.068942	–0.194276	–0.300284	–0.367176	
	±0.987971	±0.824767	±0.541042	±0.180879	
$n = 9$	–0.055097	–0.157284	–0.248553	–0.309383	–0.325687
	±0.990660	±0.861243	±0.633820	±0.336543	0.0
$n = 10$	–0.045901	–0.132519	–0.214173	–0.277405	–0.317206
	±0.992383	±0.885262	±0.694540	±0. 139646	±0.145430

After finding the derivative of the characteristic function for ω = 1, the following expression arises

$$\frac{dL_n(\omega^2)}{d\omega}_{|\omega=1} = f(C_1, C_3, \ldots, C_{n-1}) = f(\mathbf{C}) = \left\{ \sum_{i=0}^{(n-2)/2} C_{2i+1} U_{2i+1}(1) \right\}^2 \quad (6.22)$$

C will be found by maximizing the value of (6.22) under the constraint (6.9) using the Lagrange multiplier. The maximum of the following expression is to be found

Table 6.7 Coefficient of the denominator of the transfer function of the L-filters

	s^5	s^4	s^3	s^2	s^1	s^0
$n = 3$			1	1.3107	1.3590	0.577348
$n = 4$		1	1.5629	1.8879	1.2416	0.408249
$n = 5$	1	1.5515	2.2036	1.6927	0.898.340	0.223606
$n = 6$	1.7261	2.6897	2.4334	1.6331	0.679635	0.141421
$n = 7$	2.9928	2.9246	2.3322	1.2308	0.437942	0.075593
$n = 8$	3.7232	3.3477	2.1189	0.993930	0.299670	0.0451753
$n = 9$	4.2477	3.0119	1.7074	0.680434	0.1815714	0.0238093
$n = 10$	3.0847	2.1108	1.0830	0.422042	0.1077180	0.0137465
	s^{10}	s^9	s^8	s^7	s^6	
$n = 6$					1	
$n = 7$				1	1.7279	
$n = 8$			1	1.8614	3.4466	
$n = 9$		1	1.8663	3.7416	4.2490	
$n = 10$	1	1.4443	3.2653	3.5424	3.9521	

$$F(\mathbf{C},\lambda)=f(\mathbf{C})+\lambda\left\{\sum_{i=0}^{(n-2)/2} C_{2i+1}^2-1\right\}. \tag{6.23}$$

A system of nonlinear equation is obtained after differentiating and equating the derivatives with zero:

$$\frac{\partial F}{\partial C_1}=0;\quad \frac{\partial F}{\partial C_3}=0,\ldots,\frac{\partial F}{\partial C_{n-1}}=0,\quad \text{and}\quad \frac{\partial F}{\partial C_\lambda}=0. \tag{6.24}$$

After elimination of λ, one gets

$$C_{2i-1}U_{2i+1}(1)-C_{2i+1}U_{2i-1}(1)=0 \quad \text{for} \quad i=0,1,\ldots,(n-2)/2 \tag{6.25a}$$

and

$$\sum_{i=0}^{(n-2)/2} C_{2i+1}^2=1. \tag{6.25b}$$

This system of nonlinear equations is solved easily by recursion i.e. by successive substitution of (6.25a) into (6.25b). For n odd we get

$$C_{2i+1}=U_{2i+1}(1)\Bigg/\sqrt{\sum_{i=0}^{(n-2)/2} U_{2i+1}^2(1)}. \tag{6.26}$$

By analogy, for n even we get:

$$C_{2i} = U_{2i}(1) \Big/ \sqrt{\sum_{i=0}^{(n-1)/2} U_{2i}^2(1)}. \tag{6.27}$$

6.3.3 Halpern Filters

The criterion implemented in the derivation of this class of CMAC filters is maximum asymptotic attenuation. By inspection of (6.5) and (6.6) one easily comes to the conclusion that maximum asymptotic attenuation is obtained if we choose:

$$\begin{aligned} C_{2i} = 0 \qquad &\text{for} \quad i = 0, 1, \ldots, (n-4)/2 \\ &C_{n-1} = 1 \end{aligned} \tag{6.28}$$

for n even, and

$$\begin{aligned} C_{2i+1} = 0 \qquad &\text{for} \quad i = 0, 1, \ldots, (n-3)/2 \\ &C_{n-1} = 1 \end{aligned} \tag{6.29}$$

for n odd.

The values of the C constants obtained in that way are given in Table 6.8 for polynomials up to 10th order. It is followed by Table 6.9 containing the poles and Table 6.10 containing the coefficients of the transfer function of the Halpern filters.

6.3.4 LSM Filters

To produce the LSM characteristic function we perform search for the values of the components of **C** which lead to minimum area under the characteristic function in

Table 6.8 The **C** vector for the H-filters

n	$C_0(C_1)$	$C_2(C_3)$	$C_4(C_5)$	$C_6(C_7)$	$C_8(C_9)$
3	0	1			
4	0	1			
5	0	0	1		
6	0	0	1		
7	0	0	0	1	
8	0	0	0	1	
9	0	0	0	0	1
10	0	0	0	0	1

Table 6.9 Poles of the transfer function of the Halpern filters (Top are the real and bottom the imaginary parts)

$n = 3$	–0.318376	–0.470239			
	±0.980778	0.0			
$n = 4$	–0.206952	–0.447263			
	±1.008669	±0.338156			
$n = 5$	–0.134294	–0.314217	–0.328275		
	±1.019378	±0.617671	0.0		
$n = 6$	–0.099637	–0.249876	–0.330841		
	±1.019675	±0.728701	±0.226010		
$n = 7$	–0.074166	–0.192915	–0.262618	–0.2571461	
	±1.018807	±0.811295	±0.445714	0.0	
$n = 8$	–0.059154	–0.158060	–0.226951	–0.265206	
	±1.017105	±0.855494	±0.560318	±0.170818	
$n = 9$	–0.047289	–0.128618	–0.189611	–0.222291	–0.213964
	±1.015461	±0.890037	±0.657738	±0.347976	0.0
$n = 10$	–0.039430	–0.108714	–0.164096	–0.200631	–0.222735
	±1.013914	±0.911414	±0.718454	±0.453220	±0.137708

Table 6.10 Coefficient of the denominator of the transfer function of the H-filters

	s^5	s^4	s^3	s^2	s^1	s^0
$n = 3$			1	1.1070	1.3627	0.5
$n = 4$		1	1.3084	1.7449	1.0785	0.333333
$n = 5$	1	1.2253	2.0007	1.3535	0.768140	0.166666
$n = 6$	1.3607	2.3658	1.9082	1.3280	0.515370	0.100000
$n = 7$	2.6171	2.1957	1.8819	0.89546	0.327179	0.05
$n = 8$	2.7668	2.6300	1.5218	0.720899	0.202964	0.0285715
$n = 9$	3.3225	2.1019	1.2390	0.451799	0.1214301	0.0142857
$n = 10$	2.9624	1.9859	0.893745	0.3222619	0.0712828	0.0078896
	s^{10}	s^9	s^8	s^7	s^6	
$n = 6$					1	
$n = 7$				1	1.3169	
$n = 8$			1	1.4187	2.9656	
$n = 9$		1	1.3896	3.2155	3.0712	
$n = 10$	1	1.4512	3.5253	3.6016	4.1933	

the normalized passband. The area is obtained by integration the characteristic function in the passband. Since we are manipulating squares (of voltages) this quantity may be seen as power. Its minimum would mean minimum reflected power in the passband hence the importance of the class of filters.

The area is numerically obtained by calculating the following integral

$$P = \int_0^1 L_n(\omega^2)\mathrm{d}\omega. \tag{6.30a}$$

After substitution of (6.6), for n odd, it becomes

$$P = \int_0^1 \left\{ \int_0^\omega x \cdot \left[\sum_{i=0}^{(n-1)/2} C_{2i} U_{2i}(x) \right]^2 dx \right\} d\omega. \tag{6.30b}$$

By changing the order of integration we may rearrange to get

$$P = \int_0^1 dx \int_x^1 x \cdot \left[\sum_{i=0}^{(n-1)/2} C_{2i} U_{2i}(x) \right]^2 d\omega = \int_0^1 x \cdot (1 - x) \cdot \left[\sum_{i=0}^{(n-1)/2} C_{2i} U_{2i}(x) \right]^2 dx. \tag{6.30c}$$

The value of this integral should be minimal under the condition 6.9). Again, using Lagrange multiplier, for n odd we search for **C** which minimizes the function:

$$f(C_0, C_2, \ldots, C_{n-1}, \lambda) = \int_0^1 \left\{ \int_0^\omega x \left[\sum_{i=0}^{(n-1)/2} C_{2i} U_{2i}(x) \right]^2 dx \right\} \mathrm{d}\omega + \lambda \left(\sum_{i=0}^{(n-1)/2} C_{2i}^2 - 1 \right), \tag{6.31a}$$

and, similarly, for n even

$$f(C_1, C_2, \ldots, C_{n-1}, \lambda) = \int_0^1 x \cdot (1 - x) \cdot \left[\sum_{i=0}^{(n-2)/2} C_{2i+1} U_{2i+1}(x) \right]^2 dx + \lambda \cdot \left(\sum_{i=0}^{(n-2)/2} C_{2i+1}^2 - 1 \right) \tag{6.31b}$$

To get a minimum it is necessary that (for n odd only)

$$\frac{\partial f}{\partial C_{2j+1}} = 0 \quad \text{for}\, j = 0, 1, \ldots, (n/2 - 1) \tag{6.32a}$$

and

$$\frac{\partial f}{\partial \lambda} = 0. \tag{6.32b}$$

After differentiation and rearranging (what includes elimination of λ), the following systems of nonlinear equations arises:

$$\begin{gathered}C_{2j+1} \cdot \int_0^1 x^2 \left[\sum_{i=0}^{\frac{n-2}{2}} C_{2i+1} U_{2i+1}(x)\right] \cdot U_{2j-1}(x) \cdot dx \\ - C_{2j-1} \cdot \int_0^1 x^2 \left[\sum_{i=0}^{(n-2)/2} C_{2i+1} U_{2i+1}(x)\right] \cdot U_{2j+1}(x) \cdot dx = 0 \\ \sum_{i=0}^{(n-2)/2} C_{2i+1}^2 = 1 \quad \text{for}\, j = 1, 2, \ldots, (n-2)/2\end{gathered} \tag{6.33a}$$

for n even, and

$$\begin{gathered}C_{2j} \cdot \int_0^1 x^2 \left[\sum_{i=0}^{\frac{n-1}{2}} C_{2i} U_{2i}(x)\right] \cdot U_{2j-2}(x) \cdot dx \\ - C_{2j-2} \cdot \int_0^1 x^2 \left[\sum_{i=0}^{(n-1)/2} C_{2i} U_{2i}(x)\right] \cdot U_{2j}(x) \cdot dx = 0 \\ \sum_{i=0}^{(n-1)/2} C_{2i}^2 = 1 \quad \text{for}\, j = 1, 2, \ldots, (n-1)/2\end{gathered} \tag{6.33b}$$

for n odd.

These systems may look complicated but in fact they are second order polynomials in the elements of **C**. To solve these systems one needs to implement some of the methods for solving systems of nonlinear equations. Within the $\mathcal{RM}$ software the Newton-Raphson [8] iteration was implemented. Note, the **C** vector obtained in closed form for the L-filters may an excellent initial solution for this iterative process.

The values of the C constants obtained in that way are given in Table 6.11 for polynomials up to 10th order. It is followed by Table 6.12 containing the poles and Table 6.13 containing the coefficients of the transfer function of the LSM filters.

Table 6.11 *C*-constants for the LSM filters

n	$C_0(C_1)$	$C_2(C_3)$	$C_4(C_5)$	$C_6(C_7)$	$C_8(C_9)$
3	0.735595	0.677422			
4	0.816497	0.577350			
5	0.539066	0.716797	0.442277		
6	0.645810	0.661605	0.381962		
7	0.425843	0.629268	0.570416	0.311037	
8	0.529322	0.615107	0.516024	0.274191	
9	0.349642	0.547621	0.565019	0.451636	0.233753
10	0.446303	0.553681	0.531873	0.409607	0.208768

Table 6.12 Poles of the transfer function of the LSM filters (Top are the real and bottom the imaginary parts)

$n = 3$	–0.407647	–0.795875			
	±0.872900	0.0			
$n = 4$	–0.283855	–0.688579			
	±0.926542	±0.375074			
$n = 5$	–0.199170	–0.521880	–0.648323		
	±0.953073	±0.583750	0.0		
$n = 6$	–0.152964	–0.422379	–0.592000		
	±0.966473	±0.696754	±0.249779		
$n = 7$	–0.117928	–0.334199	–0.493642	–0.551073	
	±0.975225	±0.773458	±0.425329	0.0	
$n = 8$	–0.095779	–0.277145	–0.427885	–0.515928	
	±0.980565	±0.820992	±0.540215	±0.186733	
$n = 9$	–0.077963	–0.228357	–0.360859	–0.450904	–0.482181
	±0.984533	±0.856947	±0.629234	±0.332718	0.0
$n = 10$	–0.065683	–0.194412	–0.313740	–0.405969	–0.458177
	±0.987220	±0. 881412	±0.690749	±0.438095	±0.149110

6.4 Properties of the CMAC Filters

In this paragraph (mostly by diagrams and tables) the properties of the CMAC filters will be described.

Frequency domain responses are depicted in the following set of figures. Transfer functions of 8th order filters were synthesized.

Figure 6.1 depicts the passband characteristics of the 8th order CMAC filters. Being created under the criterion of maximum asymptotic selectivity, the Halpern filters may be considered as the solution introducing the largest reflected power of all. The reader is not to forget that by reducing the value of ε^2 in (6.1) one may reduce the passband distortions to any desired value. Of course, the poles and the

Table 6.13 Coefficient of the denominator of the transfer function of the LSM filters

	s^5	s^4	s^3	s^2	s^1	s^0
$n = 3$			1	1.6112	1.5770	0.738676
$n = 4$		1	1.9449	2.3357	1.6423	0.577351
$n = 5$	1	2.0904	2.9119	2.5154	1.3811	0.376838
$n = 6$	2.3347	3.6550	3.7126	2.6098	1.1704	0.262424
$n = 7$	4.1923	4.7414	3.9215	2.2696	0.853673	0.160289
$n = 8$	6.1169	5.7487	3.9801	1.9802	0.642414	0.104203
$n = 9$	7.5049	5.8983	3.5225	1.5214	0.431353	0.0611156
$n = 10$	8.6259	5.9248	3.1115	1.1937	0.301267	0.0380161
	s^{10}	s^9	s^8	s^7	s^6	
$n = 6$					1	
$n = 7$				1	2.4426	
$n = 8$			1	2.6335	4.8946	
$n = 9$		1	2.7183	5.3970	7.2412	
$n = 10$	1	2.8760	6.0631	8.7367	9.8614	

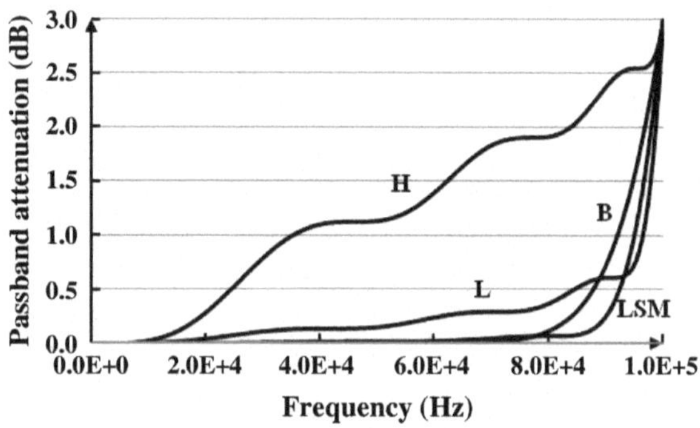

Fig. 6.1 Passband attenuation of 8th order CMAC filters

corresponding coefficients (given in the tables above) will be changed now. In addition, two other consequences will be experienced.

First, the value of the attenuation at the cut-off would be reduced too. To compensate for that (or to restore the 3 dB value at the cut-off) one may implement re-normalization in the following way.

To find the new poles of the new filter (having reduced attenuation in the passband while still exhibiting 3 dB at cut-off) one has first to solve the following equation

$$1 + \varepsilon^2 K_n(\omega^2)_{|\omega=s/j} = 0 \tag{6.34}$$

and to separate the solutions which lay in the left half of the complex frequency plane. In that way new transfer function is created which exhibits the following attenuation value at the cut-off

$$a(1) = 10 \cdot \log(1 + \varepsilon^2). \tag{6.35}$$

To re-establish the 3 dB value at the cut-off one needs first to find the frequency at which the new function reaches 3 dB. In other words one needs to solve for ω_n the following equation

$$3 = 10 \cdot \log[1 + \varepsilon^2 K(\omega_n^2)] \tag{6.36}$$

That is usually done by search of the right hand side with small granularity of ω. After that one should divide (re-normalize) all (new) poles by ω_n.

Second, the new function, whether it is original or re-normalized, will exhibit lower attenuation in the stop-band i.e. lower selectivity. The smaller the ε^2 the lower the selectivity will be.

In the next we will give a comparative study of the CMAC filters based on the criteria mentioned above. Note since only $L_n(\omega^2)$ is discussed, the table is valid for every ε^2 being simultaneously applied to all four classes without re-normalization. The first criterion for comparison will be the reflected power as depicted in Table 6.14.

We will start again with the Halpern solution. Not only it exhibits largest reflected power it also practically keeps that value (does not improve) with the rise of the order of the filter. The rest of the filters of the CMAC class reduce the reflected power with the rise of the order of the filter. Note, the 10th order LSM reflects 2.1% of the signal power only.

It is of prime interest to establish a mutual relation between the LSM and the Butterworth filters. That is because the latter are very frequently used while the first one are almost unknown to the filter design community. By inspection of Table 6.14 we easily come to the conclusion that the Butterworth filter reflects larger power in the passband which, as the order of the filter increases, becomes more than twice as large.

The Legendre filters are seen as a transitional solution between the LSM and the Halpern filters.

The stopband behaviour will be illustrated based on Fig. 6.2 and several tables.

Looking to Fig. 6.2 where the stopband attenuation is depicted, we may conclude that the Halpern filter really produce the largest attenuation at high

Table 6.14 The LSM criterion: Area under the curve $L_n(\omega^2)$

n Type	3	4	5	6	7	8	9	10
B	0.1429	0.1111	0.0999	0.0769	0.0667	0.0588	0.0526	0.0476
H	0.3714	0.3143	0.3679	0.3390	0.3663	0.348	0.3654	0.3534
L	0.1619	0.1238	0.1071	0.0894	0.0800	0.0700	0.0640	0.0574
LSM	0.1210	0.0860	0.0610	0.0470	0.0370	0.0300	0.0250	0.0210

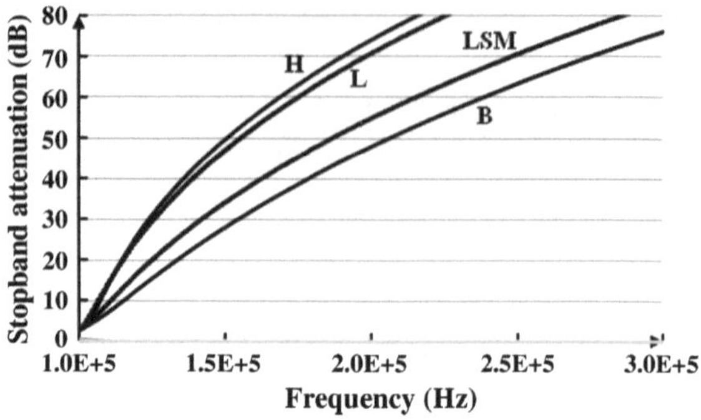

Fig. 6.2 Stopband attenuation of 8th order CMAC filters

Table 6.15 The maximum slope of $L_n(\omega^2)$ at $\omega = 1$ criterion

n Type	3	4	5	6	7	8	9	10
B	6	8	10	12	14	16	18	20
H	6	8	10	12	14	16	18	20
L	8	12	18	24	32	40	50	60
LSM	7.288	10.67	15.34	20.10	26.16	32.29	39.74	47.24

frequencies. One has to weigh that benefit with the passband distortions, however. On the other end, the Butterworth solution is by high margin the less selective one.

From Fig. 6.2 we may better appreciate the quality of the Legendre filters. Namely, their selectivity is lower than the one of the Halpern filters but by a small margin. On the other side their passband distortions are larger than the ones produced by the LSM filters but, again, by a small margin. In that way we may conclude that the Legendre filters are offering the best compromise between selectivity and passband distortion.

These conclusions may be supported by the numerical data given in Tables 6.15, 6.16, 6.17, and 6.18.

Table 6.15 illustrates the slope of the characteristic function at $\omega = 1$ for orders of the filters up to the 10th. It is surprising that the most selective (Halpern) and the least selective (Butterworth) have the same slope at the cut-off. The L-filters, as

Table 6.16 The maximum attenuation at infinity criterion. Value of $\lim_{\omega \to \infty} [L_n(\omega^2)/\omega^{2n}]$

n Type	3	4	5	6	7	8	9	10
B	1	1	1	1	1	1	1	1
H	2	3	6	10	20	35	70	126
L	1.41	2.45	4.47	7.07	13.23	22.14	42	36.37
LSM	1.35	1.73	2.65	3.82	6.22	9.60	16.36	26.30

Table 6.17 Comparisons of the attenuation characteristic of CMAC filters for $n = 5$

↓a(dB) →ω	B	L	H	LSM
0.1	0.6869	0.1726	0.0694	0.7993
0.2	0.7371	0.4661	0.1003	0.8354
0.5	0.8106	0.7005	0.1710	0.8812
1.0	0.8740	0.9221	0.3006	0.9200
30.0	1.9960	1.6144	1.5727	1.7309
50.0	3.1637	2.4300	2.3236	2.6586
70.0	5.0142	3.7700	3.5753	4.1603
100.0	10.0050	7.4417	7.0273	8.2484

Table 6.18 Comparisons of the attenuation characteristic of CMAC filters for $n = 6$

↓a(dB) →ω	B	L	H	LSM
0.1	0.7313	0.2899	0.1792	0.8448
0.2	0.7755	0.4032	0.2184	0.8728
0.5	0.8395	0.7600	0.2925	0.9085
1.0	0.8938	0.9407	0.3904	0.9386
30.0	1.7738	1.4417	1.4115	1.5336
50.0	2.6111	1.9915	1.9145	2.1639
70.0	3.8327	2.8389	2.7033	3.1173
100.0	6.0156	4.9601	4.6951	5.4018

expected are by far the steepest one while the LSM filters are, again, surprisingly steep.

Looking for the asymptotic attenuation, Table 6.16 depicts the highest order coefficient of the characteristic function. Here we may see that the Halpern filter is the best by large margin. The Butterworth solution is lagging seriously while the LSM solution is, again, surprisingly good.

We will come back to these results when discussing the filters with finite transmission zeros obtained by improving the selectivity of the above class of filters.

Finally, Tables 6.17 and 6.18 are kind of numerical representation of Figs. 6.1 and 6.2. For example, one may recognize from these tables that even for low orders of the filters the LSM filters exhibit very low attenuation in the passband e.g. less the 0.1 dB in 80% of the passband.

The phase and the group delay characteristics of the CMAC filters are depicted in Figs. 6.3 and 6.4. The group delay is easier to interpret from the phase distortion point of view since the phase characteristic is deceiving by its shape which looks as if it is linear.

We may see from Fig. 6.4 that the Halpern filters produce a high peek of the group delay at the end of the passband. That is not a favorable property if linear phase is sought in conjunction with selectivity. Namely, in such situations the

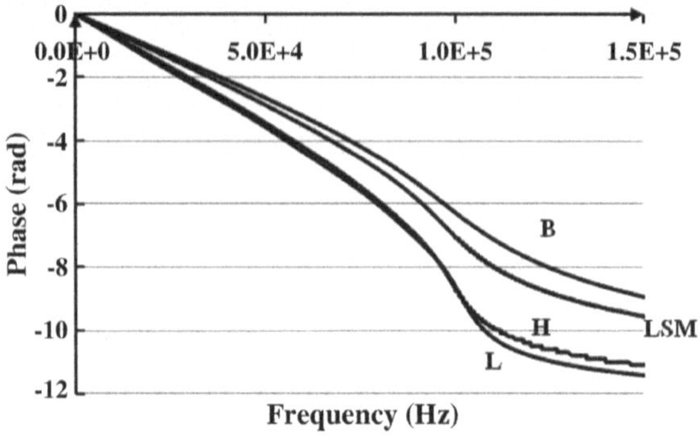

Fig. 6.3 Phase characteristics of 8th order CMAC filters

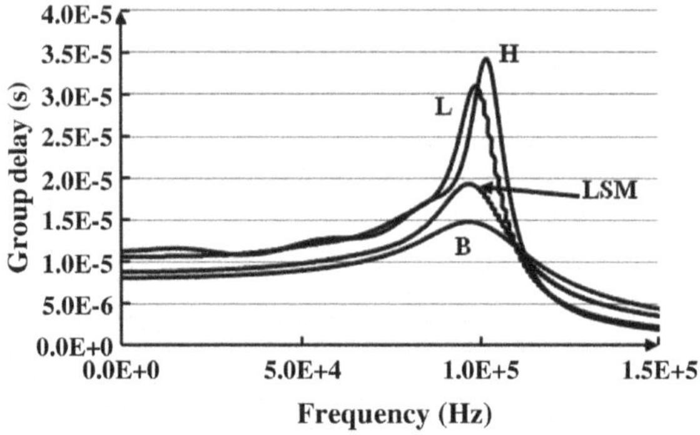

Fig. 6.4 Group delay characteristics of 8th order CMAC filters

selective filter is followed (in cascade connection) by an additional circuit named all-pass phase corrector. That circuit is supposed not to affect the amplitude characteristic while correcting the phase. Since the phases (and, consequently group delays) of two cascaded networks add to each other the goal is that the overall (the sum) group delay to become constant which means the phase characteristic will become linear. Fact is, unfortunately, that the schematic of the all-pass phase corrector is most frequently more complicated than the schematic of the selective part of the overall filter. It may be relaxed if one chooses an approximation solution with lower peak of the group delay.

Looking to Fig. 6.4 it seem that from this point of view the LSM solution will provide for all three: low passband distortions, acceptable selectivity, and low phase distortions. This claim was elaborated in [9].

The time domain responses of the CMAC filters of 8th order are depicted in Figs. 6.5 and 6.6. For now we may recognize that two pairs are formed. The first one is the Butterworth and the LSM filters while the second is made of the Legendre and Halpern filters.

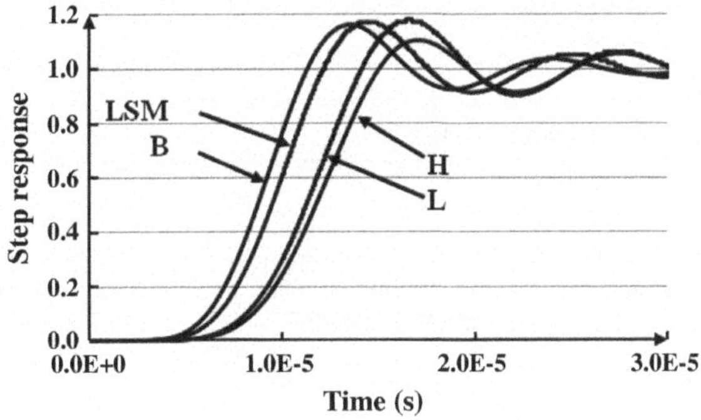

Fig. 6.5 Step responses of the 8th order CMAC filters

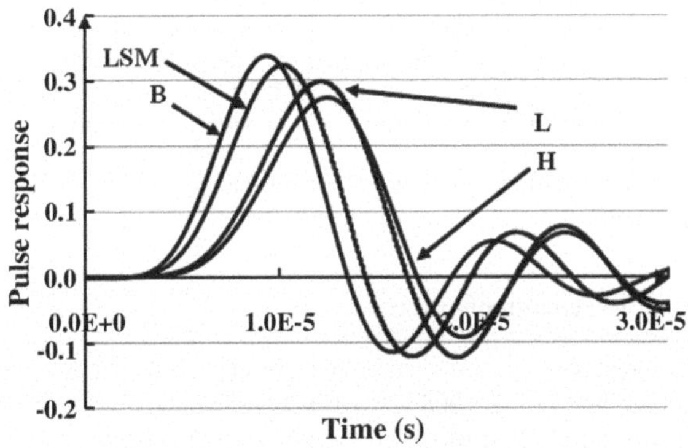

Fig. 6.6 Responses to a Dirac pulse of the 8th order CMAC filters

We will speculate here that this pairing is a consequence of the shape of the passband amplitude characteristic. Namely, the first pair is related to very small distortions in the lower part of the passband where most of the energy of the signal is located while the second pair exhibits substantial reflection of the signal power.

We will discuss these responses with more detail in the next chapter where the so called Chebyshev filters will be studied.

6.5 Design Example

In this paragraph we will show the synthesis results for an 8th order LSM filter performed by the $\mathcal{CMAC}$ program of the $\mathcal{RM}$ software.

First excerpts will be given from the transfer function synthesis program.

Start.E

```
Program CMAC                    SYNTHESIS OF CMAC FILTERS

 Project name:  lsm_8
+++++++++++++++++++++++++++++++++++++++++++++++++++++++++++++++++++++++++++++
Order of the numerator=0          Order of the denominator n=8
Function type=4          Maximum pass-band attenuation=3.01030e+000
The transfer function is re-normalized to show Anorm=3.00000e+000 dB at the pass-
band edge of fc=1.00000e+005 Hz
                    LSM-FILTERS

==============================================================================
 General information
 The constants  C :
5.861261183e-001        6.512512426e-001            4.775842446e-001
6.512512426e-002
 Area under the Ln(w**2) curve=1.067294840e-001
 Inflection points:         3.125048143e-001 6.953692630e-001
 Order of denominator=8,           Order of numerator=0
==============================================================================
The rest of calculations are performed for passband attenuation of 3.000000
[dB]

==============================================================================
 All zeros at infinity
==============================================================================
 real pole                  Imag. pole
-1.096573351e+000          1.057013372e-001
-1.096573351e+000          -1.057013372e-001
-1.271142352e-001          9.746006678e-001
-1.271142352e-001          -9.746006678e-001
-3.784561168e-001          7.554555813e-001
-3.784561168e-001          -7.554555813e-001
-6.363272680e-001          3.470964773e-001
-6.363272680e-001          -3.470964773e-001

==============================================================================
Coefficients of the squared modulus of the characteristic function (ascending) (s**2)
-1.110223025e-016     -0.000000000e+000        4.942619340e-001
3.044760292e+000      5.595749207e+000         -1.373229932e+000
-9.778986741e+000     -1.184813034e+000        5.170955242e+000
Coefficients of the squared modulus of the characteristic function (ascending) (w**2)
-1.110223025e-016      0.000000000e+000        4.942619340e-001
-3.044760292e+000      5.595749207e+000        1.373229932e+000
-9.778986741e+000      1.184813034e+000        5.170955242e+000
Coefficients of the denominator of the transfer function (ascending)
4.397588727e-001      2.441869896e+000         6.779543246e+000
1.221903753e+001      1.569948704e+001         1.467939277e+001
9.906940341e+000      4.476941941e+000         1.000000000e+000
```

Second, excerpts will be given from the transfer function analysis program (*LP_Analysis*). Note, the drawings depicted in Fig. 6.1 to Fig. 6.6 are all created based on the results produced by the *LP_Analysis* program.

```
**************************************************************************
   The cut-off frequency defined by 6.00000e+001 dB is equal to 2.14468e+005
**************************************************************************
   delay time (at 0.5)=9.47500e-006      rise      time      (0.1-0.9)=4.83509e-006
**************************************************************************
   The number of oscillations at the beginning of the step response is=1
   time=4.77465e-009      value of the extremum=2.09441e-016
**************************************************************************
   The number of oscillations at the end of the step response is=16
   time=1.44253e-005      value=1.17298e+000
   time=1.98113e-005      value=9.13266e-001
   time=2.49719e-005      value=1.05122e+000
   time=3.00695e-005      value=9.67806e-001
   time=3.51558e-005      value=1.02139e+000
   time=4.02650e-005      value=9.85644e-001
   time=4.53903e-005      value=1.00960e+000
   time=5.05231e-005      value=9.93615e-001
   time=5.56538e-005      value=1.00424e+000
   time=6.07846e-005      value=9.97186e-001
   time=6.59153e-005      value=1.00187e+000
   time=7.10481e-005      value=9.98760e-001
   time=7.61768e-005      value=1.00082e+000
   time=8.13056e-005      value=9.99454e-001
   time=8.64384e-005      value=1.00036e+000

**************************************************************************
```

End.E

6.6 Developer's Corner

It will be repeated several times throughout these texts that both transfer function synthesis and circuit synthesis needs a powerful and robust high order polynomial solver. That is not only because we need to analyze the transfer function synthesis results but for creation networks, too.

Next, we owe to the reader additional information on the subject of critical monotonic characteristic function. What we intend to describe will probably be a nice music into the ears of potential program developer who is not very familiar with orthogonal polynomials and hyper-geometric functions as such. Namely, there is an alternative method to describe the CMAC function which is involving the attenuation zeros (zeros of the characteristic functions) in place of the polynomials given above.

For odd filter orders one may write [10, 11]

$$\frac{dL_n(\omega^2)}{d\omega} = A \cdot \omega \cdot \prod_{i=1}^{(n-1)/2} (\omega^2 - \omega_i^2)^2, \tag{6.37}$$

where we choose:

$$A = 1/\int_0^1 \omega \cdot \prod_{i=1}^{(n-1)/2} (\omega^2 - \omega_i^2)^2 d\omega \tag{6.38}$$

which leads to $K(1) = 1$. The integral of (6.37) is the CMAC polynomial. Similar expression may be written for even order filters. It is obvious that instead of the **C** vector, to create a proper filter function under some criterion, one is to do with the abscissas of the inflection points $\pm\ \omega_i$.

We do not claim that any of these two representations is favorable. Simply, the one accepted in this book was more frequently elaborated in literature. Interested filter designer and skilful programmer may find the latter method more convenient and proceed that way.

In a subsequent chapter, the characteristic function of this kind will be extended with poles of attenuation to raise the selectivity. Once the polynomial (6.37) was created, however, as concerns the algorithm for finding the attenuation poles and the final results, there will be no difference at all as to which way the original polynomial characteristic function was created.

References

1. Topisirović D, Litovski V, Andrejević Stošović M (2015) Unified theory and state-variable implementation of critical-monotonic all-pole filters. Circuit Theory Appl 43(4):502–515
2. Halpern P (1969) Optimum monotonic low-pass filters. IEEE Trans Circuit Theory 16 (2):240–242
3. Butterworth S (1930) On the theory of filter amplifiers. Exp Wirel Wirel Eng 7(85):536–541
4. Papoulis A (1958) Optimum filters with monotonic response. Proc IRE 46(3):606–609
5. Papoulis A (1959) On monotonic response filters. Proc IRE 47(2):332–333
6. Fukada M (1959) Optimum filters of even orders with monotonic response. IRE Trans Circuit Theory 6(3):277–281
7. Raković BD, Litovski VB (1973) Least-squares monotonic low-pass filters with sharp cutoff. Electron Lett 9(4):75–76
8. Ralston A, Rabinowitz P (2001) A first course in numerical analysis. Dover Publications, New York
9. Andrejević Stošović M, Topisirović D, Litovski V (2019) Frequency and time domain comparison of selective polynomial filters with corrected phase characteristics. Int J Electron 106(5):770–784

10. Rabrenović D, Jovanović V (1971) L-Filters. Publikacije Elektrotehničkog fakulteta-serija ATA 56:41–46
11. Rabrenović D, Jovanović V (1974) Low-pass filters with maximally flat magnitude beyond the origin and arbitrary transmission zeros. Int J Circuit Theory Appl 2(1):83–87

Chapter 7
Chebyshev and Modified Chebyshev Filters

Chebyshev (C) filters are among the most frequently used. Their transfer function is obtained via the characteristic function (which is introduced in this chapter) to offer the most selective polynomial filters of all. In addition, their amplitude characteristic in the passband is equi-ripple. The value of the ripple (or the maximum passband attenuation) may be controlled by a single parameter assigned before the poles of the function are obtained by solving the so called Feldkeller equation. Since to accommodate the resistive terminals of the passive LC realization of even order C filters needs a transformer, we here, for the first time, introduce the Modified Chebyshev (MC) filter. The MC behaves in the passband as if the order of the filter was reduced by one while keeping the asymptotic slope of the original even order filter. In order to get a feeling about the mutual advantages and disadvantages of the CMAC and the Chebyshev filters comparisons are given of the attenuation, group delay and time domain responses as well as the sensitivity of the attenuation characteristics to variation of the circuit element values. Elaborated example of synthesis of a C filter is given, too.

7.1 Introduction

The most popular non-monotonic amplitude characteristic is the one exhibiting equi-ripple amplitude characteristic in the passband. Since it is obtained by use of Chebyshev polynomials it was named Chebyshev. For a given variation of the amplitude in the passband this characteristic exhibits the largest selectivity of all.

Since the pass-band attenuation touches the abscissa maximum number of times, one may state that this class of filters has all its attenuation zeroes distinct and laying on the axis of real frequencies.

V. Litovski, *Electronic Filters*, Lecture Notes in Electrical Engineering 596,
https://doi.org/10.1007/978-981-32-9852-1_7

The attenuation characteristic of a 9th order polynomial Chebyshev low-pass filter having 0.75 dB maximum attenuation in the passband is depicted in Fig. 7.1.

It is important to mention that for even order Chebyshev filters the amplitude characteristic does not start with unity gain (or attenuation of 0 dB). In fact, the gain at the origin is equal to the minimum passband gain. When passive LC ladder physical realization is sought, this leads to filters with unequal terminal resistances. The resistance of the load is different than the resistance of the source. To accommodate for equal resistances one would need an additional transformer.

To avoid this, the $\mathcal{RM}$ software provides with a special routine which allows for the gain to start at the unity for the price of slight reduction of the selectivity since one order of freedom is partly sacrificed so that the amplitude characteristic has an additional inflection point at the origin. If this option is used, for example, the 16th order filter will have attenuation as depicted in Fig. 7.2. It will exhibit slightly smaller selectivity than the normal 16th order Chebyshev polynomial filter while still more selective than its 15th order counterpart which have similar passband response. This comparison is illustrated in Fig. 7.3 for the case of the 8 and 7th order filters.

Figure 7.4 depicts the group delay characteristic of the 7th order Chebyshev filter. As can be seen strong distortions may be observed even deep in the passband. Note, since the area under the group delay curve is equal to $n \cdot \pi/2 \approx 11$, one would expect (for the ideal group delay characteristic approximation in the passband) the constant group delay in the passband to be approximately 11 s. In fact, it is approximately twice as small. To compensate for the group delay nonlinearities, if necessary, one is to add an all-pass corrector whose group delay occupies an area as

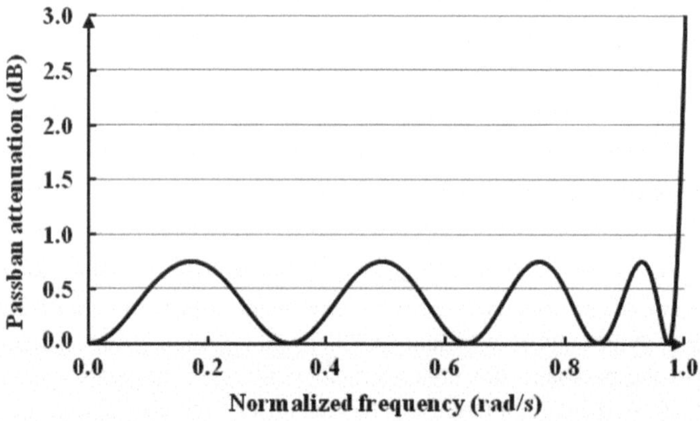

Fig. 7.1 Attenuation characteristics of the Chebyshev filter for $n = 9$ with passband attenuation of 0.75 dB. The characteristic is renormalized so as to produce 3 dB at the cut-off frequency

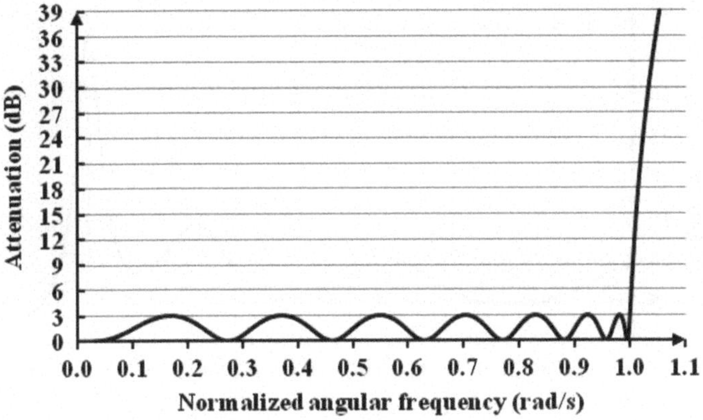

Fig. 7.2 Attenuation characteristic of a modified Chebyshev polynomial filter of 16th order. The passband attenuation is 3 dB

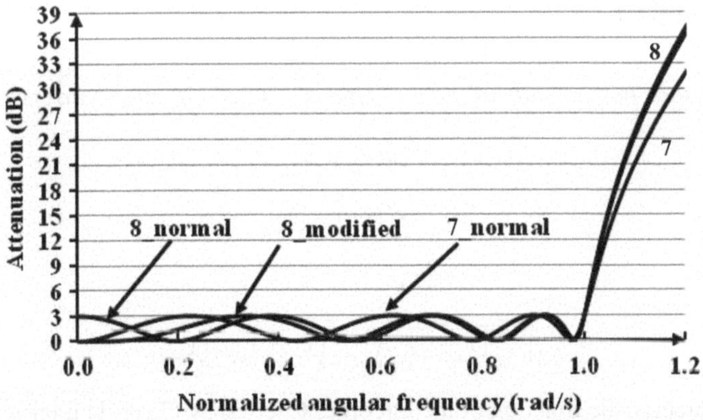

Fig. 7.3 Attenuation characteristics of original and modified equi-ripple passband polynomial filters

depicted with the dashed line rectangle in the figure. That means that (approximately) one will need a 6th order corrector which in most realizations are much more complicated a circuit than the filter itself.

We will come back to the time domain response of the Chebyshev filters and its properties.

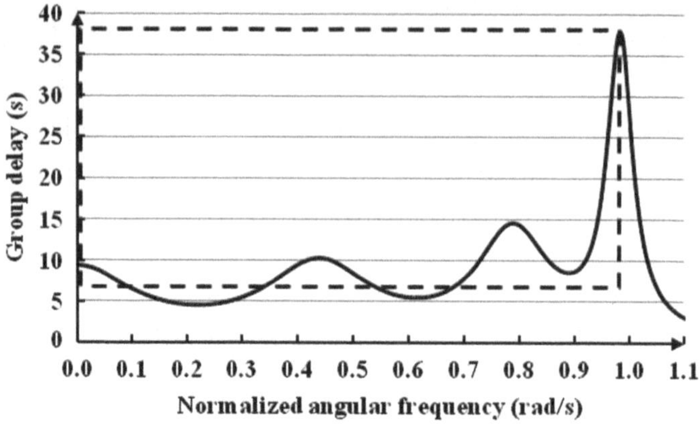

Fig. 7.4 Group delay characteristics of the Chebyshev filter for $n = 7$

7.2 The Chebyshev Polynomial and the Characteristic Function

The *Chebyshev polynomials of the first kind* are defined by the recurrence relations

$$T_0(x) = 1 \tag{7.1}$$

$$T_1(x) = x \tag{7.2}$$

and

$$T_{n+1}(x) = 2 \cdot x \cdot T_n(x) - T_{n-1}(x). \tag{7.3}$$

The characteristic function of the nth order Chebyshev (Type 1) filter is given by

$$K_n(\omega^2) = [T_n(x)]^2_{|x=\omega}. \tag{7.4}$$

Accordingly, the amplitude characteristic is given by

$$|H(\omega)| = \frac{1}{\sqrt{1+\varepsilon^2 K_n(\omega^2)}} = \frac{1}{\sqrt{1+\varepsilon^2 [T_n(\omega)]^2}} \tag{7.5}$$

where ε was defined earlier.

Of course, to find the transfer function one needs first to solve the following equation (similarly to the CMAC case)

$$1+\varepsilon^2[T_n(\omega)]^2_{|\omega=s/\mathrm{j}}=0 \tag{7.6}$$

and to separate the poles belonging to the left half of the complex frequency plane. If coefficients of the denominator of the transfer function are sought one needs to have a routine to calculate them from the poles just obtained.

Due to the nature of the function used, formulas were developed [1] for the location of the poles as follows

$$\begin{aligned} p_m = & -\sinh\left[\frac{1}{n}\cdot\ \mathrm{ar\,sinh}\left(\frac{1}{\varepsilon}\right)\right]\cdot\ \sin(\theta_m) \\ & +j\cdot\ \cosh\left[\frac{1}{n}\cdot\ \mathrm{ar\,sinh}\left(\frac{1}{\varepsilon}\right)\right]\cdot\ \cos(\theta_m) \end{aligned} \tag{7.7}$$

where $\theta_m=\frac{\pi}{2}\cdot\frac{2m-1}{n}$ and $m = 1, 2,\ldots, n$.

Despite the fact that this expression looks attractive, the $\mathcal{RM}$ software uses a polynomial solver for (7.6) so making the procedure of evaluation of the poles common for all types of denominator polynomials of the amplitude squared.

7.3 The Modified Chebyshev Characteristic Function

The modified Chebyshev polynomial may be defined as:

$$T_{n,\mathrm{mod}}(\mathbf{x},x)=x^2\cdot\frac{\prod_{j=1}^{n/2-1}\left(x^2-x_j^2\right)}{\prod_{j=1}^{n/2-1}\left(1-x_j^2\right)}. \tag{7.8}$$

In (7.8) **x** stands for the unknown zeros of the modified Chebyshev polynomial as depicted in Fig. 7.5. Normalization so as $T_{n,\mathrm{mod}}(1)=1$ is implemented. n is an even number!

To find **x** an iterative procedure is implemented in the *Chebyshev* program of the $\mathcal{RM}$ software. It runs as follows.

For a given initial solution for **x**:

STEP 1: Find the abscissas of the extrema points of $T_{n,\mathrm{mod}}(x)$. The resulting vector is $\boldsymbol{\omega}$ as depicted in Fig. 7.5.

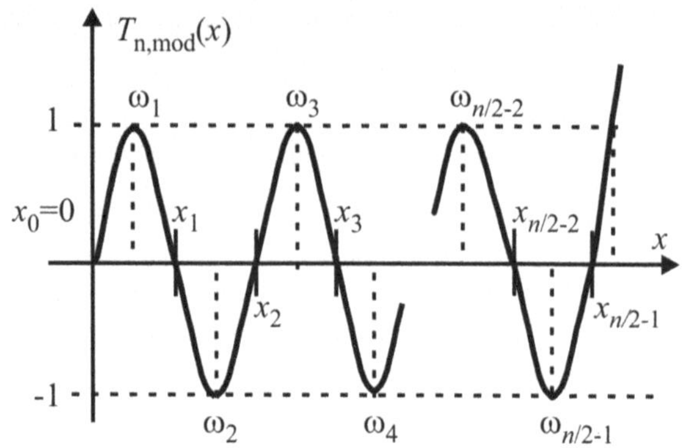

Fig. 7.5 Definition of the modified Chebyshev polynomial

STEP 2: Solve, for one step only, the following system of nonlinear equations (i.e. find the corrections of **x** and correct it only once)

$$T_{n,\text{mod}}(\mathbf{x}, \omega_i) = (-1)^{i+1}, \text{ for } i = 1, \ldots, n/2 - 1. \tag{7.9}$$

In fact, using a Newton-Raphson iterative solution procedure [2], one has to solve the following system of linear equations

$$\sum_{k=1}^{n/2-1} \frac{\partial T_{n,\text{mod}}(\mathbf{x}, \omega_i)}{\partial x_k} \cdot \Delta x_k = -T_{n,\text{mod}}(\mathbf{x}, \omega_i) + (-1)^{i+1}, \text{for } i = 1, \ldots, n/2 - 1. \tag{7.10}$$

STEP 3: Check for convergence. In this step we usually check for the norm $\|\Delta \mathbf{x}\|$ to be smaller than a previously chosen small number δ, e.g. $\delta = 10^{-6}$. If necessary continue with STEP 1.

As an initial solution one may use the roots of that Chebyshev polynomial (only the positive half) of the lover odd order (i.e. of the order $m = n$-1) which are found to be given as

$$x_k = \cos\left[\frac{\pi(k + 1/2)}{m}\right], \quad \text{for} \quad k = 0, 1, 2, \ldots, m - 1. \tag{7.11}$$

As for the derivatives involved in (7.10) one may use the following development. First we find

$$\frac{\partial T_{n,\mathrm{mod}}(\mathbf{x},x)}{\partial x_k} = x^2 \times \frac{\left\{\prod_{\substack{j=1\\ j\neq k}}^{n/2-1}\left(x^2-x_j^2\right)\cdot\right\}(-2x_k)\cdot\prod_{j=1}^{n/2-1}\left(1-x_j^2\right) - \left\{\prod_{j=1}^{n/2-1}\left(1-x_j^2\right)\right\}\cdot\left\{\prod_{\substack{j=1\\ j\neq k}}^{n/2-1}\left(1-x_j^2\right)\right\}(-2x_k)}{\left[\prod_{j=1}^{n/2-1}\left(x^2-x_j^2\right)\right]^2} \tag{7.12}$$

This may be rearranged so that

$$\begin{aligned}\frac{\partial T_{n,\mathrm{mod}}(\omega_i)}{\partial x_k} &= -2x^2x_k\cdot\frac{\prod_{j=1}^{n/2-1}\left(x^2-x_j^2\right)}{\prod_{j=1}^{n/2-1}\left(1-x_j^2\right)}\cdot\frac{\left(\frac{1}{x^2-x_k^2}-\frac{1}{1-x_k^2}\right)}{\prod_{j=1}^{n/2-1}\left(1-x_j^2\right)}\\ &= \frac{T_{n,\mathrm{mod}}(\omega_i)}{\prod_{j=1}^{n/2-1}\left(1-x_j^2\right)}\cdot\frac{2x_k(x^2-1)}{\left(x^2-x_k^2\right)\cdot\left(1-x_k^2\right)}_{|x=\omega_i}\end{aligned} \tag{7.13}$$

Similar developments will be necessary whenever we need to manipulate the characteristic function of a filter in which factored form of a polynomial appears.

As for the abscissas of the extrema points, ω_i, $i = 1, 2,\ldots, n/2-1$, one may use the modified secant method as follows.

Since, in fact, we are solving the equation

$$\frac{\mathrm{d}T_{n,\mathrm{mod}}(\mathbf{x},x)}{\mathrm{d}x} = 0 \tag{7.14}$$

$n/2$-1 times, we have to repeat the following.

STEP 1: Set $k \in (1, n/2-1)$, start with $k = 1$ and δ (to ensure accuracy after convergence, e.g. $\delta = 10^{-6}$).
STEP 2: Choose $l = 0$, $a^l = x_{k-1}$ and $b^l = x_k$
STEP 3: Compute

$$f_\mathrm{a} = \frac{\mathrm{d}T_{n,\mathrm{mod}}(\mathbf{x},x)}{\mathrm{d}x}_{|x=a^l} \tag{7.15a}$$

$$f_\mathrm{b} = \frac{\mathrm{d}T_{n,\mathrm{mod}}(\mathbf{x},x)}{\mathrm{d}x}_{|x=b^l} \tag{7.15b}$$

STEP 4: Compute

$$c^l = b^l - f_b \frac{b^l - a^l}{f_b - f_a} \tag{7.16}$$

STEP 5: Check for convergence i.e. if $\left|c^l - b^l\right| \leq \delta$.
If yes $\omega_k = c^l$; and if $k = n/2-1$, END.
If $k < n/2-1$, increment k by 1and continue with STEP 2.
Else, compute

$$f_c = \frac{\mathrm{d}T_{n,\mathrm{mod}}(\mathbf{x}, x)}{\mathrm{d}x}_{|x=c^l} \tag{7.17}$$

Increment l by 1 and
If $f_c{\cdot}f_a > 0$ select $a^l = c^{l-1}$, $f_a = f_c$ and $b^l = b^{l-1}$ and continue with STEP 4, or
If $f_c{\cdot}f_a < 0$ select $a^l = a^{l-1}$ and $b^l = c^{l-1}$, $f_b = f_c$ and continue with STEP 4.

7.4 Synthesis Example

Below is an excerpt from the textual report of the *Chebyshev* program. To save space some of the output data are edited and rearranged in a form of a table.

Start.E
Project name: CHEB_5

General information
Order of the numerator=0 Order of the denominator n=5 type of approximation=0

Requested pass-band width is fc=1.00000e+005 Hz
Requested pass-band attenuation is Amax=1.00000e+000 dB
The transfer function is re-normalized to show Anorm=3.00000e+000 dB at the pass-band edge of fc=1.e+5 Hz
++
TRANSFER FUNCTION SYNTHESYS OF A CHEBYSHEV FILTER
==

General information
Order of denominator=5, Order of numerator=0

This part of calculations is performed for passband attenuation of 1.000000 [dB]

The results about the poles, zeros and coefficients listed below are valid for frequency normalization with wc=2*pi*fc=6.28318530e+005

Normalized poles of the transfer function

Real part	Imaginary part
-2.8949334124e-001	0.0000000000e+000
-8.9458362200e-002	-9.9010711200e-001
-8.9458362200e-002	9.9010711200e-001
-2.3420503282e-001	6.1191984772e-001
-2.3420503282e-001	-6.1191984772e-001

Coefficients of the squared modulus of the CHARACTERISTIC function (ascending s**2)
0 -6.473135295 -51.78508236 -144.9982306 -165.7122635
-66.28490542

Coefficients of the denominator of the TRANSFER function (ascending power of s)
0.1228266705 0.5805341513 0.9743960731 1.688815979
0.9368201313 1

The nominal 3.00000e+000 dB cut-off frequency=1.03367e+005

===

The rest of calculations is performed for renormalized poles
Maximum attenuation inside passband= 3.000000 [dB]
Attenuation at the band-edge= 60.000000 [dB]

The results about the poles, zeros and coefficients listed below are valid for frequency normalization with
wc=2*pi*fc*fcnorm=6.49474211e+005 [rad/s]

Normalized poles of the transfer function

Real part	Imaginary part
-2.8006351516e-001	0.0000000000e+000
-8.6544385689e-002	-9.5785580763e-001
-8.6544385689e-002	9.5785580763e-001
-2.2657614327e-001	5.9198744544e-001
-2.2657614327e-001	-5.9198744544e-001

The new 3.00000e+000 dB cut-off frequency=1.000000e+005 [Hz]

Coefficients of the denominator of the transfer function:
Ascending exponent of frequency (s)
1 4.885591338 8.476311962 15.18573371 8.707464417
9.607658039
Descending exponent of frequency (s)

9.607658039 8.707464417 15.18573371 8.476311962 4.885591338
1

We do proceed now with an excerpt from the report of the *LP_analysis* program of the *RM* software.

****

delay time (at 0.5) =7.51476e-006 rise time (0.1-0.9)=4.94868e-006

****

The number of oscillations at the beginning of the step response is=0

****

The number of oscillations at the end of the step response is=13

time=1.22714e-005	value=1.10171e+000
time=1.76127e-005	value=8.62085e-001
time=2.31096e-005	value=1.01106e+000
time=2.73432e-005	value=9.60167e-001
time=3.26446e-005	value=1.03094e+000
time=3.79796e-005	value=9.71375e-001
time=4.33650e-005	value=1.01805e+000
time=4.85502e-005	value=9.86819e-001
time=5.37386e-005	value=1.00976e+000
time=5.89156e-005	value=9.92293e-001
time=6.41538e-005	value=1.00583e+000
time=6.93861e-005	value=9.95628e-001
time=7.46083e-005	value=1.00326e+000

****

End.E

The set of figures (Fig. 7.6) represent the properties of the fifth order Chebyshev filter in time and frequency domain.

The first question of interest which may pose a beginner in filter design is the following. Suppose one reduces the maximum passband attenuation of a Chebyshev filter while keeping the attenuation at the cut-off at 3 dB, will the Chebyshev filter's attenuation characteristic converge to the Butterwort solution? To answer that we made an experiment the result of which are depicted in Fig. 7.7.

Figure 7.7 depicts four Chebyshev filters of the 7th order. The maximum passband attenuation was swept from 1 dB, via 0.1 dB and 0.01 dB, to 1 mdB. By inspection of the results we easily come to the conclusion that for all practical values of the maximum passband attenuation, the Chebyshev filter keeps its main

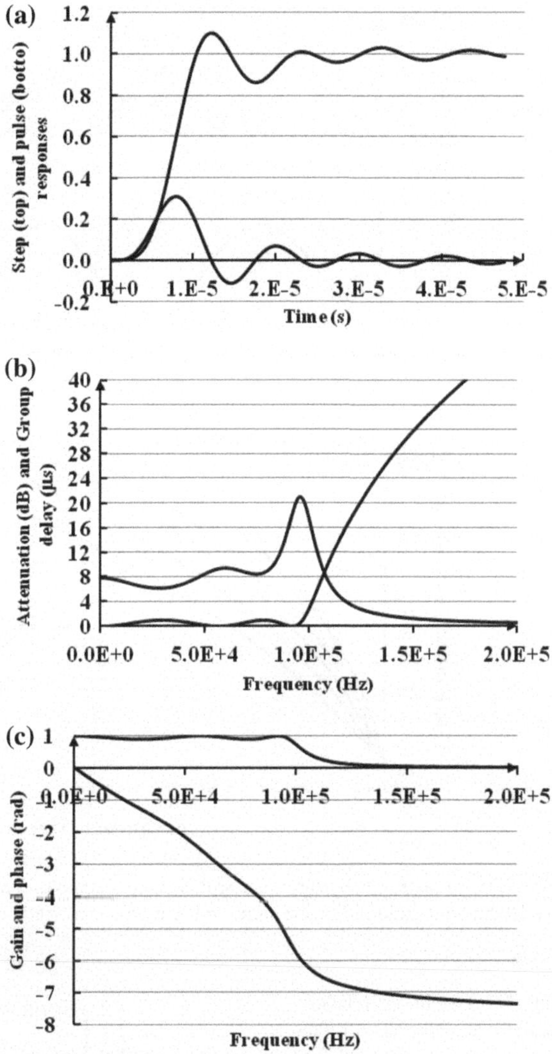

Fig. 7.6 Chebyshev 5th order, **a** time domain responses **b** overall attenuation and group delay and **c** phase and gain

property to be the most selective one. As for the passband, one may say that when the Chebyshev filter goes below 0.1 dB it becomes preferable too.

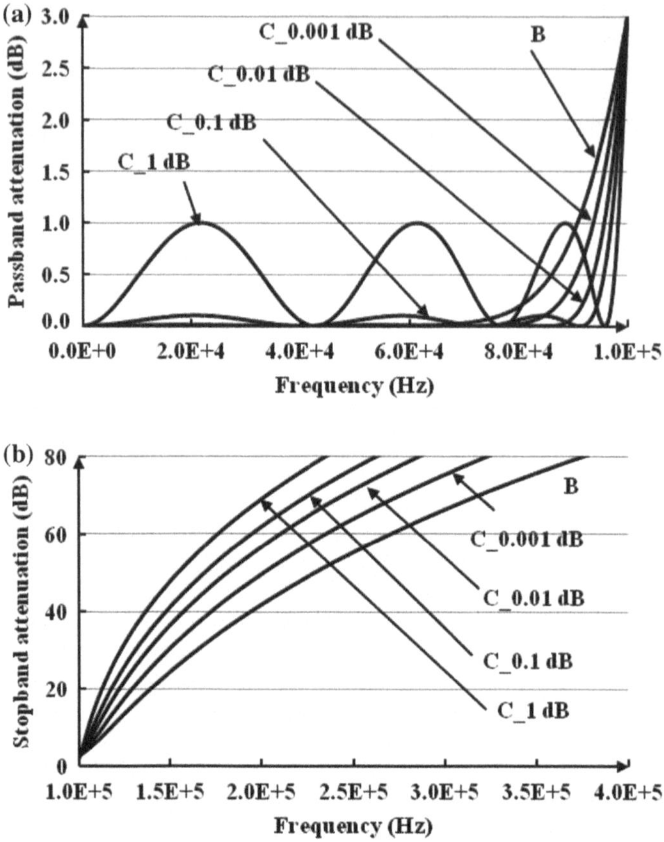

Fig. 7.7 7th order Chebyshev filter with various maximum passband attenuation compared with the Butterworth solution of the same order, **a** passband attenuation and **b** stopband attenuation

Fig. 7.8 Definitions related to the step response of a filter

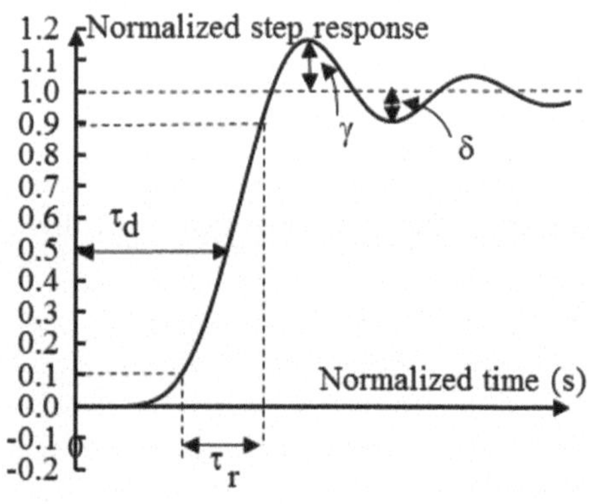

7.5 Discussion and Some Comparisons

Let us now consider a parallel between the Chebyshev and the CMAC filters [3]. Unless otherwise stated the comparisons will be based on the 5th order filter. Chebyshev filter exhibiting passband attenuation of 1 dB but renormalized to 3 dB at the cut-off, will be used.

We will start with the time domain responses. The parameters of the step and pulse responses of the CMAC and Chebyshev filters, as defined by Fig. 7.8, are quantified in Table 7.1. For illustration purposes Figs. 7.9 and 7.10 represent the step and pulse responses.

The LSM filter (followed by the Butterworth) is seen as fastest since it exhibits the smallest delay (t_d) and rise (t_r) time. On the other end is the Chebyshev filter exhibiting largest delay and rise times. As for the overshoot (γ) and undershoot (δ) the Butterworth filter (followed by the LSM) has the smallest ringing about its final value. Note, according to this the LSM filter would produce the largest opening of the eye diagram since it has the smallest undershoot. The Chebyshev filter is seen to have the largest undershoot in the step response as already mentioned. Finally, looking to the pulse response, the Butterworth filter (followed by the LSM) have the largest value of the main lobe as depicted in the last column of Table 7.1. All together, from the time domain prospective the LSM and the Butterworth filters would be compared favorably to all others (Table 7.2).

Let consider the comparisons of the attenuation characteristics. Figure 7.11 depicts the passband attenuation while Fig. 7.12 the stopband attenuation of the group of filters under consideration. Looking to the passband, we have here a Chebyshev filter whose maximum passband attenuation is kind of a compromise between the Halpern solution, having the largest distortions, and the LSM solution having the lowest distortion. One is to notice the presence of a lobe at low frequencies for the Chebyshev filter and ever large attenuation in the passband of the Halpern filter. We think that these values are related to the undershoot of the step

Table 7.1 Parameters of the time domain responses of the 5th order Chebyshev and CMAC filters

Type	Step response				Dirac response
	Normalized delay time t_d (s)	Normalized rise time t_r (s)	Over-shoot γ (%)	Under-shoot δ (%)	Main lobe
Butterworth	3.494	2.561	12.777	4.349	0.365
Papoulis	4.252	2.867	13.282	7.208	0.333
Halpern	4.453	3.121	4.883	9.286	0.307
LSM	3.207	2.482	11.466	3.751	0.376
Chebyshev (1 dB)	4.722	3.109	10.171	13.792	0.309

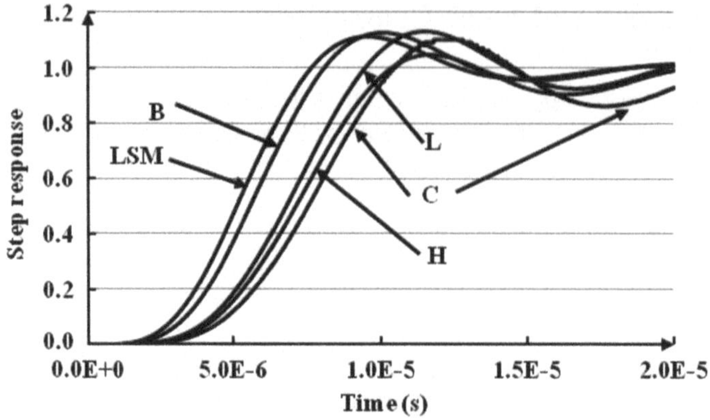

Fig. 7.9 Step responses of the CMAC and a Chebyshev filter

Fig. 7.10 Pulse responses of the CMAC and a Chebyshev filter

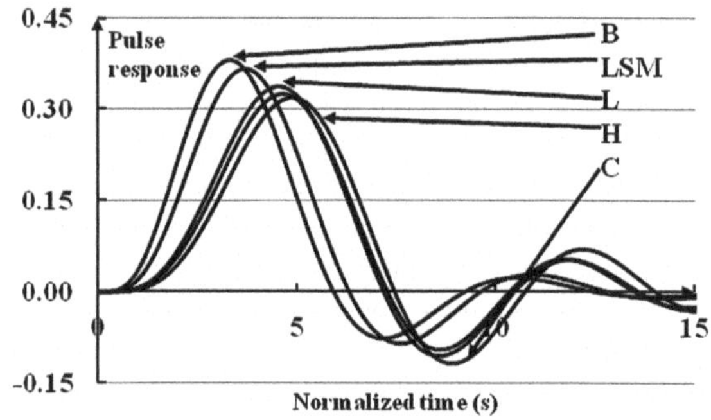

Table 7.2 Parameters of the time domain responses of the 7th order Chebyshev and CMAC filters

Filter type	Step response				Dirac response
	Normalized delay time t_d (s)	Normalized rise time t_r (s)	Over-shoot γ (%)	Under-shoot δ (%)	Main lobe
Butterwort	4.831	2.788	1.154	0.935	0.345
Papoulis	6.146	3.208	1.163	0.907	0.307
Halpern	6.438	3.480	1.078	0.895	0.280
LSM	5.527	3.005	1.170	0.918	0.325
Chebyshev (1 dB)	6.669	3.424	1.121	0.843	0.288

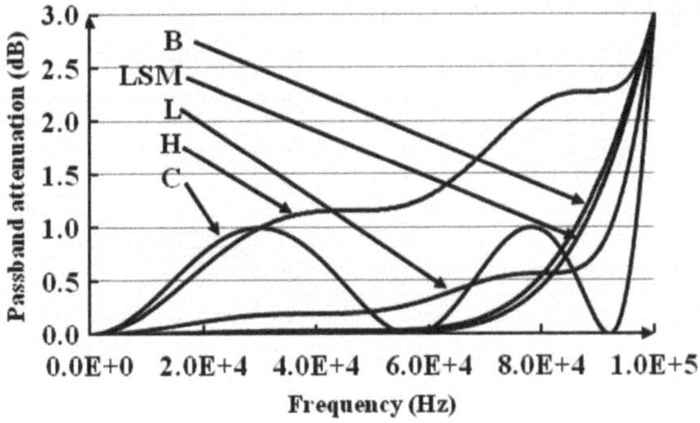

Fig. 7.11 Passband attenuation of the CMAC and a Chebyshev filter

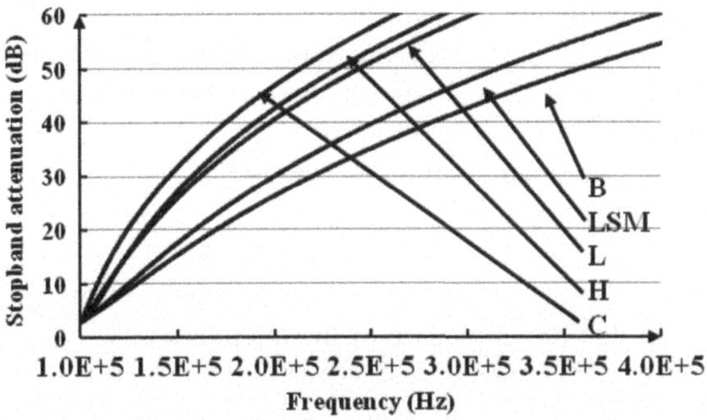

Fig. 7.12 Stopband attenuation of the CMAC and a Chebyshev filter

response since they "suffocate" part of the spectrum of the input signal carrying large energy. With no doubt the LSM response in the passband is preferable. The Papoulis (Legendre) filters may be considered as a good transition between Butterworth and LSM from one side and Halpern and Chebyshev on the other.

As for the stopband attenuation the situation is opposite. Chebyshev is unapproachable and it is followed by the Halpern filter. On the other side are LSM and Butterworth the latter being unapproachable, too. Here again, the Papoulis filters may be considered as a transition.

The group delay of the Chebyshev filter was already mentioned. As can be seen from Fig. 7.13 it has the largest distortions of all. We say "distortions" since the group delay is derivative of the phase and when it is not constant the phase looses linearity. Note, here the passband attenuation is 1 dB as opposed to Fig. 7.4 where it was 3 dB. Not only it has large main lobe but also has large lobes within the passband which makes the synthesis of a phase corrector circuit extremely difficult which we will discuss later on. Correcting the phase of the Legendre, LSM and Butterworth filters is not such a difficult task since the passband group delay is practically monotonic. Of course, the Butterworth filter is the easiest to correct since even its main lobe is small. We will come back to this issue later on when corrector synthesis discussed.

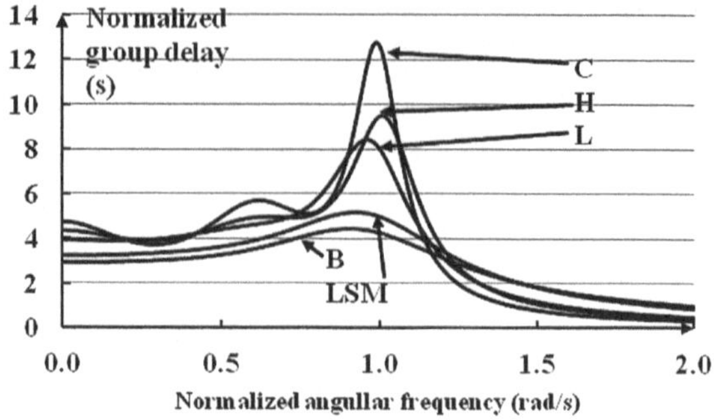

Fig. 7.13 Group delay of 5th order monotonic (CMACs) and Chebyshev filters

To complete the comparisons the derivative of the passband amplitude characteristic will be considered. It will be used as a measure for the sensitivity of the attenuation characteristic on the variation of the circuit parameters. To get that measure we use the following [4].

The attenuation and the characteristic function are connected by the relation:

$$a = 10 \cdot \mathbf{log}\left[1 + K\left(\omega^2\right)\right], \tag{7.18}$$

So that the derivative of the attenuation is

$$\frac{\partial a}{\partial \omega} = \frac{10}{\mathbf{ln}(10)} \cdot \frac{1}{1 + K(\omega^2)} \cdot \frac{\partial K(\omega^2)}{\partial \omega}. \tag{7.19}$$

Now, as shown in [4] for the partial increment of the attenuation (Δa_i) being consequence of a change of a single circuit element change (Δx_i), one may write

$$\Delta a_i = \frac{\Delta x_i}{x_i} \cdot \omega \cdot \frac{\partial a}{\partial \omega} = \frac{\Delta x_i}{x_i} \cdot F(\omega) \tag{7.20}$$

where x_i is the value of the ith parameter of the circuit.

Figure 7.14a depicts the function $F(\omega)$ for a 7th order monotonic (Halpern) and Chebyshev (with 3 dB minimum passband attenuation) filters. It is obvious that despite the large value of the passband attenuation produced by the Halpern filter, due to its critical monotonicity, the sensitivity may be stated as incomparably smaller than the corresponding Chebyshev filter in the whole passband.

The next example goes even further. Figure 7.14b represents the value of $F(\omega)$ for 9th order LSM and 7th order Chebyshev (with 1 dB maximum passband attenuation but renormalized to produce 3 dB at the cut-off) filters. The LSM filter is chosen to be of higher order in order to get almost equal (slightly more selective) stopband attenuation characteristic as the Chebyshev one. It is easy to notice that even in 95% of the passband the LSM filter produces smaller absolute value of $F(\omega)$ than the Chebyshev one.

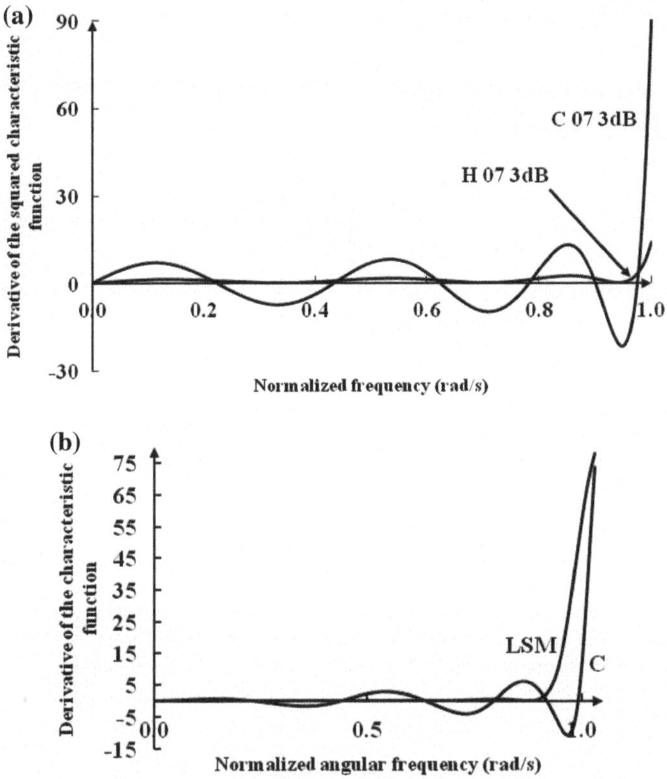

Fig. 7.14 Derivative of the characteristic function multiplied by the angular frequency (The function $F(\omega)$) of the **a** 7th order monotonic (Halpern) and Chebyshev (3 dB) filters and **b** 9th order LSM and 7th order Chebyshev (1 dB) filters

To further exemplify, suppose that the relative tolerance of the ith parameter be $100 \cdot \Delta x_i / x_i = 1\%$. From Fig. 7.14b at the normalized angular frequency of $\omega = 0.85$ rad/s for the LSM filter we read $F(\omega) = 0.092$ so that the partial increment is $(\Delta a_i)_{LSM} = 0.01 \cdot 0.092 = 9 \cdot 2 \cdot 10^{-4}$ dB. At the same normalized angular frequency for the Chebyshev filter we read $F(\omega) = 3.43$, so that $(\Delta a_i)_C = 0.01 \cdot 3.43 = 0.343$ mdB which is 35 times larger. Note that the increments due to separate elements get added to produce much larger overall increment. In a complex system that would mean the advantage of CMAC filters is substantial.

7.6 Developer's Corner

Having explicit expression for its zeros (Eq. 7.11) one may generate the Chebyshev polynomial in a factored form

$$T_m(x, \mathbf{x}) = \frac{\prod_{i=0}^{m-1}(x - x_i)}{\prod_{i=0}^{m-1}(1 - x_i)}. \tag{7.21}$$

This expression may become useful as an initial solution in subsequent developments where the characteristic function becomes rational.

References

1. Dimopoulos H (2012) Analog electronic filters theory design and synthesis. Springer, Heidelberg
2. Ralston A, Rabinowitz P (2001) A first course in numerical analysis. Dover Publications, New York
3. Topisirović D, Litovski V, Andrejević Stošović M (2015) Unified theory and state-variable implementation of critical-monotonic all-pole filters. Int J Circuit Theory Appl 43(4):502–515
4. Geher K (1971) Theory of network tolerances. Akademiai Kiadó, Budapest

Chapter 8
Low-Pass Selective Filters with Increased Selectivity

The characteristic function of the polynomial (CMAC and C) filters available, when requirements for increased selectivity is to be satisfied, a question arises as to what is better from the circuit complexity point of view: to increase the order of the filter (the order of the polynomial) or to introduce transmission zeros on the ω-axis so making the transfer function rational. This dilemma is regularly avoided in the literature and no solutions are offered which would help the designer to come to "the cheapest" physical system. Here we propose an algorithm for introduction of arbitrary (up to the maximum possible) number of transmission zeros located on the ω-axis based on the minimum stopband attenuation. It is shown by examples that by introduction of transmission zeros benefits are collected in both the passband and the stopband compared to polynomial solutions extended to the same circuit complexity. In the stopband the selectivity is raised while in the passband the reflected power is decreased (the attenuation in all but the upper end of the passband is reduced). Special extension of this chapter is related to the transfer functions with multiple transmission zeros including the ones exhibiting maximally flat attenuation in the origin.

8.1 Introduction

Having the characteristic function of a selective polynomial filter one may produce even more selective one by adding transmission zeros. In order not to significantly disturb the original characteristic function i.e. in order to preserve (as much as possible) its passband behavior, the "amplitude correction" is performed in the characteristic function domain. In that way the newly synthesized function will have new poles and zeros as compared with the original one.

V. Litovski, *Electronic Filters*, Lecture Notes in Electrical Engineering 596,
https://doi.org/10.1007/978-981-32-9852-1_8

As may be seen from simple analysis (and the examples) given below the method implemented by the *Chang_1* program of the *RM* software will lead to a new attenuation characteristic having lower attenuation in the passband than the original one while introduction transmission zeros being responsible for dramatic increase of selectivity.

In this chapter we will first describe a procedure for transfer function synthesis of amplitude corrected selective low-pass filters with distinct transmission zeros. That will be followed by analysis of the effects of the correction. Then, a special case will be described related to transfer functions with multiple transmission zeros. Discussion of the properties of such a new paradigm will be given, too. Design examples will be included.

8.2 A Procedure for Amplitude Characteristic Correction of Selective Polynomial Filters

The amplitude characteristic of a filter may be expressed as

$$|H(\mathrm{j}\omega)| = 1/\sqrt{1+\varepsilon^2 K(\omega)}. \tag{8.1}$$

Here $K(\omega)$ is the characteristic function while ε is a normalization constant

$$\varepsilon = \sqrt{10^{\frac{a_{max}}{10}} - 1} \tag{8.2}$$

where $a_{\max}$ is the maximum passband attenuation reached at the cut-off which, in turn, hereafter is normalized to unity.

The characteristic function of a polynomial filter is given by

$$L_{\mathrm{n}}(\omega^2) = \sum_{k=0}^{n} a_k \omega^{2\cdot k}. \tag{8.3}$$

This will be used as a prototype low-pass polynomial solution. The values of the coefficients a_k are given in Tables 8.1, 8.2, 8.3, 8.4 for the Halpern, Papoulis, LSM and Chebyshev filters, respectively. Note, for Butterworth filters one has

$$a_k = \begin{cases} 0 & \text{for} \quad k < n \\ 1 & \text{for} \quad k = n \end{cases}. \tag{8.4}$$

Given a prototype polynomial filter, to get a "corrected" characteristic function exhibiting higher selectivity one may use the following

Table 8.1 Coefficients of the characteristic function of the Halpern filters ($a_0 = 0$)

n	ω^2	ω^4	ω^6	ω^8	ω^{10}
3	2.000000000	–4.800000000	3.428571429		
4	0.000000000	6.400000000	–13.71428571	8.000000000	
5	3.333333333	–24.00000000	68.57142857	–80.00000000	32.72727273
6	0.000000000	2.160000000	–123.4285714	272.0000000	–261.8181818
7	4.666666667	–67.20000000	408.0000000	–1182.222222	1756.363636
8	0.000000000	51.20000000	–548.5714286	2453.333333	–5643.636364
9	6.000000000	–144.0000000	1491.428571	–7760.000000	22647.27273
10	0.000000000	100.0000000	–1714.285714	12666.66667	–50909.09091
n	ω^{12}	ω^{14}	ω^{16}	ω^{18}	ω^{20}
6	92.30769231				
7	1292.307692	373.3333333			
8	7015.384615	–4480.000000	1152.941176		
9	38769.23077	38640.00000	–20752.94118	4642.105263	
10	121476.9231	–176960.0000	154494.1176	–74273.68421	15120.00000

Table 8.2 Coefficients of the characteristic function of the Legendre filters ($a_0 = 0$)

n	ω^2	ω^4	ω^6	ω^8	ω^{10}
3	0.6666666667	–2.400000000	2.571428571		
4	0.000000000	2.400000000	–6.857142857	5.333333333	
5	0.6666666667	–6.400000000	24.00000000	–35.55555556	18.18181818
6	0.000000000	4.800000000	–34.28571429	93.33333333	–109.0909091
7	0.6666666667	–12.00000000	90.00000000	–315.5555556	559.0909091
8	0.000000000	8.000000000	–102.8571429	546.6666667	–1476.363636
9	0.6666666667	–19.20000000	236.5714286	–1443.555556	4887.272727
10	0.000000000	12.00000000	–240.0000000	2053.333333	–9469.090909
n	ω^{12}	ω^{14}	ω^{16}	ω^{18}	ω^{20}
6	46.15384615				
7	484.6153846	163.3333333			
8	2132.307692	–1568.000000	461.1764706		
9	9614.769231	10923.73333	–6640.941176	1671.157895	
10	25716.92308	–42336.00000	41505.88235	–22282.10526	5040.000000

Table 8.3 Coefficients of the characteristic function of the LSM filters ($a_0 = 0$)

n	ω^2	ω^4	ω^6	ω^8	ω^{10}
3	0.8478357666	−2.816621905	2.784877484		
4	0.000000000	3.581397291	−9.246042864	6.498047462	
5	0.07747767909	−0.3793896394	0.3482011173	0.6497128872	0.2007808143
6	0.000000000	−0.6222363543	−1.373596059	−0.2380818415	1.344616757
7	0.08341747199	−0.5896867915	1.490946443	−0.6050760313	−2.332466319
8	0.000000000	0.3972917906	−2.622217778	4.997676713	1.254333484
9	0.003913063684	−0.1664833221	2.351072607	−6.523278680	0.1154617063
10	0.000000000	0.2639234081	−3.297888191	14.51964825	−19.07312888
n	ω^{12}	ω^{14}	ω^{16}	ω^{18}	ω^{20}
6	0.5663048608				
7	1.313035871	1.583411876			
8	−9.069726500	1.111089490	4.889948439		
9	18.87490168	−14.01375771	−14.63589818	14.95759189	
10	26.49858854	80.46058524	−20.69152999	−73.42835671	48.71901359

Table 8.4 Coefficients of the characteristic function of the Chebyshev filters

n	ω^0	ω^2	ω^4	ω^6	ω^8	ω^{10}
3	0	8.95736083	−23.8862955	15.9241970		
4	0.99526231	−15.9241970	79.620985	−127.393576	63.6967881	
5	0	24.8815578	−199.05246	557.346896	−636.967881	254.787152
6	0.99526231	−35.8294433	418.010172	−1783.51006	3439.62656	−3057.44583
7	0	48.7678534	−780.285654	4681.71393	−13376.3255	19618.6107
8	0.99526231	−63.6967881	1337.63255	−10701.0604	42039.8801	−89685.0777
9	0	80.6162475	−2149.7666	−22070.9371	−113507.676	327911.065
10	0.99526231	−99.526231	3284.36563	−42039.8801	273259.221	−1020167.75
n	ω^{12}	ω^{14}	ω^{16}	ω^{18}	ω^{20}	
6	1019.14861					
7	−14268.080	4076.59444				
8	105991.455	−65225.511	16306.3777			
9	−556455.14	550340.249	−293514.79	65225.5110		
10	2318563.08	−3261275.5	2772084.22	−1304510.2	260902.044	

$$K(\omega,\mathbf{x}) = \frac{\prod_{i=1}^{m/2}(\omega_0 - x_i^2)^2}{\prod_{i=1}^{m/2}(\omega^2 - x_i^2)^2} \cdot L_n(\omega^2) \tag{8.5}$$

where m is the (even) order of the numerator of the new transfer function and x_i^2, $i = 1, 2, \ldots, m/2$, are the unknown abscissas of the attenuation poles at the ω-axis in the stopband. Note the analogies: transmission zeros are equivalent to transfer function zeros and consequently to attenuation poles. On the opposite side are attenuation zeros which are zeros of the characteristic function and do not appear explicitly in the transfer function. **x** stands for the vector representing the unknown attenuation poles. ω_0 is the cut-off angular frequency which, from now on, will be considered equal to unity. Note that the numerator is chosen such as the characteristic function to be equal to unity at the cut-off as $L_n(\omega^2)$ is already equal to unity.

Since, in the passband, $\omega < \omega_0$, every difference in de denominator of the new characteristic function will be larger than the difference in the numerator, one may conclude that the pass-band attenuation will be reduced in whole pass-band except for $\omega = \omega_0$ no matter which prototype is used.

With respect to Fig. 8.1, to find the values of x_i, $i = 1, 2, \ldots, m/2$, one has to solve iteratively the following system of equations

$$\varepsilon^2 \cdot K(y_j,\mathbf{x}) = \varepsilon^2 \cdot K_{\min} \quad \text{for } j = 1,2,\ldots,m/2. \tag{8.6}$$

In the above y_j stands for the abscissas of the minimums of the characteristic function in the stopband, and $\varepsilon^2 \cdot K_{\min}$ is the required maximum obtained as a solution of

$$10 \cdot \log[1 + \varepsilon^2 K_{\min}] = a_{\min} \tag{8.7a}$$

or

$$\varepsilon^2 \cdot K_{\min} = 10^{a_{\min}/10} - 1 \tag{8.7b}$$

where $a_{\min}$ is the required minimum *stopband* attenuation.

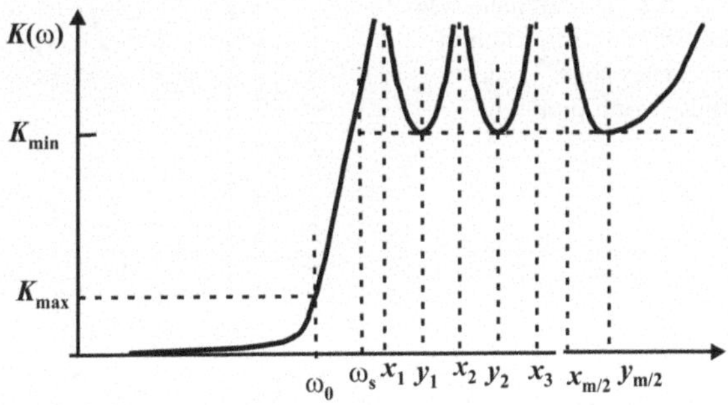

Fig. 8.1 Definition of the optimization problem

There were some attempts to produce this kind of characteristic function reported in the literature. We will group them in two categories: the ones which use nonlinear frequency transformations [1, 2], and the others related to specific prototype polynomial [3–6].

To avoid any limitation we will here propose an iterative procedure which will allow to produce new selective characteristic function based on any prototype, being of arbitrary order (n), and having arbitrary number of transmission zeros (m). Of course no restrictions will be imposed to the maximum passband attenuation $a_{\min}$ and minimum stopband attenuation $a_{\max}$.

Both **x** and **y** are unknown at the beginning. The procedure starts with a guess for **x**. Then it calculates **y** and solves (for one corrective step only) for the increments of **x**. That allows for the iterative process to proceed. To get a robust algorithm working with any kind of polynomial characteristic functions of any order n, any number of attenuation poles m, and any value of $a_{\min}$ and $a_{\max}$, one needs to implement a restrictive iteration damping of the Newton-Raphson process. In the next details about the procedure and some results will be given.

To start with we need a procedure to extract the abscissas of the maximums of the characteristic function in the stopband. To that end the derivative of the characteristic function (8.5) is found to be

$$\frac{dK(\omega^2)}{d\omega} = K(\omega^2) \cdot \left[\frac{1}{L_n(\omega^2)} \cdot \frac{L_n(\omega^2)}{d\omega} - 2 \cdot \sum_{j=1}^{m} \frac{\omega}{\omega^2 - x_j^2}\right]. \tag{8.8}$$

Here the multiplicative constant ε^2 is omitted for convenience. Now, the zeros of the characteristic function, if real, all belong to the passband so that the zeros of its derivative in the stopband may be found as the zeros of the following equation

$$\frac{1}{L_n(\omega^2)} \cdot \frac{L_n(\omega^2)}{d\omega} - 2 \cdot \sum_{j=1}^{m} \frac{\omega}{\omega^2 - x_j^2} = 0. \tag{8.9}$$

Figure 8.2 depicts the value of the left hand side of this equation as a function of frequency for 8th order Butterworth prototype. In this case $m = 3$ was used and

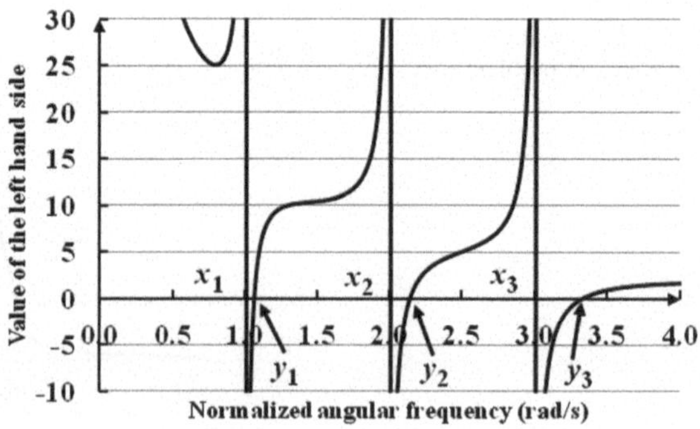

Fig. 8.2 The left hand side of Eq. (8.9) as a function of frequency for 8th order Butterworth prototype

$x_1 = 1$, $x_2 = 2$ and $x_3 = 3$ were chosen for illustration purposes only. From the figure we may see that the zeros of (8.9) belong to the intervals

$$y_j \in (x_j, x_{j+1}) \quad \text{for} \quad j = 1, 2, \ldots, m \tag{8.10}$$

with $x_{m+1} = \infty$. Having that, one may use the modified secant method described in the previous chapter to solve for the locations of the maximums of the characteristic function **y**.

Having the abscissas of the maximums of the characteristic function **y**, we may proceed with solving of (8.6) for **x**.

Since it is a system of nonlinear equations with respect **x**, we apply the Newton-Raphson linearization to get the following system of linear equations

$$\sum_{k=1}^{m/2} \frac{\partial K(y_j, \mathbf{x})}{\partial x_k} \cdot \Delta x_k = -K(y_j, \mathbf{x}) + K_{\min} \quad \text{for } j = 1, 2, \ldots, m/2. \tag{8.11}$$

This is to be solved once and the new values for the locations of the transmission zeros calculated:

$$\mathbf{x} = \mathbf{x} + \Delta\mathbf{x}. \tag{8.12}$$

If $\|\Delta\mathbf{x}\|$ is not satisfactorily small (e.g. smaller that 10^{-6}) we repeat the calculation of **y** and so close the iterative loop.

For the formulation of (8.11) one may use the following

$$\frac{\partial K(y_j, \mathbf{x})}{\partial x_k} = 2 \cdot K(y_j, \mathbf{x}) \cdot \frac{x_k \cdot (1 - y_j^2)}{(y_j^2 - x_k^2)(1 - x_k^2)}. \tag{8.13}$$

Good initial solutions are normally necessary for the iterative process to converge. If good enough these may accelerate the process, too. It was our experience that the following is satisfactory

$$x_k = \omega_s \cdot \left(1 + 1.5 \cdot \frac{k}{m/2}\right) \quad \text{for } k = 1, 2, \ldots, m/2. \tag{8.14}$$

where ω_s is the desired frequency at which the characteristic function reaches (looking from the passband side it rises to) $\varepsilon^2 \cdot K_{\min}$ for the first time.

Note, Newton-Raphson has quadratic convergence which means it produces the largest increment at the very beginning of the iterative process. That may prevent convergence. The reader is advised to use

$$\mathbf{x} = \mathbf{x} + \beta \cdot \Delta\mathbf{x} \tag{8.15}$$

instead of (8.12), with $\beta < 1$ or even $\beta \ll 1$ at the beginning of the iterations.

8.3 Properties of Amplitude Corrected Selective Polynomial Filters

In this paragraph we will demonstrate results of implementation of the procedure described above. The examples will include all four CMAC classes and the Chebyshev filters. The value of m will be varied to produce a notion on its influence to the overall solution. Note, the special case when Butterworth characteristic function is used as a prototype and the number of transmission zeros is largest possible (i.e. $m = n - 1$ for n odd and $m = n - 2$ for n even) is widely known as Inverse Chebyshev filter.

The first analyses will be related to the solutions where maximum number of transmission zeros are introduced. Some results are depicted in Figs. 8.3 and 8.4 as well as in Tables 8.5 and 8.6. Filters of 7th and 8th order were used as prototypes to be improved. Note, the Chebyshev prototype has 3 dB maximum attenuation all over the passband.

Let begin with discussion of the effects on the selectivity. First of all, one may notice in both Figs. 8.3 and 8.4 that the mutual relation of the prototypes was not changed. That, of course was expected. By comparing the first two rows in Tables 8.5 and 8.6 one may conclude that substantial rise of the selectivity was

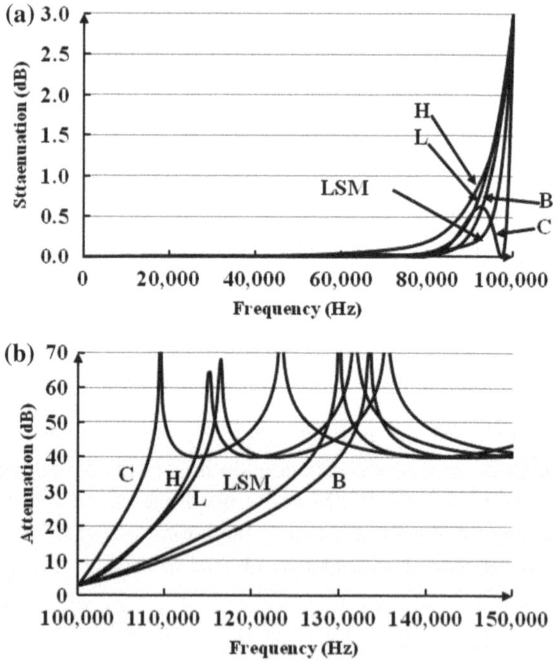

Fig. 8.3 **a** Passband and **b** stopband attenuation characteristic of amplitude corrected selective polynomial filters of 7th order

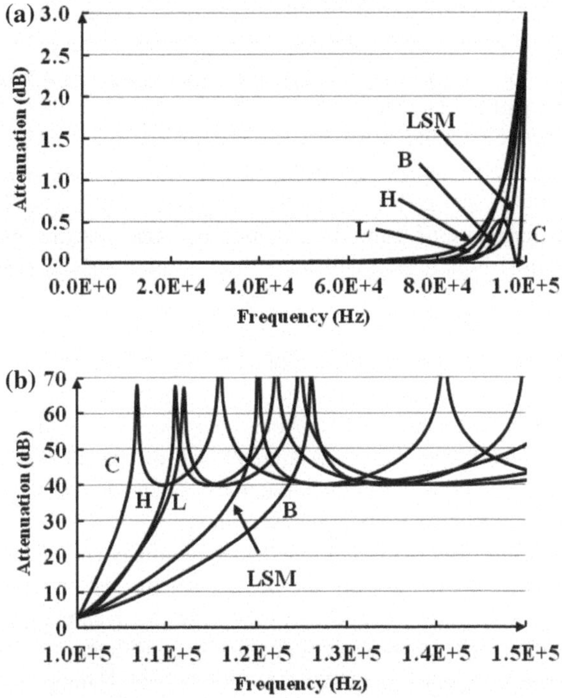

Fig. 8.4 Passband (top) and stopband (bottom) attenuation characteristic of amplitude corrected selective polynomial filters of 8th order, **a** passband and **b** stopband attenuation

Table 8.5 Normalized upper edge of the transition band (ω_s) for 7 and 10th order polynomial and 7th order corrected selective filters with maximum number of transmission zeros

Type	Chebyshev	Butter-worth	Halpern	Papoulis	LSM
Polynomial 7th order	1.4217	1.9313	1.4569	1.4966	1.8456
Corrected 7th order	1.0824	1.3006	1.1358	1.1460	1.2683
Polynomial 10th order	1.1636	1.5852	1.2309	1.2505	1.3871

Table 8.6 Normalized upper edge of the transition band (ω_s) for 8th order polynomial and corrected selective filters with maximum number of transmission zeros

Type	Chebyshev	Butterworth	Halpern	Papoulis	LSM
Polynomial 8th order	1.2595	1.7788	1.3563	1.3872	1.6215
Corrected 8th order	1.0586	1.2358	1.0998	1.1066	1.1814

obtained in all cases. For example, in the case of 7th order Chebyshev filter the transition region is reduced almost four times. Similar is the case with the LSM filters. The order of the prototype polynomial increased (Table 8.6) even larger relative improvement is obtained.

To fully verify the benefits of the introduction of transmission zeros we made an experiment. Namely, we raised the order of the 7th order prototype by three to equalize the number of freedoms with the improved functions. The results are depicted in the third row of Table 8.5. It may be seen that the improvement are almost halved in comparison with the case where transmission zeros were introduced.

Let consider now the effects on the passband attenuation. As already mentioned the corrective function keeps the attenuation at the cut-off untouched while reducing it as the frequency decreases toward the origin. That may be observed in both Figs. 8.3 and 8.4. The function which benefitted the most is the Chebyshev one since it has large values at the lower frequencies of the passband as shown in Fig. 7.7. The next function to benefit the most is the Halpern while the LSM function is the one which benefits almost nothing.

Based on the results described here one has to seriously consider the issue of the complexity (the value of n and m) of the filter no matter which technology will be used. Namely, it seems that the Chebyshev and the Halpern filter are not so repulsive any more as the passband attenuation is considered.

To complete the story one more example will be given. Namely, the algorithm for evaluation the locations of the transmission zeros described above is not affected by the number of zeros as long as it satisfies $m < n$. In cases when the asymptotic slope of the attenuation is of importance, however, lower values of m will be preferred (The asymptotic slope is calculated as $S = 6 \cdot (n - m)$ dB/oct). Such would be the case when our signal to be filtered contains large high-order harmonics which may distract the neighbouring channels if not eliminated at the proper spot.

One such result is depicted in Fig. 8.5 where the Butterwort prototype was corrected by single new transmission zero. Of course, improvements are obtained

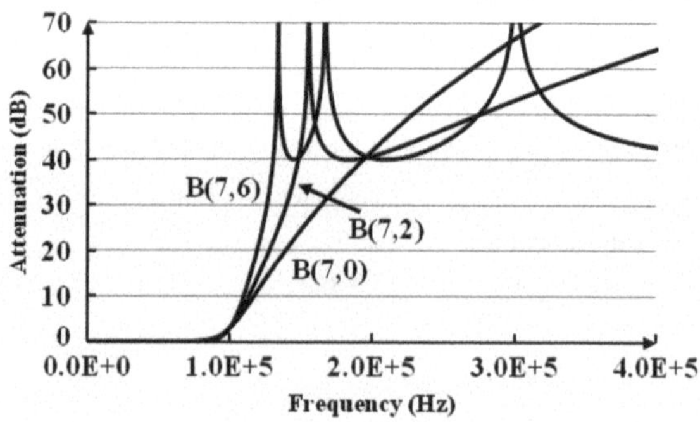

Fig. 8.5 Butterworth filter of 7th order B(7,0), The same extended by one pair of transmission zeros B(7,2), and the same extended by three pairs of transmission zeros (B7,6) (or Inverse Chebyshev)

Table 8.7 The normalized upper edge of the transition band for Butterworth and extended filters

Type	ω_s
B(7, 0)	1.93134
B(7, 2)	1.47667
B(7, 6)	1.30066
B(10, 0)	1.58526

according the price paid. We used this example to stress that the algorithm described in this chapter considers the Inverse Chebyshev filters an instance of improved Butterworth filters.

An additional important conclusion may be drawn from Table 8.7, however. Namely, we added the upper frequency of the stopband for the 10th order Butterworth to be compared with the case of 7th order with single transmission zero. It comes that one zero is more valuable than three poles. The transition region width calculated as

$$\omega_t = \omega_s - \omega_0 \tag{8.16}$$

is significantly smaller for the case B(7, 2) the relative difference being

$$100 \cdot \left[\omega_{B(10,0)} - \omega_{B(7,2)}\right] / \omega_{t,B(10,0)} = 19\% \tag{8.17}$$

which is significant. Note, in any technology a function being of higher order by 2 would need one more second order cell.

Having in mind the increment of the values of ω_s from B(7, 0) to B(10, 0) and result represented by (8.17), one may conclude that even higher order polynomial solutions may be substituted with the one with transmission zeros. Hence the immense importance of this chapter.

8.4 Design Example of Corrected Selective Filters

In this section excerpt will be given of the report produced by the *Chang_1* program of the *RM* software describing part of the results related to the synthesis of amplitude corrected LSM filter of 7th order [7, 8]. The attenuation characteristic of this filter is depicted in Fig. 8.4.

Start.E

Project name: LSM_8
Order of the numerator n=6 Order of the denominator m=8

Polynomial of the characteristic function given as input
t[1]=-1.110223025e-016 t[2]=0.000000000e+000
t[3]=0.000000000e+000 t[4]=0.000000000e+000
t[5]=4.942619340e-001 t[6]=0.000000000e+000
t[7]=-3.044760292e+000 t[8]=0.000000000e+000
t[9]=5.595749207e+000 t[10]=0.000000000e+000
t[11]=1.373229932e+000 t[12]=0.000000000e+000
t[13]=-9.778986741e+000 t[14]=0.000000000e+000
t[15]=1.184813034e+000 t[16]=0.000000000e+000
t[17]=5.170955242e+000 t[18]=0.000000000e+000

++

SOLUTION

Order of the Numerator=6,
Order of the denominator=8,

The results about the poles, zeros and coefficients listed below are valid for frequency normalization with
wc=2*pi*fc=6.28318530e+005

The set of results below are related to normalization so that at the edge of the pass-band is Amax=3.00000e+000

Real parts of the poles:

-8.422599750e-002	-8.422599750e-002	-3.291283913e-001
-3.291283913e-001	-8.225961467e-001	-8.225961467e-001
-1.485394193e+000	-1.485394193e+000	

Imaginary parts of the poles:

9.990882071e-001	-9.990882071e-001	9.781760610e-001
-9.781760610e-001	8.703755444e-001	-8.703755444e-001
4.139521385e-001	-4.139521385e-001	

Coefficients of the denominator of the transfer function:
3.651551335e+000 1.161949862e+001 2.362830468e+001
3.290652279e+001 3.409638874e+001 2.627459938e+001
1.469687003e+001 5.442689457e+000 1.000000000e+000
Real parts of the zeros:
0.000000000e+000 0.000000000e+000 0.000000000e+000
0.000000000e+000 0.000000000e+000 0.000000000e+000
Imaginary parts of the zeros:
1.202323383e+000 -1.202323383e+000 1.408406545e+000
-1.408406545e+000 2.171489364e+000 -2.171489364e+000
Coefficients of the numerator of the transfer function:
1.535715627e+000 2.968018381e+000 5.583102423e+000
5.615508679e+000 4.975622756e+000 2.471901071e+000
1.000000000e+000

The attenuation reaches 4.00000e+001 dB at the frequency of 1.18143e+005 rad/s
End.E

8.5 Filters with Multiple Transmission Zeros

The idea of multiple transmission zeroes is related to the need for a frequency interval in the stopband in which extremely high attenuation occurs. In that way a component of the spectrum (or several of them) which are of high value may be fully eliminated before disturbing the neighbouring channel. Attempts were reported to produce this kind of transfer function [9, 10] restricted to maximally flat solutions. We will here first demonstrate a general method for synthesis of selective amplitude characteristic with multiple transmission zeros of any multiplicity. Then we will elaborate [11] a method for creating maximally flat characteristic functions with multiple transmission zeros in a closed form.

The *RM* software for filter design has a routine named *Chang_MZ* which is performing the synthesis of transfer functions with multiple transmission zeros. Its structure and its way of working is similar to the *Chang_1* program so in the next, no design example will be give.

8.5.1 General Prototype

The solution offered here (for the general case) is not much different from the one just described. Simply, instead of distinct values all zeros have the same value to produce the following characteristic function

$$K(\omega,\mathbf{x}) = \frac{\left(\omega_0 - x_{\mathrm{A}}^2\right)^m}{\left(\omega^2 - x_{\mathrm{A}}^2\right)^m} \cdot L_n(\omega^2) \tag{8.18}$$

where x_{A} is the abscissa of the attenuation pole. $L_n(\omega^2)$ is the given polynomial of the characteristic function to be improved.

To find the value of x_{A} one has to solve iteratively the following

$$\frac{\left(\omega_0 - x_{\mathrm{A}}^2\right)^m}{\left(y^2 - x_{\mathrm{A}}^2\right)^m} \cdot L_n(\omega^2) = K_{\min} \tag{8.19}$$

with reference to (8.7a), (8.7b). Here y stands for the (unknown) abscissa of the only minimum of the characteristic function in the stopband.

The solution may proceed exactly as for the case of distinct zeros with proper accommodation of the expressions for the derivatives.

Figure 8.6 depicts the attenuation characteristics of the 7th order Chebyshev filter (C(7, 0)) with 3 dB passband attenuation, the corrected Chebyshev (C(7, 6)) with three distinct zeros, and the one with a triple zero (C(7, 6) MZ). Note, MZ is used as abbreviation for multiple zeros.

Looking to the passband, the C(7, 6) MZ case is less improved than the C(7, 6) one. Similarly, in the stopband, the C(7, 6) solution is more selective while both are noticeable better than the C(7, 0) case. So, generally speaking, no improvements are to be expected except for the interval of very large attenuation in part of the stopband.

The benefits are clearly noticeable from Fig. 8.7 where enlarged part of the stopband attenuation characteristic for C(7, 6) and C(7, 6) MZ is depicted. As can be seen there exist a wide frequency interval (more than 45% of the width of the passband) in which the attenuation is for at least 10 dB higher than the nominal minimum stopband attenuation of both examples.

8.5.2 Maximally Flat Prototype

We will introduce here the term Maximally Flat with maximum number of transmission zeros with Maximum Multiplicity (MFMM) [11]. In solutions described in the previous paragraphs a restriction was imposed to the number of transmission zeros of the form "$n > m$". That restriction is obvious for odd n. For even n,

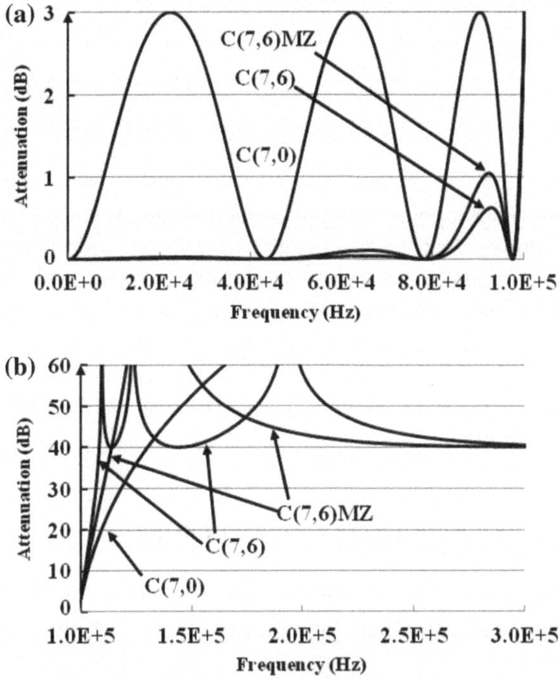

Fig. 8.6 Attenuation characteristics of the 7th order Chebyshev filter (C(7,0)) with 3 dB passband attenuation, the corrected Chebyshev (C (7,6)) with three distinct zeros, and the one with a triple zero (C(7,6)MZ), **a** passband and **b** stopband

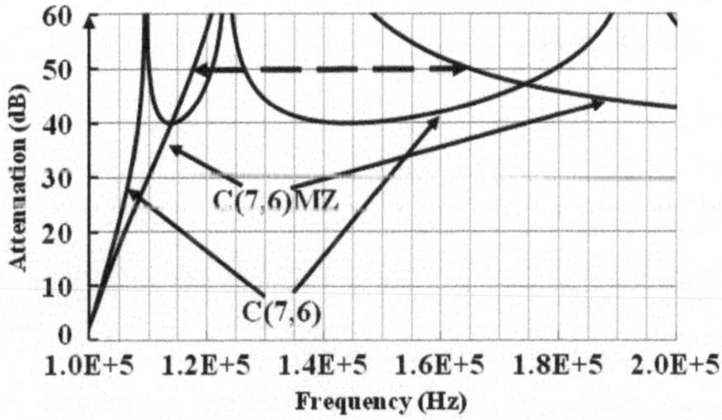

Fig. 8.7 Enlarged part of Fig. 8.6

however, there is no justification for that restriction since filters with zero valued asymptotic slopes are possible. Simply, in the next $n = 2m$ will be allowed.

Having that in mind in the next we will try to accomplish the following. First, we will give closed-form expressions for generation of the MFMM amplitude characteristic, and then we will give new MFMM characteristic functions for band-pass, band-stop, and high-pass cases. Based on these, we will recommend a simple design procedure, too.

The characteristic function of the low-pass MFMM filter may be written as:

$$K(\omega^2) = \frac{(1-\omega_A^2)^{2\cdot\lfloor n/2\rfloor}}{(\omega^2-\omega_A^2)^{2\cdot\lfloor n/2\rfloor}}\cdot\omega^{2n}, \tag{8.20}$$

where $\lfloor.\rfloor$ denotes the flooring function. ω_A denotes the transmission zero (or the attenuation pole) of the filter. Its multiplicity is $\lfloor n/2\rfloor$. In the above expression it is supposed that the bandwidth of the filter is normalized to unity.

The first $2n$–1 derivatives of this function with respect to ω in the origin are equal to zero hence it is maximally flat.

There is only one available requirement that may be fulfilled by a function of this kind. It is the maximum gain (or, in other words, the minimum attenuation) in the stop-band, A_{max}. Having that in mind we may start the development of the design procedure.

8.5.2.1 Even-Order Low-Pass Filters

For even n the value of the gain at infinity is finite and equal to A_{max}, so a simple design equation may be written:

$$1/\left[1+\varepsilon^2\cdot K(\omega^2)_{|\omega\to\infty}\right] = A_{max}^2 \tag{8.21}$$

which becomes:

$$(1-\omega_A^2)^n = \alpha \tag{8.22}$$

where

$$\alpha = (1/A_{max}^2 - 1)/\varepsilon^2 \tag{8.23}$$

Combining (8.22) and (8.23) one gets:

$$(1-\omega_A^2) = -\sqrt[n]{\alpha} \tag{8.24}$$

The negative sign comes since n is even.
Finally,

$$\omega_A = \sqrt{1+\sqrt[n]{\alpha}}. \tag{8.25}$$

Table 8.8 contains the values of ω_A for several values of the minimum attenuation in the stop-band (a_{min}) for ε^2 = 1 and for n = 6. The following was used:

Table 8.8 Transmission zero and the stopband edge frequency for $n = 6$

n	a_{min} (dB)	ω_A	ω_s
6	20	1.775	1.255
6	25	1.900	1.343
6	30	2.040	1.442
6	35	2.198	1.554
6	40	2.375	1.6795
6	45	2.574	1.820
6	50	2.795	1.976

$$a_{min} = 20 \cdot \log_{10}(1/A_{max}). \tag{8.26}$$

Table 8.8 contains also the value of ω_s which is the frequency at which the amplitude characteristic reaches A_{max} for the first time. It is a measure of the selectivity. For n even it may be calculated from

$$\omega_s = \omega_A/\sqrt{2} \tag{8.27}$$

For further illustration, if $n = 8$, for $A_{max} = 0.01$ ($a_{min} = 40$ dB) one gets $\omega_A = 2.0401563$ and $\omega_s = 1.4426084$. The amplitude characteristic so obtained is depicted in Fig. 8.8.

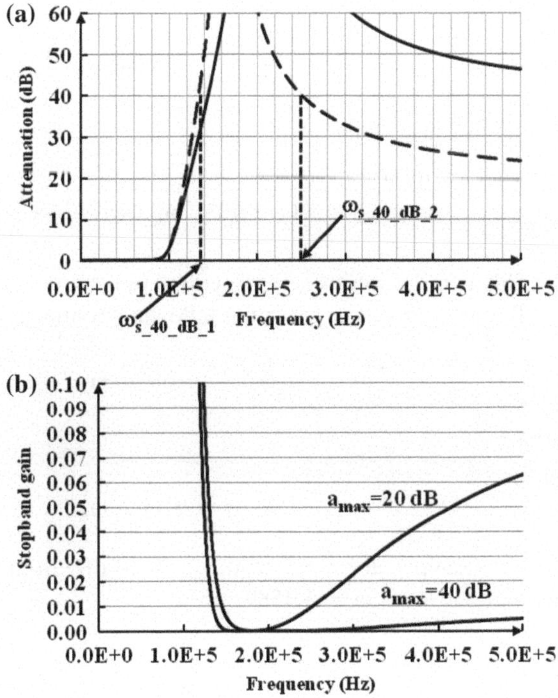

Fig. 8.8 **a** MFMM low-pass attenuation characteristic for $n = 8$ and $a_{min} = 40$ dB (full line) and for $n = 8$ and $a_{min} = 20$ dB (dashed line), and **b** stopband gain of the same

To appreciate this property we will give a numerical example. Let $n = 8$ and $a_{min} = 20$ dB ($\omega_{A\_20dB} = 1.6661472$, $\omega_{s\_20dB} = 1.17814404$, as depicted in Fig. 8.8a by doted lines.), the attenuation of this new MFMM function is higher than 40 dB in the frequency band from $\omega_{s\_40dB\_1} = 1.3332837$ to $\omega_{s\_40\_dB\_2} = 2.5165120$ which is $\Delta\omega = 1.1832283$ i.e. for approximately 20% broader than the pass-band width of the neighboring channel so preventing inter-channel interference. This filter, while designed for $a_{min} = 20$ dB, has for 8.2% smaller transition band at 40 dB than the one designed to have $a_{min} = 40$ dB. So, by careful choice of the value of a_{min}, when using MFMM function, one may get much better selectivity than in a possible original design.

8.5.2.2 Odd-Order Low-Pass Filters

The evaluations for n odd, here given for completeness, have to be a bit different. Namely, in this case the maximum of the gain in the stop-band is not at infinity.

At ω_B (the frequency of the minimum) the first derivative of $K(\omega^2)$ has to be equal to zero. After differentiation of (8.20) and equating the derivative to zero we obtain

$$\omega_B = \omega_A \cdot \sqrt{n} \tag{8.28}$$

Now, from

$$1/\left(1 + \varepsilon^2 \cdot K(\omega^2)_{|\omega=\omega_B}\right) = A_{max}^2 \tag{8.29}$$

and (8.20), we obtain

$$\left(1 - \omega_A^2\right)^{n-1} \omega_A^2 = \alpha \cdot \frac{(1 - 1/n)^{n-1}}{n} \tag{8.30}$$

This is to be solved iteratively to get ω_A.

To get ω_s one is to solve the following nonlinear equation:

$$\omega_s^{2n} - \gamma \cdot (\omega_s^2 - \omega_A^2)^{n-1} = 0 \tag{8.31}$$

where

$$\gamma = \alpha/(1 - \omega_A^2)^{n-1}. \tag{8.32}$$

For example, for $n = 7$, and $a_{min} = 40$ dB, from (8.30) and (8.31), one gets: $\omega_A = 1.8307$ and $\omega_s = 1.4589$.

8.5.2.3 Design Procedure for Low-Pass MFMM Filters

The design task in this case may be stated as: assuming the maximum allowed passband loss to be a_{max} dB (i.e. $\varepsilon^2 = 10^{a_{max}/10} - 1$) and given ω_s and a_{min}, one needs to find n and ω_A.

To solve this problem for n-even from (8.27) we have: $\omega_A = \omega_s\sqrt{2}$. Then, substituting in (8.24) we get $n = \log(\alpha)/\log(2\omega_s^2 - 1)$.

For odd n we have to solve a pair (system) of nonlinear Eqs. (8.30) and (8.31) for n and ω_A.

To illustrate we will elaborate the following example. Let $\omega_s = 1.6$, and $a_{min} = 40$ dB. Find n and ω_A.

If even n was presumed as initial solution one would obtain $\omega_A = 2.262741699$ and $n = 6.50508164$. So, we choose $n = 7$, for which we already calculated $\omega_A = 1.8307$. Now, from (8.31), the corresponding $\omega_s = 1.4589$ satisfies the imposed requirement.

The above procedure may be recommended as general. Since the even n case is simple to manipulate, one is first to start with it. If unsatisfactory, at least n will be known (now odd valued) and only (8.30) will be needed to be solved for ω_A.

8.5.2.4 The Transfer Function of an MFMM Filter

After substitution (8.20) into (8.1), and after some rearrangements, the amplitude squared may be written in the form:

$$\left|H(s)_{|s=j\omega}\right|^2 = \frac{\prod_{i=1}^{\lfloor n/2 \rfloor}\left(\omega^2 - \omega_A^2\right)^2}{\prod_{i=1}^{\lfloor n/2 \rfloor}\left(\omega^2 - \omega_A^2\right)^2 + \varepsilon^2\left(1 - \omega_A^2\right)^{2\cdot\lfloor n/2 \rfloor} \cdot \omega^{2n}} \tag{8.33}$$

From here we get the poles of the transfer function after:

- 1 substituting ω^2 by $-s^2$;
- 2 solving the denominator polynomial for s; and
- 3 separating the left-half plane zeros.

These will be used to create the denominator (Hurwitz) polynomial of the transfer function.

The reader will notice that we could explain how the poles are obtained much earlier in this book but he/she should appreciate the fact that due to the simplicity of the characteristic function this is a good opportunity to exploit.

8.5.2.5 Band-Pass MFMM Filters

A band-pass amplitude characteristic may be obtained from a low-pass one by substituting the proper LP-to-BP transformation (4.1b) which is here rewritten for convenience

$$\frac{\omega}{\omega_0} = \frac{1}{B_r}\left(\frac{\Omega}{\omega_c} - \frac{\omega_c}{\Omega}\right) \tag{8.34}$$

where ω_c is the central frequency of the new band-pass filter, and $B_r = BW/\omega_c$ is the relative bandwidth. ω_h and ω_l are the upper and lower cut-off frequencies of the band-pass filter. The absolute bandwidth is given by $BW = \omega_h - \omega_l$, while, since geometrical symmetry is sought, we have $\omega_0^2 = \omega_h\omega_l$.

In the next we will assume normalization is performed in both the low-pass and the band-pass domain meaning that $\omega_0 = 1$ rad/s and $\omega_c = 1$ rad/s. If so (8.34) becomes

$$\omega \leftarrow \frac{1}{B_r \cdot \Omega}\left(\Omega^2 - 1\right) \tag{8.35}$$

After implementation of (8.35) into (8.20) and substituting Ω with ω we get:

$$K(x) = \frac{\left(1-\omega_A^2\right)^n\left(\omega^2-1\right)^{2n}}{\left[\omega^4 - \left(2+B_r^2\omega_A^2\right)\omega^2+1\right)]^n} \quad \text{for } n \text{ even} \tag{8.36}$$

and

$$K(x) = \frac{\left(1-\omega_A^2\right)^{n-1}\left(\omega^2-1\right)^{2n}}{B_r^2\omega^2\left[\omega^4 - \left(2+B_r^2\omega_A^2\right)\omega^2+1\right)]^{n-1}} \quad \text{for } n \text{ odd.} \tag{8.37}$$

To our knowledge this is the only characteristic function written in closed-form directly as band-pass.

The design procedure in the case of a band-pass filter starts again with the known ε^2. Assuming normalization performed i.e. $\omega_0 = 1$, to continue, one needs the value of ω_s which is obtained in the following way. From the requirements of the band-pass case one has ω_{sl} and ω_{sh} standing for the lower and upper edge of the transition band of the filter we use. By substitution of ω_{sh} into (8.35) one gets ω_s. The rest is to get n and ω_A as described earlier. In other words ω_A is calculated from (8.27) or (8.31) depending on the value of n. Of course, the resulting filter will be of order $2n$. For example, if $\omega_{sh} = 1.09$ and $B_r = 0.1$, from (8.35) $\omega_s = 1.7256880$ is obtained. Now, if $a_{min} = 40$ dB, we see from Table 8.8 that a sixth order MFMM prototype filter will suffice and $\omega_A = 2.375186618$. The final filter order will be 12 but in (8.36) and (8.37) we use $n = 6$.

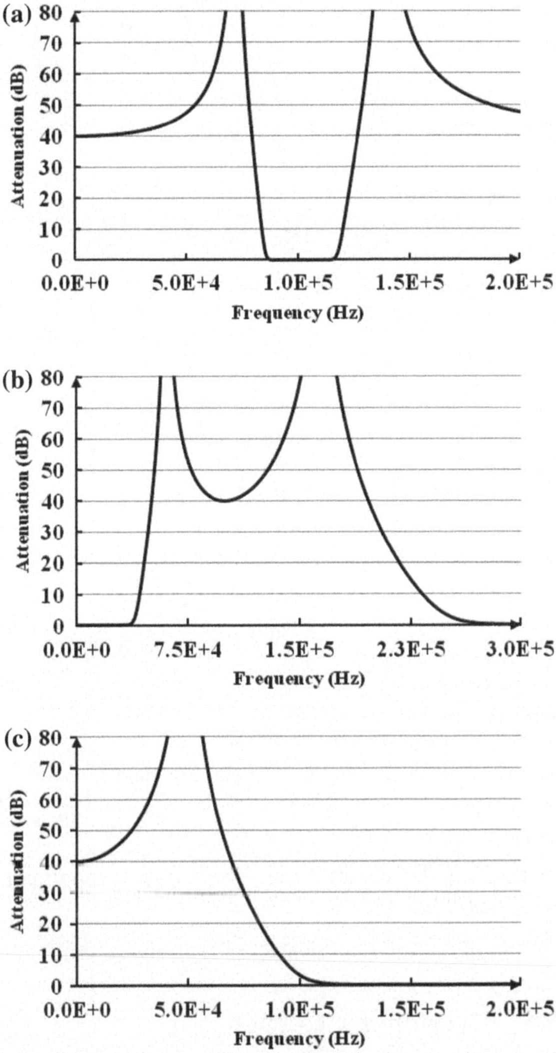

Fig. 8.9 Attenuation characteristics of the transformed 8th order MFMM filter. **a** band-pass, **b** band-stop, and **c** high-pass

Figure 8.9a illustrates the transformed band-pass 8th order MFMM filter with $B_r = 0.3$.

8.5.2.6 Band-Stop MFMM Filters

When normalization is implemented so that $\omega_0 = 1$, after implementation of (4.16b)–(8.20), and after proper rearrangements for the ban-stop MFMM characteristic function one obtains:

$$K(\omega^2) = \frac{(1-\omega_A^2)^n \cdot B_r^{2n} \cdot \omega^{2n}}{\left[\omega_A^2\omega^4 - \left(B_r^2 + 2\omega_A^2\right)\omega^2 + \omega_A^2\right)\right]^n} \quad \text{for } n \text{ even,} \tag{8.38}$$

and

$$K(\omega^2) = \frac{(1-\omega_A^2)^{n-1} \cdot B_r^{2n} \cdot \omega^{2n}}{\left[\omega_A^2\omega^4 - \left(B_r^2 + 2\omega_A^2\right)\omega^2 + \omega_A^2\right)\right]^{n-1}(\omega^2-1)^2} \quad \text{for } n \text{ odd.} \tag{8.39}$$

To our knowledge this is the only characteristic function written in closed-form directly as band-stop. Figure 8.9b depicts the transformed amplitude characteristics for $n = 8$, $a_{\min} = 40$ dB, and $B_r = 0.3$.

The design procedure is analogous to the one described for the band-pass filters.

8.5.2.7 High-Pass MFMM Filters

For the high-pass MFMM characteristic function one easily gets

$$K(\omega^2) = \frac{(1-\omega_A^2)^n}{(1-\omega_A^2\omega^2)^n} \quad \text{for } n \text{ even} \tag{8.40}$$

and

$$K(\omega^2) = \frac{(1-\omega_A^2)^{n-1}}{(1-\omega_A^2\omega^2)^{n-1}\omega^2} \quad \text{for } n \text{ odd.} \tag{8.41}$$

Figure 8.9c depicts the transformed amplitude characteristics for $n = 8$, $a_{\min} = 40$ dB.

8.6 Developer's Corner

We will repeat here the necessity for a good polynomial solver as well as a good solver of a system of linear equations.

The software developer, who is not in a position to write a program creating the characteristic functions of the polynomial filters, may memorize the tables given and instantiate a proper polynomial at will.

One should spend some effort and time before taking a decision on the order of the filter and the number of transition zeros. This may be of crucial importance for the complexity of the physical system. For example, an 8th order polynomial filter in cascade active RC technology will ask for 4 second order cells. A 6th order filter of the same approximation origin, having one or two transmission zeros will ask for three cells while producing higher selectivity. Of course, there are additional criteria to be considered such as phase distortions and sensitivity to tolerances but still, reducing the complexity of an 8th order filter by one second order cell means reducing it for 25%.

References

1. Temes G, Gyi M (1967) Design of filters with arbitrary pass-band and Chebyshev stopband attenuation. IEEE Int Conv. Rec. Pt 5:2–12
2. Rakovic B (1974) Designing monotonic low-pass filters - comparison of some methods and criteria. Int J Circuit Theory Appl 2(3):215–221
3. Chang C (1968) Maximally flat amplitude low-pass filters with an arbitrary number of pairs of real frequency transmission zeros. IEEE Trans Circuit Theory CT 15(4):465–467
4. Budak A, Aronhime P (1970) Maximally flat low-pass filters with steeper slopes at cutoff. IEEE Trans Audio Electroacoust AU 18(1):63–66
5. Dutta-Roy S (1971) On maximally flat sharp cutoff low-pass filters. IEEE Trans Audio Electroacoust AU 19(1):58–63
6. Harris F (1972) On maximally flat low-pass filters with ripple in the stopband. IEEE Trans Audio Electroacoust AU 18(5):345–352
7. Rakovich BD, Litovski VB (1974) Monotonic passband low-pass filters with Chebyshev stopband attenuation. IEEE Trans Acoust Speech Signal Process ASSP 22(1):39–44
8. Litovski VB, Raković BD (1974) Attenuation characteristics and element values of least-squares monotonic passband filters with an arbitrary number of transmission zeros. Publications of Electrical Engineering Faculty, University of Belgrade, Series ETA 97–101:67–101
9. Dutta-Roy SC (1971) On maximally flat sharp cutoff low-pass filters. IEEE Trans Audio Electroacoust 19(1):58–63
10. Wei CH (1968) An alternative approach for designing maximally flat low-pass filters with multiple order imaginary-axis zeros. IEEE Trans Circuits Syst 33(6):642–644; and Wei CH (1987) Correction to an alternative approach for designing maximally flat low-pass filters with multiple order imaginary-axis zeros. IEEE Trans Circuits Syst 34(1):109–109
11. Andrejević Stošović MV, Živanić JM, Litovski VB (2014) Maximally flat filter functions with maximum number of transmission zeros having maximal multiplicity. IEEE Trans Circuits Syst II: Express Briefs 61(10):778–782

Chapter 9
Modified Elliptic Low-Pass Filters

The ultimate selective amplitude characteristic is the one which approximates the gain in both the passband and in the stopband in equi-ripple manner. The passband attenuation is limited by a maximum value (a_{max}) while the stopband attenuation is limited by a minimum value (a_{min}). The mathematical apparatus allowing for such a solution is based on elliptic integrals. Since not everyone is familiar with this part of mathematics, tables are available to read the zeros and poles of the transfer function of the now elliptic filters. Elliptic filters as such have two main limitations: the values of a_{min} and a_{max} are coupled and generated from a single constant to be supplied by the designer; and the number of transmission zeros is restricted to its maximum. In this chapter we describe an iterative algorithm which leads to a solution based on the same equi-ripple requirements having no restrictions as above. We refer to the solution described here as the Modified elliptic filters. After a design example a short comparison with the filters obtained by extending the polynomial function with zeros will be given. Finally, advice is given as to how to select the overall complexity of the filter.

9.1 Introduction

The Elliptic or, by the author, Zolotarev filters were introduced in the second half of the 19th century [1–4]. Being the best solution to the amplitude characteristic problem, however, they attracted much attention in the next period [5–12]. These are about mini-max approximation of the attenuation both in the passband and in the stopband. For that purpose elliptic functions are used in place of the characteristic function of the filter, hence the name. The main limitations coming from the properties of the elliptic functions is that the number of attenuation poles is maximal and the values of the maximum passband attenuation (a_{max}) and minimum stopband attenuation (a_{min}) are not fully independent meaning that not any combination of

V. Litovski, *Electronic Filters*, Lecture Notes in Electrical Engineering 596,
https://doi.org/10.1007/978-981-32-9852-1_9

these two is possible. Of course, elliptic functions are not a simple matter to manipulate and need skills and serious amount of computational efforts to produce the final results.

Here we propose an iterative procedure which has no limitations. The number of attenuation poles is arbitrary (from 1 to the maxim) and the values of the maximum passband and minimum stopband attenuations are arbitrary and unrelated. Having that in mind we introduced a new name "Modified Elliptic" and the results contained in the output file produced by the program *Elliptic* of the *RM* software is entitled Elliptic_mod.

In the sequel we will first elaborate the algorithm we are proposing. It will be followed by a design example produced by the *Elliptic* program. Finally, to appreciate the mutual position, we will try to put together the Modified Elliptic and the Chebyshev filters having improved selectivity.

9.2 The Algorithm

The amplitude characteristic of a filter may be expressed as

$$|H(\mathrm{j}\omega)| = 1/\sqrt{1+\varepsilon^2 K(\omega)}. \tag{9.1}$$

here $K(\omega)$ is the characteristic function while ε is a normalization constant.

Given the design data (maximum passband attenuation, $a_{\max}$, and minimum stopband attenuation, $a_{\min}$) and the complexity of the filter (orders of the numerator, m, and denominator, n, polynomials) the characteristic function which may be optimized to produce highest selectivity of all may be written in the following form

$$K(\omega,\mathbf{x},\mathbf{y}) = \omega^{2(n-2\cdot\lfloor n/2\rfloor)} \frac{\prod_{i=1}^{m/2}\left(\omega_0^2 - x_i^2\right)^2}{\prod_{i=1}^{m/2}\left(\omega^2 - x_i^2\right)^2} \cdot \frac{\prod_{i=1}^{\lfloor n/2\rfloor}\left(\omega^2 - y_i^2\right)^2}{\prod_{i=1}^{\lfloor n/2\rfloor}\left(\omega_0^2 - y_i^2\right)^2} \tag{9.2}$$

where m is the (even) order of the numerator of the new transfer function and x_i^2, i = 1, 2, ..., $m/2$, are the unknown abscissas of the attenuation poles (in the stopband) at the ω-axis; $\mathbf{x}$ is the corresponding vector; n is the order of the denominator of the transfer function and y_i^2, i = 1, 2, ..., $\lfloor n/2\rfloor$, are the unknown abscissas of the attenuation zeros (in the passband) at the ω-axis; and $\mathbf{y}$ is the corresponding vector. ω_0 is the cut-off angular frequency which, from now on, will be considered equal to unity. The symbol $\lfloor . \rfloor$ denotes the flooring function.

To find the values of x_i, i = 1, 2, ..., $m/2$, and y_i, i = 1, 2, ..., $\lfloor n/2\rfloor$, one has to solve iteratively the following system of equations

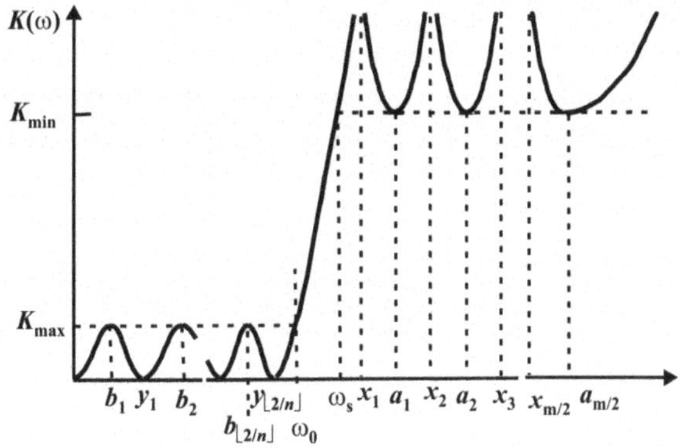

Fig. 9.1 Definition of the mini-max characteristic function

$$K\left(a_j, \mathbf{x}, \mathbf{y}\right) = \varepsilon^2 \cdot K_{\min}, \quad \text{for } j = 1, 2, \ldots, m/2. \tag{9.3}$$

$$K\left(b_j, \mathbf{x}, \mathbf{y}\right) = \varepsilon^2 \cdot K_{\max}, \quad \text{for } j = 1, 2, \ldots, \lfloor n/2 \rfloor. \tag{9.4}$$

In the above a_j stand for the abscissas of the minimums of the characteristic function in the stopband, b_j stand for the abscissas of the maximums of the characteristic function in the passband. $\varepsilon^2 \cdot K_{\max}$ is the required maximum obtained as a solution of (see Fig. 9.1)

$$10 \cdot \log\left[1 + \varepsilon^2 K_{\max}\right] = a_{\max} \tag{9.5a}$$

or

$$\varepsilon^2 \cdot K_{\max} = 10^{a_{\max}/10} - 1 \tag{9.5b}$$

while for $\varepsilon^2 \cdot K_{\min}$ we have

$$10 \cdot \log\left[1 + \varepsilon^2 K_{\min}\right] = a_{\min} \tag{9.5c}$$

$$\varepsilon^2 \cdot K_{\min} = 10^{a_{\min}/10} - 1. \tag{9.5d}$$

Since usually $\omega_0 = 1$ (rad/s) and $K_{\max}\omega_0 = 1$, one has

$$\varepsilon^2 = 10^{a_{\max}/10} - 1. \tag{9.6}$$

where $a_{\max}$ is usually reached at $\omega = \omega_0$.

Both **x** and **y** are unknown at the beginning. So are **a** and **b**. The procedure starts with a guess for **x** and **y**. Then it calculates **a** and **b** and solves (for one corrective step only) for the increments of **x** and **y**. That allows for the iterative process to proceed. To get a robust algorithm working with any kind of polynomial characteristic functions of any order *n*, any number of attenuation poles *m*, and any value of $a_{\min}$ and $a_{\max}$, one needs to implement a very restrictive iteration damping.

The derivative of the characteristic function which is necessary to find the locations of extrema a_j and b_j may be expressed as follows

$$\frac{1}{K(\omega)}\cdot\frac{dK(\omega)}{d\omega}=\frac{2}{\omega}(n-2\cdot\lfloor n/2\rfloor) +4\cdot\omega\cdot\left(\sum_{k=1}^{\lfloor n/2\rfloor}\frac{1}{\omega^2-y_k^2}-\sum_{k=1}^{m}\frac{1}{\omega^2-x_k^2}\right). \tag{9.7}$$

It is depicted in Fig. 9.2 for the case of a 7th order filter with two transmission zeros at the positive half of the imaginary axis. From the figure we may see that the zeros of (9.7) belong to the intervals

$$a_j\in\left(y_j,y_{j+1}\right)\text{ for }\quad j=0,2,\ldots,\lfloor n/2\rfloor \tag{9.8a}$$

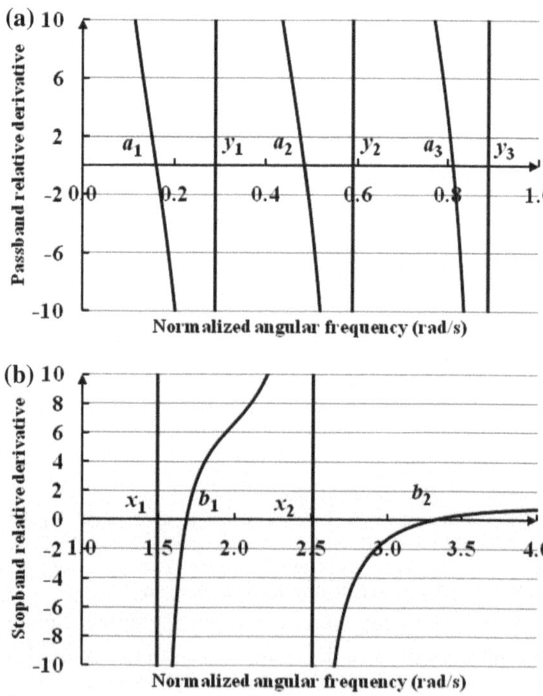

Fig. 9.2 The relative derivative expressed by (9.7) for a 7th order filter with two transmission zeros at the positive ω axis, **a** passband and **b** stopband

$$b_j \in \left(x_j, x_{j+1}\right) \text{ for } \quad j = 1, 2, \ldots, m/2 \tag{9.8b}$$

with $y_0 = 0$ and $x_{m/2+1} = \infty$. Having that, one may use the modified secant method described in Chap. 7 to solve for the locations of the maximums of the characteristic function **a** and **b**.

Having the abscissas of the maximums of the characteristic function **a** and **b**, we may proceed with solving of (9.7) for **x** and **y**.

Since it is a system of nonlinear equations with respect **x** and **y**, we apply the Newton-Raphson linearization to get the following system of linear equations

$$\sum_{k=1}^{m/2} \frac{\partial K\left(a_j, \mathbf{x}, \mathbf{y}\right)}{\partial x_k} \cdot \Delta x_k = -K\left(a_j, \mathbf{x}, \mathbf{y}\right) + K_{\min} \text{ for } \quad j = 1, 2, \ldots, \lfloor n/2 \rfloor. \tag{9.9a}$$

$$\sum_{k=1}^{m/2} \frac{\partial K\left(b_j, \mathbf{x}, \mathbf{y}\right)}{\partial y_k} \cdot \Delta y_k = -K\left(b_j, \mathbf{x}, \mathbf{y}\right) + K_{\max} \text{ for } \quad j = 1, 2, \ldots, m/2. \tag{9.9b}$$

This is to be solved as a single system once and the values of the locations of the transmission zeros updated as:

$$\mathbf{x} = \mathbf{x} + \Delta\mathbf{x} \tag{9.10a}$$

and

$$\mathbf{y} = \mathbf{y} + \Delta\mathbf{y}. \tag{9.10a}$$

If $\|\Delta\mathbf{x}\| + \|\Delta\mathbf{y}\|$ is not satisfactorily small (e.g. smaller that 10^{-6}) we repeat the calculation of **a** and **b** and so close the iterative loop.

The derivatives necessary for formulation of the system (9.9) may be calculated from

$$\frac{1}{K\left(a_j, \mathbf{y}, \mathbf{x}\right)} \frac{\partial K\left(a_j, \mathbf{y}, \mathbf{x}\right)}{\partial x_k} = 4 \cdot x_k \cdot \left(\frac{1}{\omega_0^2 - x_k^2} - \frac{1}{a_j^2 - x_k^2}\right) \tag{9.11a}$$

$$\frac{1}{K\left(b_j, \mathbf{y}, \mathbf{x}\right)} \frac{\partial K\left(b_j, \mathbf{y}, \mathbf{x}\right)}{\partial y_k} = -4 \cdot y_k \cdot \left(\frac{1}{\omega_0^2 - y_k^2} - \frac{1}{a_j^2 - y_k^2}\right). \tag{9.11b}$$

Good initial solutions are normally necessary for the iterative process to converge. If good enough these may accelerate the process, too. It was our experience that the following is satisfactory

$$x_j = \omega_s \cdot \left(1 + 1.5 \cdot \frac{j}{m/2}\right) \quad \text{for} \quad j = 1, 2, \ldots, m/2. \tag{9.12a}$$

$$y_j^2 = (2 \cdot j + 0.5)/n \quad \text{for} \quad j = 1, 2, \ldots, \lfloor n/2 \rfloor \tag{9.12b}$$

where ω_s is the *desired* frequency at which the characteristic function reaches (looking from the passband side it rises to) $\varepsilon^2 \cdot K_{\min}$ for the first time, as depicted in Fig. 9.1.

Note, Newton-Raphson has quadratic convergence which means it produces the largest increment at the very beginning of the iterative process. That may prevent convergence. The reader is advised to use

$$\mathbf{x} = \mathbf{x} + \beta \cdot \Delta\mathbf{x} \tag{9.13a}$$

$$\mathbf{y} = \mathbf{y} + \beta \cdot \Delta\mathbf{y} \tag{9.13a}$$

instead of (9.10), with $\beta < 1$ at the beginning of the iterations.

9.3 Synthesis Example

The example given below was produced by the *Elliptic* program of the *RM* software. It was chosen to be simple enough for the details to be recognizable and complex enough for the properties of the Modified Elliptic filters to be pronounced. The filter is of 7th order with maximum ($m = 6$) number of transmission zeros at the ω axis. The passband attenuation was chosen to be small ($a_{\max} = 0.1$ dB only) while the stopband attenuation is moderate ($a_{\min} = 40$ dB).

Figure 9.3 depicts the responses of this filter in the frequency and time domain. It is followed by excerpts from the reports related to this function generated by the *Elliptic* and the *LP_Analysis* programs. Note, the transfer function was renormalized after synthesis to produce 3 dB attenuation at cut-off.

Except for the extreme selectivity (which is unreachable by any other approximation criterion) there are not much to note as a difference between the polynomial Chebyshev and the Modified Elliptic. Of course, the pick of the group delay is much more pronounced and so is the undershoot of the step response (despite the fact that only 0.1 dB lobes are present in the passband attenuation characteristic).

Here are excerpts from the reports produced by the *RM* software. First goes the report produced by the transfer function synthesis program *Elliptic*. To reduce the length of the text the report was edited and some data are transformed into tables.

Start.E

++

Project name: allip_7_6_01_40

Order of the filter=7 Number of transmission zeros on the omega axis=3

++

FINAL SOLUTION

Input requirements:
Pass-band attenuation=0.1000
Stop-band attenuation=40.0000
Order of the filter=7 Order of the numerator=6

The results about the poles, zeros and coefficients listed below are valid for frequency normalization with
wc=2*pi*fc=6.28318530e+005

Locations of the m=6 zeros/poles (rad/s) of attenuation:
5.835793889e-001 8.892378953e-001 9.899571711e-001
1.115674154e+000 1.242040655e+000 1.892578271e+000
Transmission zeros (rad/s): 1.115674154e+000 1.242040655e+000
1.892578271e+000
Locations of the extrema (rad/s):
3.188193805e-001 7.712737337e-001 9.564762354e-001
1.154727763e+000 1.431874338e+000 3.464206700e+000
Values of the extrema [dB]:
9.999999999e-002 9.999999997e-002 9.999999999e-002
4.000000004e+001 4.000000009e+001 4.000000000e+001

Zeroes of the transfer function (rad/s)

0.	1.1156741539
0.	-1.1156741539
0.	1.2420406552
0.	-1.2420406552
0.	1.8925782707
0.	-1.8925782707

POLYNOMIAL P(w**2) = squared denominator of the characteristic function of order=7
0.000000000e+000 6.965258708e-002 -7.273571624e-001
3.013314854e+000 -6.346643256e+000 7.213688859e+000
-4.222648276e+000 1.000000000e+000

POLYNOMIAL S(w**2)= squared numerator of the characteristic function of order=6
4.730511835e+001 -1.637515742e+002 2.293244824e+002
-1.653977286e+002 6.437576480e+001 -1.273849264e+001
1.000000000e+000

Poles of the transfer function

-2.901174e-002	-1.018648e+000
-2.901174e-002	1.018648e+000
-3.728252e-001	-7.016109e-001
-3.728252e-001	7.016109e-001
-5.940394e-001	0.000000e+000
-1.311660e-001	9.557825e-001
-1.311660e-001	-9.557825e-001

Coefficients of the numerator of the transfer function:
Ascending exponent of frequency (s)
1.000000000e+000 0.000000000e+000 1.730801866e+000
0.000000000e+000 9.260490232e-001 0.000000000e+000
1.453938154e-001
Descending exponent of frequency (s)
1.453938154e-001 0.000000000e+000 9.260490232e-001
0.000000000e+000 1.730801866e+000 0.000000000e+000
1.000000000e+000

Coefficients of the denominator of the transfer function:
Ascending exponent of frequency (s)
1.000000000e+000 3.202337248e+000 6.593176484e+000
1.008612954e+001 1.021966562e+001 9.622992181e+000
4.580123065e+000 2.759035073e+000
Descending exponent of frequency (s)
2.759035073e+000 4.580123065e+000 9.622992181e+000
1.021966562e+001 1.008612954e+001 6.593176484e+000
3.202337248e+000 1.000000000e+000
The resulting function will be re-normalized so that at the cut-off to have 3.00000e+000

Frequency at which the attenuation of 40.000 dB
is reached for the first time is f=1.07850e+005 [Hz]

The next is part of the report produced by the *LP_Analysis* program.

****
delay time (at 0.5)=5.88920e-006 rise time (0.1-0.9)=5.60833e-006
****
The number of oscillations at the beginning of the step response is=0
****

The number of oscillations at the end of the step response is=9

time=1.15767e-005	value=1.18235e+000
time=1.75214e-005	value=8.83966e-001
time=2.31182e-005	value=1.07682e+000
time=2.84568e-005	value=9.39599e-001
time=3.37136e-005	value=1.05054e+000
time=3.89203e-005	value=9.56861e-001
time=4.40825e-005	value=1.03702e+000
time=4.92042e-005	value=9.68046e-001
time=5.42931e-005	value=1.02779e+000

****

End.E

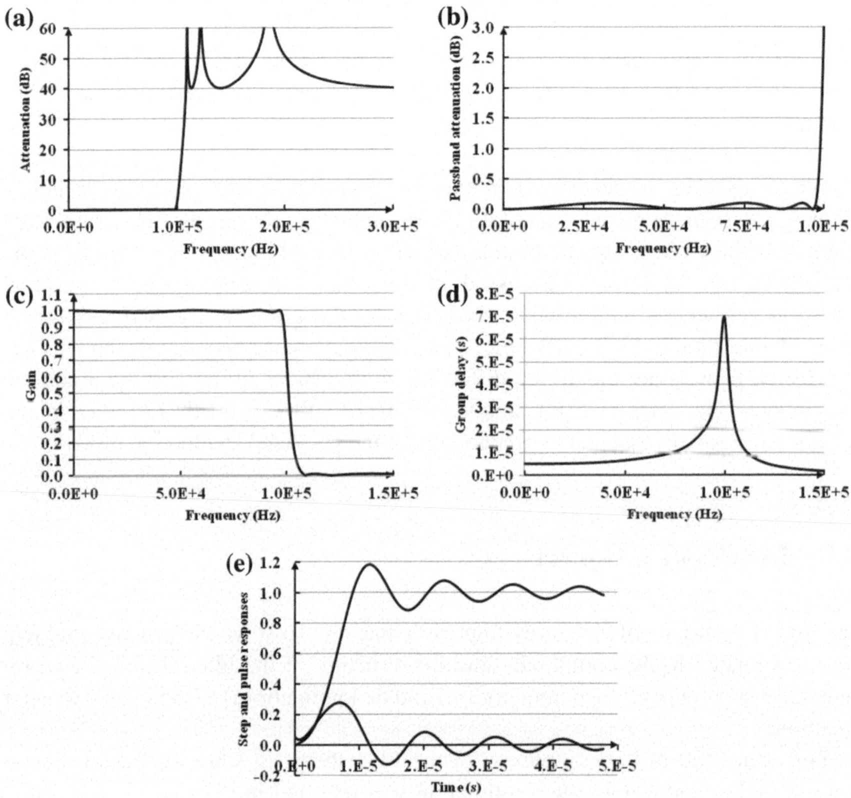

Fig. 9.3 The responses of a 7th order Modified Elliptic filter in frequency and time domain, **a** overall attenuation, **b** passband attenuation, **c** overall gain, **d** group delay, and **e** time domain responses

9.4 Some Comparisons

In this paragraph we will give a comparison between the Modified Elliptic (in the figures below denoted by ME) and the Chebyshev filter with improved selectivity introduced in the previous chapter (below denoted by C). The goal is to create information enabling appreciations of the advantages and disadvantages of ME over C.

For comparison, the above described 7th order ME (now denoted as ME(7,6)) and the one described in the previous chapter (now denoted by C(7,6)) will be considered. To make them comparable the C(7,6) was developed from 3 dB passband attenuation of Chebyshev prototype while the ME(7,6) was synthesized with 0.7 dB of passband attenuation.

From the part of Fig. 9.4 depicting the passband attenuation (Fig. 9.4a) we may conclude that the maximum passband attenuation is almost the same for both filter which, in conjunction with fact that the stopband attenuation is the same (Fig. 9.4b) as is the complexity of the filters, allows for fair comparison.

From the figure depicting the stopband attenuation and Table 9.1 we may conclude that the transition region of the ME(7,6) is almost half of the one of C(7,6) which is the main advantage of ME(7,6). That, of course, was to be expected but not so pronounced as it is.

What was to be expected is the mutual relation of the peaks of the group delay. Namely, as seen from the part of Fig. 9.4c depicting the group delay and from Table 9.1, the peak of the group delay of ME(7,6) is almost twice as big as is the one of C(7,6). As already discussed in the chapter describing the polynomial Chebyshev filters, in cases when correction of the group delay characteristic is planned, one should think twice before making final choice between the above two.

Finally, there is not much of a difference in the values of the overshoot and the undershoot. The later is a bit smaller for C(7,6) than for ME(7,6) due to the presence the (small) lobe at low frequencies in the passband attenuation of ME(7,6).

9.5 Developer's Corner

Having no catalogue of Zolotarev filters one has first so solve the primary problem which is related to the complexity and the structure of the filter. These are represented by the orders of the numerator (m) and denominator (n) of the filter's transfer function.

For estimation of the complexity of the filter there are some suggestions as to how to find m and n. One such estimation was reported in [5]:

$$\left(n+\frac{m}{2}\right)_{\min} = t_{\min}\cdot\left(\frac{5}{4}\right)^{\frac{7-t_{\min}}{6}}, \tag{9.14a}$$

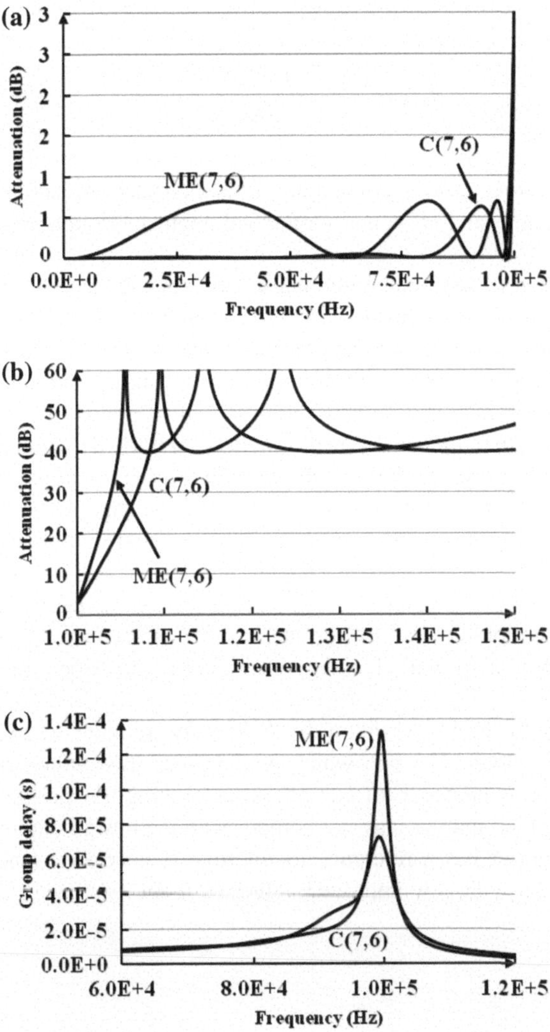

Fig. 9.4 Comparison of ME(7.6) and C(7.6) in the frequency domain. **a** Passband attenuation, **b** stopband attenuation, and **c** group delay

Table 9.1 Comparison of ME(7.6) and C(7.6) in the frequency domain and time domain

Quantity	C(7,6)_3 dB	ME(7,6)_0.7 dB
Width of the transition region $f_s - f_0$ (kHz)	8.244	4.720
Peak value of the group delay τ_{d_max} (μs)	72.752	132.94
Overshoot (%)	18.765	14.176
Undershoot (%)	11.028	13.840

where

$$t_{\min} = \frac{a_{\min}/10 + \log(1+\varepsilon^2)}{3 \cdot \log(\omega_s)}. \tag{9.14b}$$

$a_{\min}$ is the minimum stopband attenuation, ε^2 is a constant reflecting the maximum passband attenuation and ω_s is the normalized angular frequency at which $a_{\min}$ is first reached.

After evaluating (9.14) small additional effort is needed to separate the values of n and m. A bigger problem, however, is the fact that the result of (9.14) being real number from one side and not fully accurate on the other, will not necessarily produce final acceptable values for n and m. For example, if from (9.14) one gets $(n + m/2) = 10$, one may choose between ($n = 7$, $m = 6$) and ($n = 8$, $m = 4$). Even if both functions satisfy the requirements on the selectivity a dilemma is still present as to which one is better. Unfortunately, not always the solution is optimal in the sense that it may be less or even much more selective than necessary.

While considering this issue one has to have in mind that choice of maximum value of m, as shown in the previous chapter, will lead to optimal solution from the complexity of the system point of view. If, however, some restrictions are present related to the asymptotic slope of the attenuation characteristic, one may choose smaller m which is a special advantage of the above described algorithm and the *RM* software.

To cope with this problem, within the *RM* software there are two programs for synthesis of ME filters. The one which produces its own values of n and m according to (9.14a) is named *Zolotarev*. After running this program the designer gets a solution which may or may not comply with the design requirements. In any case however he/she has a reference to improve if necessary (and now run the *Elliptic* program with new improved values of n and m).

References

1. Zolotarev EI (1877) Application of elliptic functions to problems about functions with least and greatest deviation from zero. Zap Imp Akad Nauk St Petersburg 30(5) (in Russian). www.math.technion.ac.il/hat/fpapers/zol1.pdf. Last visited May 2019
2. Cauer W (1958) Synthesis of linear communication networks. McGraw-Hill, New York
3. Orfanidis JS (2006) Lecture notes on elliptic filter design. Department of Electrical & Computer Engineering Rutgers University, Piscataway, NJ. http://www.ece.rutgers.edu/~orfanidi/ece521/notes.pdf. Last visited May 2019
4. Zverev AI (2005) Handbook of filter synthesis. Wiley-Blackwell, New York
5. Chen X, Parks TW (1986) Analytic design of optimal FIR narrow-band filters using Zolotarev polynomials. IEEE Trans Circuits Syst CAS 33(11):1065–1071
6. Lawden DF (1989) Elliptic functions and applications. Springer, New York
7. Levy R (1970) Generalized rational function approximation in finite intervals using Zolotarev functions. IEEE Trans Microw Theory Tech MTT 18(12):1052–1064

8. McClellan JH, Parks TW, Rabiner LR (1974) A computer program for designing optimum FIR linear phase digital filters. IEEE Trans Audio Electro-Acoust 21(6):506–526
9. Selesnick IW, Burrus CS (1996) Exchange algorithms for the design of linear phase fir filters and differentiators having flat monotonic pass bands and equi-ripple stop-bands. IEEE Trans Circuits Syst II 43(9):671–675
10. Todd J (1984) Applications of transformation theory: a legacy from Zolotarev (1847–1878). In: Singh SP, Burry JWH, Watson B (eds) Approximation theory and spline functions. D. Reidel Publishing Company, New York, pp 207–245
11. Vlček M, Unbehauen R (1999) Zolotarev polynomials and optimal FIR filters. IEEE Trans Signal Process 47(3):717–729
12. Corral AC, Lindquist CS (2001) Optimizing elliptic filter selectivity. Analog Integr Circ Sig Process 28(1):53–61

Chapter 10
Linear Phase Low-Pass Filters with Improved Selectivity

Synthesis of constant group delay low-pass polynomial filters in a maximally flat manner (having all derivatives in the origin equal to zero) is straightforward. One uses the Bessel polynomials with the complex variable introduced instead of the usual real x. That is described in this chapter and tables of coefficients and poles of the first ten Bessel polynomials are given. A design example is given too. The next issue visited is approximation of constant group delay by a polynomial filter in equi-ripple manner. An iterative algorithm is proposed allowing for this problem to be solved for arbitrary values of the order of the filter and the maximum group delay error assigned in percentage of the nominal one. Influence of the approximation error to the properties of the filter is studied. Both, maximally flat and equi-ripple group delay approximants exhibit poor selectivity. To improve that an algorithm is described and an example is given as to how transmission zeros at the ω-axis may be introduced. Comparisons are made (in the frequency and time domain) between the two linear phase polynomial solutions being extended with transmission zeros. Finally, advice is given as to how to synthesize low-pass all-pass functions.

10.1 Introduction

In this chapter a new paradigm will be introduced. Instead of insisting on the selectivity as a prime criterion for filter synthesis, the phase linearity will be considered as the most important. The filtering functions so obtained will have, unfortunately, pure selectivity as it was the case with the phase linearity of selective filters. An interesting change happens in the time domain. Namely, the linear phase polynomial filters, generally, have a monotonic step response meaning there is practically no overshoot and undershoot. That makes them attractive for pulse transmission since the eye diagram is highly open.

From the approximation point of view there are two approaches to the problem. The first one is to implement maximally flat criterion to the group delay in the

V. Litovski, *Electronic Filters*, Lecture Notes in Electrical Engineering 596,
https://doi.org/10.1007/978-981-32-9852-1_10

origin while the second would be to implement equi-ripple error (or mini-max) criterion. Both will be implemented in this chapter. Where appropriate, complete algorithms will be given.

To improve selectivity while keeping the phase linearity is an interesting designer's challenge. A method will be described below which allows for introduction of transmission zeros on the ω axis in a way not to disturb the phase linearity. From selectivity point of view, the results obtained, while bringing serious improvement, are far behind the solutions described in the previous chapters.

Examples and comparisons will be given produced by proper programs of the *RM* software.

10.2 Thomson (Bessel) Filters

The most popular monotonic approximation of low-pass filters exhibiting constant group delay is the so called Thomson solution [1–4]. It uses Bessel polynomials in the denominator of the transfer function directly putting the normalized complex frequency as the independent variable. The main property of the approximant is that the group delay is maximally flat at the origin.

We are here promoting a program *Thomson* which produces the poles of the Thomson filters along with the coefficients of the denominator of the transfer function and the coefficients of the squared modulus of the denominator at the ω-axis.

10.2.1 The Bessel Polynomials

The transfer function of this class of filters may be written as

$$H(s) = P(0)/P(s) \tag{10.1}$$

where

$$P(s) = \sum_{k=0}^{n} a_k s^k, \tag{10.2}$$

with

$$a_k = \frac{(2n-k)!}{2^{n-k} k!(n-k)!} \tag{10.3}$$

Table 10.1 Coefficients of the Bessel transfer function

n	s^0	s^1	s^2	s^3	s^4	s^5
3	15	15	6	1		
4	105	105	45	10	1	
5	945	945	420	105	15	1
6	10395	10395	4725	1260	210	21
7	135135	135135	62370	17325	3150	378
8	2027025	2027025	945945	270270	9450	6930
9	34459425	34459425	16216200	4729725	945945	135135
10	654729075	654729075	310134825	91891800	18918900	2837835
	s^6	s^7	s^8	s^9	s^{10}	
6	1					
7	28	1				
8	630	36	1			
9	13860	990	45	1		
10	315315	25740	1485	55	1	

Here, n is the filter order and s is the normalized complex angular frequency. The coefficients of the denominator of the transfer function of the Bessel filter are listed in Table 10.1 for n up to 10.

The transfer function so obtained is stated as "delay normalized" which means that it approximates a constant equal to 1 s at the origin. For implementation purposes the poles obtained by solving

$$P(s) = 0, \tag{10.4}$$

are re-normalized so that the attenuation produces 3 dB at the normalized frequency $\omega_0 = 1$ rad/s.

10.2.2 Design Example of Thomson Filters

The example described here is related to a 9th order Thomson filter. Figure 10.1 depicts the characteristics of this filter in frequency and time domain while in the sequel excerpts of the reports of the transfer function synthesis program *Thomson* and transfer function analysis program, *LP_Analysis* are given.

We will start the analysis by the phase and group delay characteristics. The cut-off frequency being 100 kHz, we may conclude that both the phase and group delay are linear far beyond the passband of the filter. That does not mean that the linear phase approximation extends to the frequency band where high attenuations are exhibited. The problem is, as it can be seen from the attenuation characteristic, the very poor selectivity. Here, a 9th order filter has difficulties to produce an

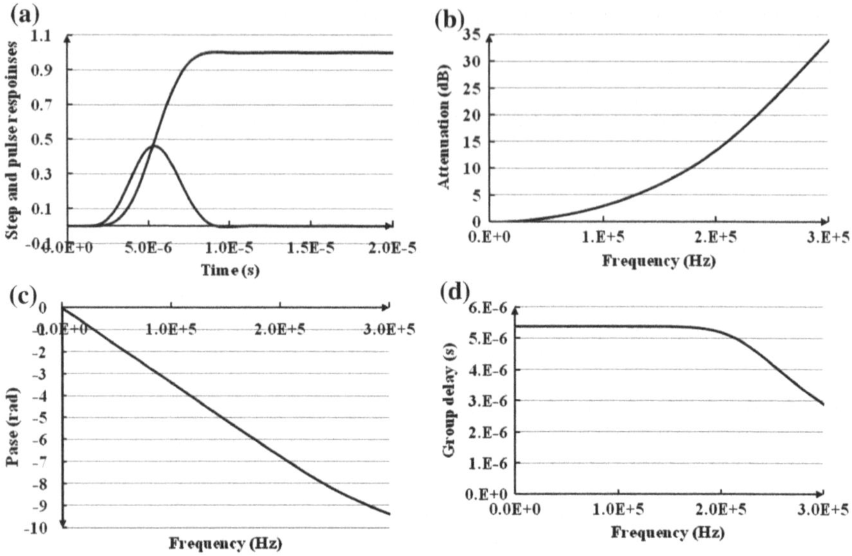

Fig. 10.1 Time and frequency domain responses of a 9th order Thomson filter, **a** time domain responses, **b** attenuation, **c** phase, and **d** group delay

attenuation of 35 dB at frequencies deep in the stopband (three times the cut-off). The reader is advised to go back to the CMAC and Chebyshev filters and to compare the stopband attenuation characteristics.

Still, having in mind the expression for the phase of a filter (3.21), it is difficult to relate the phase characteristic depicted in Fig. 10.1c to such a complex transcendental expression.

As for the time domain responses we may conclude that the step response is practically monotonic. From the data obtained by *LP_Analysis* the first overshoot is only 0.2% of the finite value. That should be opposed to values larger than 10% obtained for selective filters.

It is especially interesting that the pulse response may be found to be bell-shaped and highly symmetrical with respect to the time instant when its maximum occurs. That symmetry is convenient from the so called inter-symbol interference point of view since, in a pulse train, pulses (symbols) do not overlap if symmetrical (or one does not need to spend extra time in waiting for the asymmetrical side of the pulse to decay enough as is the case with the responses in Fig. 7.10).

Here are excerpts from the reports produced by the *RM* software. First goes the report produced by the transfer function synthesis program *Thomson*. To reduce the length of the text the report is partly edited and tables are introduced where necessary.

Start.E

Project name: Bessel9

Thomson-Bessel

Order of denominator=9, Order of nominator=0

The following calculations are performed without any transformation

The results about the poles, zeros and coefficients listed below are valid for frequency normalization with
wc=2*pi*fc=6.28318530e+005

real pole	imag. pole
-6.2970191817	0.0000000000
-6.1293679043	-1.7378483835
-6.1293679043	1.7378483835
-5.6044218195	-3.4981569179
-5.6044218195	3.4981569179
-2.9792607982	-7.2914636883
-2.9792607982	7.2914636883
-4.6384398872	-5.3172716754
-4.6384398872	5.3172716754

Coefficients of the denominator of the transfer function

3.445942500e+007 3.445942500e+007 1.621620000e+007 4.729725000e+006
9.459450000e+005 1.351350000e+005 1.386000000e+004 9.900000000e+002
4.500000000e+001 1.000000000e+000

Coefficients of the squared modulus of the characteristic function (variable=w**2)

1.187451971e+015 6.985011596e+013 2.191376187e+012 4.916549137e+010
8.939180250e+008 1.418917500e+007 2.079000000e+005 2.970000000e+003
4.500000000e+001 1.000000000e+000

The attenuation reaches 4.00000e+001 dB at the frequency of 1.11110e+006 Hz

The following calculations are performed with normalization so that attenuation (1 rad/s)=3.000e+000 dB

Order of denominator=9, Order of nominator=0

real pole	imag. pole
-1.8597045118	0.0000000000
-1.8101919046	-.51324037394
-1.8101919046	.51324037394
-1.6551590908	-1.0331139251
-1.6551590908	1.0331139251
-.87986785306	-2.1533947297
-.87986785306	2.1533947297
-1.3698747513	-1.5703547726
-1.3698747513	1.5703547726

Coefficients of the denominator of the transfer function

5.889635022e+002 1.994249327e+003 3.177690468e+003 3.138263900e+003
2.125252464e+003 1.028024725e+003 3.570179202e+002 8.634815283e+001
1.328989171e+001 1.000000000e+000

Coefficients of the squared modulus of the characteristic function (variable=w**2)

3.468780069e+005 2.339429634e+005 8.414763927e+004 2.164549518e+004
4.512185849e+003 8.211614875e+002 1.379456659e+002 2.259394999e+001
3.924916038e+000 1.000000000e+000

===

After renormalization so that the attenuation at 1.90000e+005 Hz is 3.00000e+000 dB
The attenuation reaches 4.00000e+001 dB at the frequency of 3.28145e+005 Hz

The next is part of the report produced by the *LP_Analysis* program.

****

delay time (at 0.5)=5.38089e-006 rise time (0.1-0.9)=3.44888e-006

****

The number of oscillations at the beginning of the step response is=1
time=3.18310e-009 value of the extremum=-9.04962e-015

****

The number of oscillations at the end of the step response is=20
time=9.20657e-006 value=1.00217e+000
time=1.08469e-005 value=9.98767e-001
time=1.29676e-005 value=1.00043e+000
time=1.52157e-005 value=9.99861e-001

```
time=1.75393e-005        value=1.00004e+000
time=1.98630e-005        value=9.99988e-001
time=2.22185e-005        value=1.00000e+000
time=2.45421e-005        value=9.99999e-001
***********************************************************************
```

End.E

10.3 Synthesis of Linear Phase (Equi-Ripple) Polynomial Filters

As an alternative to the best known Bessel low-pass filters in cases when constant group delay approximation is sought one may use polynomial filters approximating constant group delay in equi-ripple manner. These solutions may be used to advantage to the Bessel ones since they may broaden the approximation interval while full control of the error is possible.

This issue was discussed in [5] where all-pass networks were considered.

We are here promoting a program which produces the poles of the filters approximating constant group delay with a prescribed error along with the coefficients of the denominator of the transfe funcrion and the coefficients of the squared modulus of the denominator at the ω-axis.

Special feature to the algorithm implemented is the possibility for the group delay characteristics of even order filters to start with a maximum.

10.3.1 The Algorithm

This problem will be treated as a nonlinear optimization problem [2] which will here be solved by a variant of the Remez algorithm [6].

Given the transfer function of a filter

$$H(s) = \frac{H_0}{\prod\limits_{k=1}^{n} (s - z_k)} \tag{10.5}$$

where

$$H_0 = \prod_{k=1}^{n} (-z_k), \tag{10.6}$$

the group delay is calculated from

$$\tau(\omega) = \sum_{k=1}^{n} \frac{-\alpha_k^2/\alpha_k}{(\omega - \beta_k)^2 + \alpha_k^2} \tag{10.7}$$

for n even. After substituting $\beta_k = |\beta_k|$ it may be rewritten as

$$\tau_d(\omega) = \sum_{k=1}^{\lfloor n/2 \rfloor} T_k(\omega) \tag{10.7}$$

where

$$T_k(\omega) = T_{ka}(\omega) + T_{kb}(\omega) \tag{10.8}$$

with

$$T_{ka}(\omega) = \frac{-\alpha_k}{\alpha_k^2 + (\omega - \beta_k)^2} \tag{10.9a}$$

and

$$T_{kb}(\omega) = \frac{-\alpha_k}{\alpha_k^2 + (\omega + \beta_k)^2}. \tag{10.9b}$$

When n is odd, after denoting the real pole as $z = \alpha_0$, one uses

$$\tau_d(\omega) = T_0(\omega) + \sum_{k=1}^{\lfloor n/2 \rfloor} T_k(\omega) \tag{10.10}$$

with

$$T_0(\omega) = \frac{-\alpha_0}{\omega^2 + \alpha_0^2}. \tag{10.11}$$

In the above $\lfloor . \rfloor$ is denoting the flooring function.

With reference to Fig. 10.2 the goal of the algorithm is to describe a procedure which will lead to a group delay error characteristic having equal (in absolute values) maximum deviations from the nominal delay. The prescribed (acceptable) error is here denoted as δ and expressed in percentage.

In fact, according to Fig. 10.2 (which is valid for a polynomial filter of even order), the following system of nonlinear equations is to be solved iteratively:

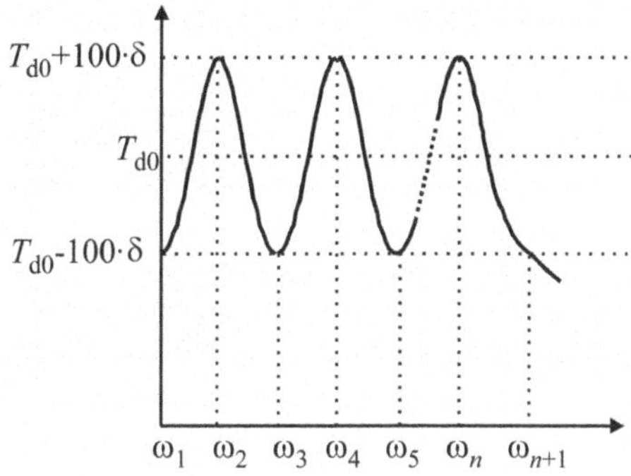

Fig. 10.2 Definition of the group delay approximation problem (Case: n-even)

$$\tau_d(\mathbf{p}, \omega_i) = T_{d0} + 100 \cdot \delta \cdot (-1)^{r+i}, \quad \text{for } i = 1, 2, \ldots, n+1 \tag{10.12}$$

where $\mathbf{p}$ is a vector of unknown coordinates of the poles (two per a complex pole and one additional real if n-odd), T_{d0} is the unknown average value of the passband group delay while $r = 1$ for n-odd and $r = 0$ for n-even.

Note, since ω_i, $i = 1, 2, \ldots, n + 1$ are unknown, in every iteration intended to produce corrections of $\mathbf{p}$, one has to first recalculate their values which will be addressed later on. A proper initial solution for $\mathbf{p}$ is necessary which, in general, may be an obstacle to the implementation of the algorithm. If, however, the so called iteration damping is implemented [6], i.e. if at the beginning of the iterative process only fractions of the corrections are added to $\mathbf{p}$, as an initial solution for the complex poles one may simply use

$$p_l = \alpha_l + j\beta_l = -\frac{1}{2 \cdot n} \pm j\frac{2 \cdot l - 1}{n+1}, l = 1, 2, \ldots, \lfloor n/2 \rfloor \tag{10.13}$$

while for the real pole (if n-odd) one may use $\alpha_0 = p_0 = -1/(2 \cdot n)$. A glimpse to the results for the poles given later on in Paragraph 10.3.2 will allow the reader to recognize that in cases of constant group delay approximation the poles are aligned almost exactly on a line parallel to the ω-axis. That property was used in creation of the real part of the poles in (10.13). In addition, as initial value, one may use $T_{d0} = (3 \cdot n) \cdot \pi/2$.

Going into details of the implementation of the Newton-Raphson iterative procedure for solving for the pole's coordinates, we first linearize (10.12):

$$\sum_{j=1}^{n} \frac{\partial \tau_d(\mathbf{p}, \omega_i)}{\partial p_j} \Delta p_j - \Delta T_{d0} = -\tau_d(\mathbf{p}, \omega_i) + T_{d0} + 100 \cdot \delta \cdot (-1)^{r+i} \tag{10.14}$$

for $i = 1, 2, \ldots, n + 1$.

For the partial derivatives one may use

$$\frac{\partial T_{ka}(\omega)}{\partial \alpha_k} = \frac{\alpha_k^2 - (\omega - \beta_k)^2}{\left[\alpha_k^2 + (\omega - \beta_k)^2\right]^2} \tag{10.15a}$$

$$\frac{\partial T_{kb}(\omega)}{\partial \alpha_k} = \frac{\alpha_k^2 - (\omega + \beta_k)^2}{\left[\alpha_k^2 + (\omega + \beta_k)^2\right]^2} \tag{10.15b}$$

$$\frac{\partial T_{ka}(\omega)}{\partial \beta_k} = \frac{-2 \cdot \alpha_k \cdot (\omega - \beta_k)}{\left[\alpha_k^2 + (\omega - \beta_k)^2\right]^2} \tag{10.16a}$$

$$\frac{\partial T_{kb}(\omega)}{\partial \beta_k} = \frac{2 \cdot \alpha_k \cdot (\omega + \beta_k)}{\left[\alpha_k^2 + (\omega + \beta_k)^2\right]^2} \tag{10.16b}$$

and

$$\frac{\partial T_0(\omega)}{\partial \alpha_0} = \frac{\left(\alpha_0^2 - \omega^2\right)}{\left(\omega^2 + \alpha_0^2\right)^2}. \tag{10.17}$$

What is still needed are the abscissas of the extremal points of the group delay denoted as ω_i, $i = 1, 2, \ldots, n + 1$. To find them one needs to solve the following nonlinear equation

$$\frac{\mathrm{d}\tau(\omega)}{\mathrm{d}\omega} = 0 \tag{10.18a}$$

or

$$\frac{dT_0(\omega)}{\mathrm{d}\omega} + \sum_{k=1}^{\lfloor n/2 \rfloor} \frac{dT_k(\omega)}{\mathrm{d}\omega} = 0 \tag{10.18b}$$

for n odd. One is to substitute in (10.18a, 10.18b) the following

$$\frac{dT_0(\omega)}{\mathrm{d}\omega} = \frac{2 \cdot \alpha_0 \cdot \omega}{\left(\omega^2 + \alpha_0^2\right)} \tag{10.19a}$$

and

$$\frac{dT_k(\omega)}{d\omega} = \frac{2\alpha_k(\omega - \beta_k)}{\left[\alpha_k^2 + (\omega - \beta_k)^2\right]^2} + \frac{2\alpha_k(\omega + \beta_k)}{\left[\alpha_k^2 + (\omega + \beta_k)^2\right]^2}. \tag{10.19b}$$

In our opinion there is no need for a sophisticated procedure to find the zeros of (10.18a, 10.18b). If sophisticated, one may search the ω-axis with a coarse step to identify the intervals where the zeros are located. That should enable implementation of the modified secant method as described earlier in this book. Otherwise, one may set a fine step for ω (say $\Delta\omega = 10^{-6}$ rad/s) and look for a change of the sign.

Since an iterative algorithm is implemented and since initial solutions are to be provided for any order of the filter, any value of the approximation error, and any type of the even order approximation, strong iteration damping should be implemented leading to a relatively long time needed for the program to run.

Note, the attempt described in [7] proposing procedure for approximation of constant group delay in equi-ripple manner, being related to polynomial coefficients and not to the poles of the transfer function, may be considered as an alternative to the method presented here.

To demonstrate the properties of possible solutions obtained by the procedure described here a 9th order filter was chosen. Figure 10.3a depicts the frequency dependence of the approximation error for three values: 1, 5 and 10%. By careful examination one may conclude that increase of the error for 10 times (from 1 to 10%) leads to increase of the approximation interval for only 12.68% (from 293.75 kHz to 331 kHz, respectively). This is far of a compromise between error and approximation interval. The reason for such a result is the fact which was already mentioned. Namely, the area under the group delay curve is fixed and by changing the error (raising the maximums and decreasing the minimums) we change the average delay for a small amount meaning that the approximation interval changes for similar amount, too.

Figure 10.3b and c represent the attenuation characteristics of the mentioned approximants. It is obvious that the rise of the error deteriorates the passband attenuation which has and additional consequence: an undershoot in the step response and closure of the eye diagram. Furthermore, the stopband attenuation characteristics does not differ significantly especially near to the passband and asymptotically.

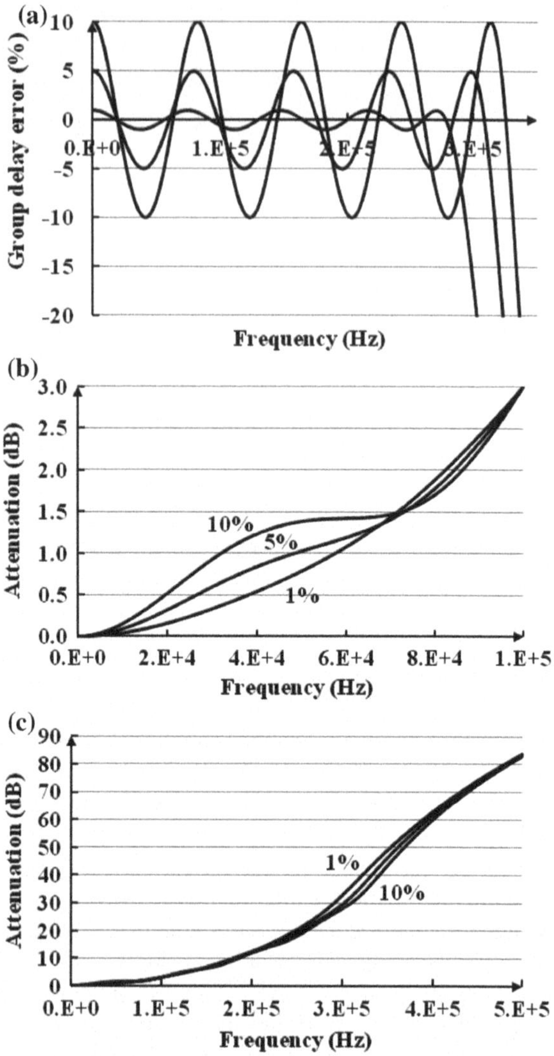

Fig. 10.3 Equi-ripple group delay polynomial filter of 9th order. **a** approximation error, **b** passband attenuation and **c** stopband attenuation

We will discuss shortly here a modification to the above algorithm which is available within the *Equ-rip-td* program. Namely, the designer is allowed to choose the group delay of an even order filter to start with a maximum. In such a case the resulting transfer function will have doubled real pole instead of a complex pair.

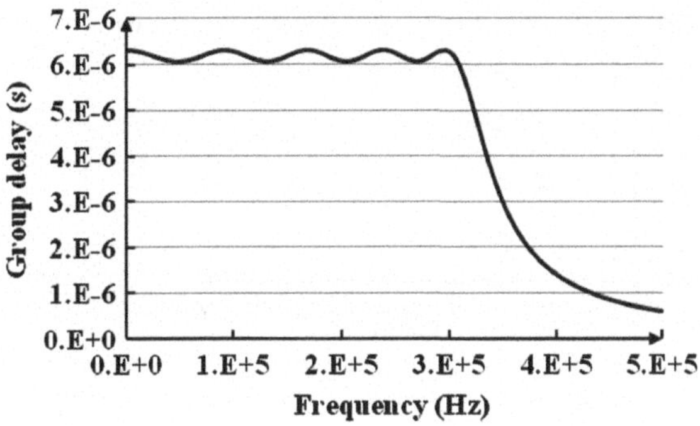

Fig. 10.4 A 10th order filter with modified equi-ripple group delay

Table 10.2 Poles of the 10th order approximant

Real poles	Imag. poles
−0.8331635529	0.000000000
−0.8331635529	0.000000000
−0.5936695213	0.9350251928
−0.5936695213	−0.9350251928
−0.5501291068	1.697514403
−0.5501291068	−1.697514403
−0.5092921115	2.406544320
−0.5092921115	−2.406544320
−0.4032678973	3.044323478
−0.4032678973	−3.044323478

One example of this kind of approximation is depicted in Fig. 10.4 with the poles of the transfer function given in Table 10.2. Note, in some physical realization procedures one needs the residues in poles which, in this case, imposes implementation of formulas for second order poles as given i Chap. 3 by (3.29). The same stands for the transient analysis.

10.3.2 *Design Example of Linear Phase (Equi-Ripple) Polynomial Filters*

The example given below was produced by the *equi-rip-td* program of the *RM* software. The filter is of 9th and 5% error of the group delay was chosen. The passband attenuation was chosen to be a_{max} = 3 dB.

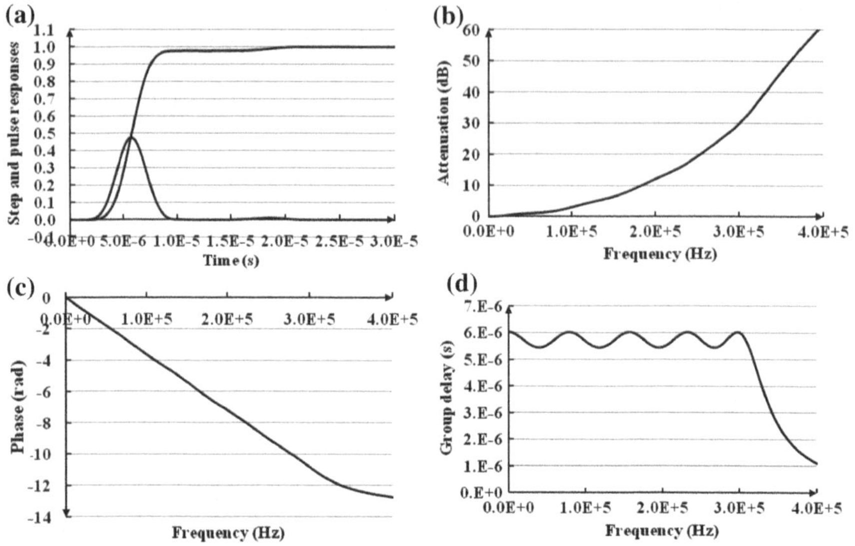

Fig. 10.5 Time and frequency domain responses of a 9th order filter approximating constant group delay in equi-ripple manner with a relative error of 5%. **a** Step and pulse responses, **b** attenuation, **c** phase, and **d** group delay

Figure 10.5 depicts the responses of this filter in the frequency and time domain. It is followed by excerpts from the reports related to this function generated by the *equi-rip-td* and the *LP_Analysis* programs.

Looking to the pulse response we may conclude, as we did with the Thomson filters, that symmetry is achieved which is important for pulse transmission from the inter-symbol interference point of view. Here again, one may note the large amplitude of the pulse response which is of order of 0.5. That is larger by significant margin than the ones given in Table 7.1 which are valid for selective polynomial filters.

The step response while exhibiting short rise time of only 3.44 μs (for a cut-off frequency of 100 kHz), due to the large value of attenuation at low frequencies in the passband, has significant undershoot which, in this type of response, lasts for a long period. It is as if it cuts the response from the top for about 2%.

The attenuation characteristic and the error were already discussed. We will mention here the linearity of the phase which is extended to an interval 3 times larger than the passband of the filter.

Here are excerpts from the reports produced by the *RM* software. First goes the report produced by the transfer function synthesis program *equi_rip_td*. To reduce the length of the text the report is partly edited and tables are introduced where necessary.

Start.E

++

Program: EQUI-RIPPLE-TD

++

Project name: Poly_td_equi_9_5

==

General information

Order of the numerator=0 Order of the denominator n=9

Requested passband relative group delay error =5.00000e+000 %

Requested attenuation at the end of the passband after renormalization is =3.00000e+000 dB

++

POLYNOMIAL FILTER: EQUI-RIPPLE APPROXIMATION OF CONSTANT GROUP DELAY

denominator: n=9 numerator: m=0

relative group delay error=5.000e+000 type of approximation=0

==

The results about the poles, zeros and coefficients listed below are valid for frequency normalization with

wc=2*pi*fc=6.28318530e+005

==

Original solution

average delay=14.137167 (s) Amin=40.000000 (dB) w(Amin)=0.848022 (rad/s)

Denormalized obtained w(Amin)=5.32828e+005 (Hz) Requested w(Amin)=1.90000e+005(Hz)

Poles of the transfer function

Real part	Imaginary part
-1.220054212e-001	0.000000000e+000
-1.215258278e-001	2.025463689e-001
-1.215258278e-001	-2.025463689e-001
-1.197349729e-001	4.028132496e-001
-1.197349729e-001	-4.028132496e-001
-1.147587102e-001	5.968241937e-001
-1.147587102e-001	-5.968241937e-001
-9.564420619e-002	7.745905825e-001
-9.564420619e-002	-7.745905825e-001

Coefficients of the denominator of the transfer function (ascending in s):
2.7046904784e-004 4.0148624701e-003 2.5388303327e-002 1.1553524718e-001
3.1159586320e-001 7.6484474453e-001 1.0673311408e+000 1.6262619389e+000
1.0253328555e+000 1.0000000000e+000

Solution after normalization so that w(Amax)=1 rad/s
average delay=1.41372e+001 (s) Amin=4.00000e+001 (dB) w(Amin)=8.48022e-001 (rad/s)

Poles of the transfer function

Real parts	Imaginary parts
-4.775355190e-001	0.000000000e+000
-4.756583657e-001	7.927769475e-001
-4.756583657e-001	-7.927769475e-001
-4.686488671e-001	1.576631860e+000
-4.686488671e-001	-1.576631860e+000
-4.491715180e-001	2.336000714e+000
-4.491715180e-001	-2.336000714e+000
-3.743563621e-001	3.031787540e+000
-3.743563621e-001	-3.031787540e+000

Coefficients of the denominator of the transfer function (ascending in s):
5.8311811484e+001 2.2114803400e+002 3.5728891066e+002 4.1540699488e+002
2.8623612205e+002 1.7950621588e+002 6.3999832069e+001 2.4914008044e+001
4.0132057449e+000 1.0000000000e+000

Denormalized obtained w(Amin)=3.31919e+005 (Hz) Requested w(Amin)=1.90000e+005(Hz)

The next is part of the report produced by the *LP_Analysis* program.

****

delay time (at 0.5) =5.77998e-006 rise time (0.1-0.9)=3.43583e-006

****

The number of oscillations at the beginning of the step response is=1
time= 4.77465e-009 value of the extremum= 5.22648e-017

****

The number of oscillations at the end of the step response is=22
time=1.12135e-005 value=9.77366e-001
time=1.26121e-005 value=9.77084e-001
time=1.41599e-005 value=9.77307e-001
time=1.47249e-005 value=9.77290e-001
time=2.16163e-005 value=9.99529e-001
time=2.27781e-005 value=9.99441e-001

time=2.42424e-005 value=9.99503e-001
time=2.57702e-005 value=9.99467e-001
****

End.E

10.4 Synthesis of Selective Low-Pass Filters Based on the Transfer Function

Having the transfer function of a non-selective (say linear phase) polynomial or "extended polynomial" filter one may produce a "selective" one by adding transmission zeros. We will introduce the definition of the "extended polynomial" soon. In order not to disturb the original phase characteristic, "amplitude correction" is performed so that the numerator of the transfer function is extended by adding transmission zeros at the imaginary axis of the complex frequency plane. In that way the newly synthetized function will have the original poles and new zeros as compared with the prototype with additional zeros just added. Due to the (relatively minor) influence of the added zeros to the pass-band attenuation characteristic, however, renormalization is to be implemented so that in the final solution both poles are zeros are new.

One has to have in mind that, usually, the polynomial transfer functions that approximate constant group delay (being of maximally-flat or equi-ripple type) cover a bandwidth much larger than the 3 dB bandwidth. In other words, in the prototype function selectivity is lost in favor to constant group delay far outside the pass-band. As a consequence, the improvements obtained by the "amplitude correction" have limitations in the sense that the resulting filter is still not a "selective" one. Nevertheless, reduction of the transition region between the pass-band and the stop-band is achieved which allows for the neighboring transmission channel to be nearer than previously. Since the frequency is kind of a natural resource one may say that in this way a frequency band is saved or that rising the selectivity has an "ecological effect".

There are special approximation procedures which perform simultaneous approximation of the phase and amplitude characteristic [8, 9] so that they control the poles of the function to get selectivity, and control the complex zeros of the transfer function to get phase (or amplitude) correction in the pass-band [5, 10, 11]. Here, however, we will not disturb the prototype be it polynomial or extended polynomial. We will simply extend the numerator by a new polynomial having zeros on the ω-axis which will not affect the group delay characteristic while improving the amplitude characteristic. The program *Chang_ph* allows for such prototype functions to be used and produces the amplitude corrected solution.

10.4.1 The Algorithm

The transfer function of an “extended polynomial” [11] filter may be stated as

$$H_e(s) = \frac{P(0)}{P(s)} \cdot \frac{Q(s)}{Q(0)}. \qquad (10.20)$$

Here *P*(s) is the known denominator (of order *n*) of an extended polynomial filter (usually) approximating constant group delay. *Q*(*s*) (being of order *l*) is also known but contains only complex zeros including zeros at the real axis of the complex frequency plane. This polynomial may arise when instead to increase the order of the polynomial approximating linear phase, one adds right-half plane zeros to the polynomial transfer function. Of course, an approximation algorithm may be necessary dealing simultaneously with numerator and the denominator to get linear phase.

Otherwise, one may use already synthesized polynomial linear phase functions in the following way. Take several (say a pair or more) poles of the transfer function and convert them into right half plane zeros as it is done in Table 10.3 for the 9th order Thomson filter described in Paragraph 10.2. By doing that one creates a 7th order transfer function with two complex zeros. Since, however, the zeros are mirrored poles, the phase and group delay characteristics are remaining the same. The benefit of this transaction is in *reduction of the complexity* of a, say, cascaded system since the number of cells is reduced by one. The drawback is in the deterioration of the attenuation characteristic. That may be observed in Fig. 10.6. Not only the asymptotic slope is reduced by 24 dB/oct but the passband attenuation became non-monotonic. In addition, the number of possible attenuation poles at the ω-axis to be added to the extended polynomial function is now reduced (in the example above from 4 to 3 pairs.

Now, since the slope of the attenuation in the transition region of the original function is small the loss at infinity does not much harm. In fact, if the attenuations depicted in Figs. 10.1b and 10.6 are compared, the later shows that the extended

Table 10.3 The poles and zeros of the extended polynomial thomson transfer function

Real zero	Imag. zero
1.8101919046	−0.51324037394
1.8101919046	0.51324037394
Real pole	Imag. pole
−1.8597045118	0.0000000000
−1.6551590908	−1.0331139251
−1.6551590908	1.0331139251
−0.87986785306	−2.1533947297
−0.87986785306	2.1533947297
−1.3698747513	−1.5703547726
−1.3698747513	1.5703547726

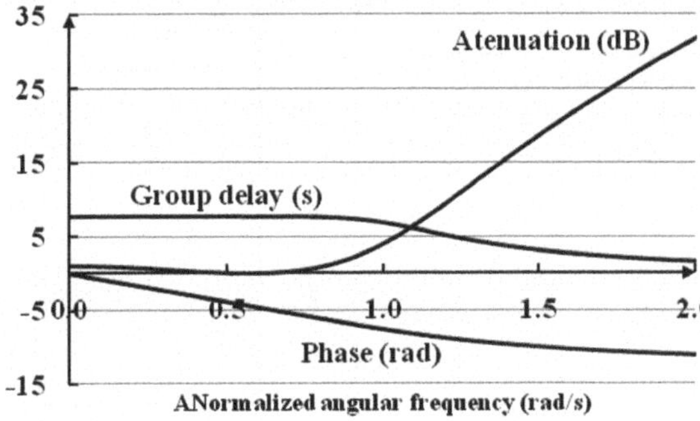

Fig. 10.6 Attenuation, phase and group delay characteristic of the extended polynomial Bessel filter

polynomial filter has higher attenuation at ω = 2 rad/s then the original Thomson by remarkable margin. As for the passband one should remember that the extension to the transfer function which was introduced to improve selectivity has a beneficial effect to the passband too. That may be exploited.

Having all that in mind (even in cases where simple not extended polynomial transfer functions are to be processed) we will proceed with algorithm of creating a new rational transfer function, now, having zeros at the ω-axis. To keep the phase characteristic unchanged while improving selectivity, new transfer function is to be created in the following way

$$H(s,\mathbf{x}) = H_e(s) \cdot \frac{\prod_{i=1}^{m/2}\left(-s^2 - x_i^2\right)}{\prod_{i=1}^{m/2}\left(-x_i^2\right)}. \tag{10.21a}$$

or, on the $j\omega$-axis,

$$H(j\omega,x) = H_e(j\omega) \cdot \frac{\prod_{i=1}^{m/2}\left(\omega^2 - x_i^2\right)}{\prod_{i=1}^{m/2}\left(-x_i^2\right)} \tag{10.21b}$$

where m is the (even) number of transmission zeros x_i^2, i = 1, 2,…, m/2, which are unknown.

With reference to Fig. 10.7, to find the values of x_i, i = 1, 2,…, m/2, one has to solve iteratively the following system of nonlinear equations

$$|H(jy_k,\mathbf{x})| = H_{\max} \quad \text{for } k = 1,2,\ldots,m/2. \tag{10.22}$$

In the above y_k stands for the abscissas (on the ω-axis) of the maximums of the amplitude characteristic, and $H_{\max}$ is the required maximum obtained as a solution of

$$-20 \cdot \log|H_{\max}| = a_{\min} \tag{10.23}$$

where $a_{\min}$ is the required minimum *stop-band* attenuation.

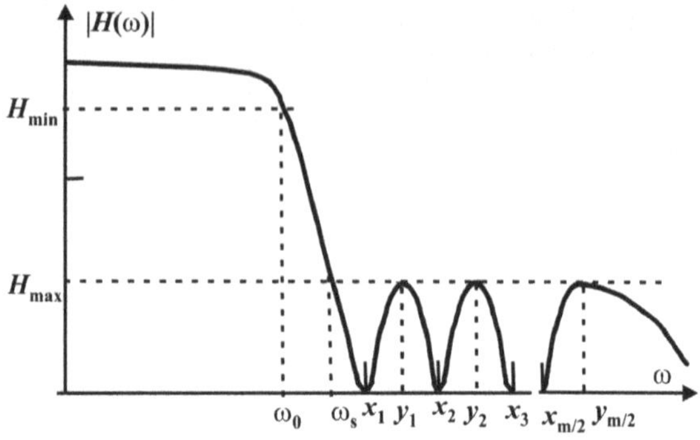

Fig. 10.7 Amplitude characteristic of a monotone passband filter having finite transmission zeros in the stopband

Both **x** and **y** are unknown at the beginning. The procedure starts with a guess for **x**. Then it calculates **y** and solves (for one corrective step only) for the increments of **x**. That allows for the iterative process to proceed. To get a robust algorithm working with any kind of polynomial characteristic functions of any order n, any number of complex zeros l, any number of attenuation poles m, and any value of $a_{\min}$ and $a_{\max}$, one need to implement a restrictive iteration damping.

The final solution is obtained by renormalization of all poles and zeros so that

$$-20 \cdot \log|H(1, \mathbf{x})| = a_{\max} \tag{10.24}$$

where $a_{\max}$ is the maximum *pass-band* attenuation.

To solve the system of nonlinear Eq. (10.22) one may, again, implement the Newton-Raphson linearization which leads to the following system of linear equations

$$\sum_{l=1}^{m/2} \frac{\partial |H(\mathrm{j}y_k, \mathbf{x})|}{\partial x_l} \cdot \Delta x_l = H_{\max} - |H(\mathrm{j}y_k, \mathbf{x})|. \tag{10.25}$$

The partial derivatives needed for formulation of this system are as follows

$$\frac{1}{|H(\mathrm{j}y_k, \mathbf{x})|} \frac{\partial |H(\mathrm{j}y_k, \mathbf{x})|}{\partial x_l} = -2 \cdot \left(\frac{x_l}{\omega^2 - x_l^2} - \frac{1}{x_l} \right). \tag{10.26}$$

To find the locations of the maximums one may go slowly or fast. Slowly would mean that one chooses a very small time step (say 10^{-6} rad/s) and tabulates the current amplitude characteristic to find its maximums. Fast means to implement the modified secant method to the derivative of the amplitude characteristic which may be expressed as

$$\frac{1}{|H(\mathrm{j}y_k, \mathbf{x})|}\frac{\partial|H(\mathrm{j}y_k, \mathbf{x})|}{\mathrm{d}\omega} = \frac{1}{|H_\mathrm{e}(\omega)|}\cdot\frac{d|H_\mathrm{e}(\omega)|}{\mathrm{d}\omega} + \sum_{k=1}^{m/2}\frac{4\cdot\omega}{\omega^2 - x_k^2}. \tag{10.27}$$

In the above we have

$$\frac{1}{|H_\mathrm{e}(\omega)|}\cdot\frac{d|H_\mathrm{e}(\omega)|}{\mathrm{d}\omega} = \frac{1}{2}\cdot\left(\frac{1}{Q_2}\cdot\frac{dQ_2}{\mathrm{d}\omega} - \frac{1}{P_2}\cdot\frac{dP_2}{\mathrm{d}\omega}\right), \tag{10.28a}$$

$$P_2 = P(s)\cdot P(-s)_{|s=j\omega} \tag{10.28b}$$

and

$$Q_2 = Q(s)\cdot Q(-s)_{|s=j\omega}. \tag{10.28c}$$

Here ends the description of the iterative algorithm.

10.4.2 Design Examples of Selective Low-Pass Filters Based on the Transfer Function

The example given below was produced by the *Chang_ph* program of the *RM* software. Two filters of 9th order with maximum (m = 6) number of transmission zeros at the ω axis were synthesized, one starting with a Thomson prototype the other with equi-ripple group delay prototype. The passband attenuation was chosen to be a_{max} = 3 dB while the stopband attenuation was a_{min} = 40 dB.

Figure 10.8 depicts the responses of these filters in the frequency domain. ER is used to denote the equi-ripple prototype while T denotes Thomson prototype.

One may consider these results as a continuation of what we have seen in Fig. 10.3. Namely, looking to the passband attenuation we see that the Thomson prototype leads to smaller distortions which are in accordance with Fig. 10.3 if the Thomson case is considered as an equi-ripple approximation with zero valued error. That maps itself into the stopband too. The Thomson prototype leads to a more selective solution than the equi-ripple one.

These claims are firmly supported by the data given in Table 10.4 where widths of the transition regions of the polynomial prototypes (9,0) and amplitude corrected functions (9,8) are given.

As for the group delay approximation interval, the equi-ripple solution is broader by large margin. In the case of the phases however, one may say that the phase linearity is abruptly destroyed at the abscissa of the first transmission zero.

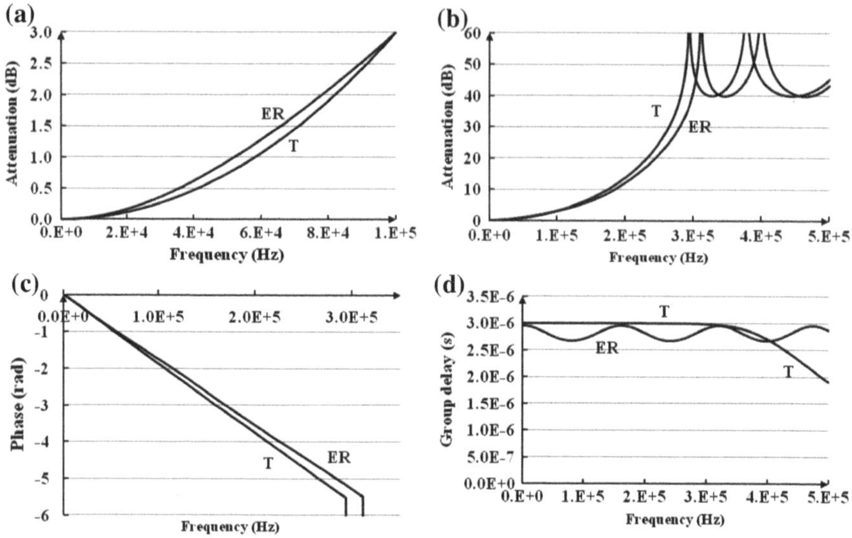

Fig. 10.8 Frequency domain responses of the 9th order filters with maximum number of transmission zeros at the ω-axis. T stands for Thomson prototype and ER stands for equi-ripple prototype, **a** passband attenuation, **b** overall attenuation, **c** phase, and **d** group delay

Table 10.4 Transition regions of the example filters

Type	T(9,0)	T(9,8)	ER(9,0)	ER(9,8)
f_s–f_0 (kHz)	228.145	181.572	231.919	198.378

Fig. 10.9 Step responses of the T(9,8) and ER(9,8) transfer functions

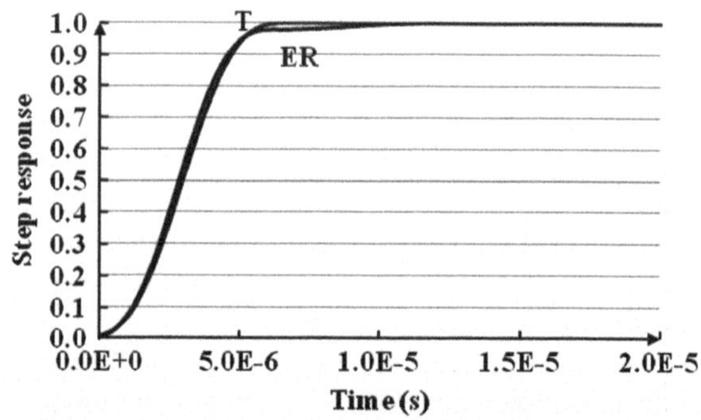

Finally, Fig. 10.9 depicts the step responses of the filter functions whose frequency responses are depicted in Fig. 10.6. A flat region (undershoot) is noticeable for the ER(9,8) filter.

10.5 Developer's Corner

The reader should consider as a serious challenge the task of finding the ω-axis transmission zeros (for a given stop-band attenuation and order of the numerator) of the function represented by Table 10.2.

It is worth mentioning that by synthesis of polynomial low-pass linear phase filters one is in fact in a position to create all-pass filters approximating constant group delay at low frequencies. These are frequently referred to as delay lines and will be addressed in more detail later one. Here we want to mention that an all-pass transfer function is easy to be obtained by simply extending the polynomial one with complex transmission zeros being in fact the poles mirrored against the ω-axis.

References

1. Thomson WE (1949) Delay networks having maximally flat frequency characteristics. Proc Inst Electr Eng Part III 96(44):487–490
2. Starch L (1954) Synthesis of constant-time-delay ladder networks using Bessel poly-nomials. Proc IRE 42:1666–1675
3. Dimopoulos H (2012) Analog electronic filters theory design and synthesis. Springer, New York
4. Bond C (2003) Bessel filters polynomials, poles and circuit elements. http://www.crbond.com/papers/bsf2.pdf. Last visited May 2019
5. Litovski VB (1979) Synthesis of monotonic passband sharp cutoff filters with constant group delay response. IEEE Trans Circuits Syst CAS 26(8):597–602
6. Remez EY (1934) Sur la détermination des polynômes d'approximation de degré donnée. Comm Soc Math Kharkov 10(41):41–63
7. Humpherys DV (1963) Rational function approximation of polynomials with equiripple error. Report R-159, Coordinated Science Laboratory University of Illinois, Urbana, Contract Da-36-039-Sc-85122 D/A Sub-Task 3-99-01-002
8. Skwirzinski J, Zdunek J (1964) Design of networks with prescribed delay and amplitude characteristics. Marconi Rev Fourth Quart 1964:162–187
9. Carlin H, Wu J (1976) Amplitude selectivity versus constant delay in minimum-phase lossless filters. IEEE Trans Circuits Syst 23(7):447–455
10. Gregorian R, Temes GC (1978) Design techniques for digital and analog all-pass circuits. IEEE Trans Circuits Syst CAS 25(12):981–988
11. Hibino M, Ishizaki Y, Watanabe H (1968) Design of chebyshev filters with flat group-delay characteristics. IEEE Trans Circuit Theory 15(4):316–325

Chapter 11
Low-Pass and Band-Pass Group Delay Correctors

To get a transfer function exhibiting both high selectivity and constant group delay in the passband one may use the so called phase correctors. The property that when two transfer functions are multiplied the amplitudes are multiplied but the phases (and group delays) are added to each other is exploited in this case. In that, for a given selective function, additional all-pass function is created which is consequently multiplied with the first one to keep the amplitude characteristic unchanged while adding a correction to the group delay. Both low-pass and band-pass approximants are visited in this chapter and detailed descriptions of the implemented algorithms are given. In both cases equi-ripple group delay approximation is implemented. Influence of the order of the corrector and the value of the group delay to be approximated to the realizability and the properties of the final solution is studied. Examples are given of a wide variety of cases. Results of a study are reported where corrected Chebyshev and LSM polynomial filters of the same complexity were compared. It was shown that while with almost equal selectivity the LSM solution is favourable from both group delay and passband attenuation approximation point of view.

11.1 Introduction

It is well known that selective filter functions produce large phase (group delay) distortions in the passband. In many applications these are unwanted or even make the solution unacceptable. The most frequent solution to get high selectivity and in the same time linear phase (constant group delay) in the passband is to cascade a new all-pass filter called phase corrector. Its role is to compensate for the phase distortions while not affecting the attenuation characteristic.

Here we promote two programs which, within the physical limitations of the problem, produce equi-ripple solutions for any type of pass-band attenuation and any final value of the group delay error.

V. Litovski, *Electronic Filters*, Lecture Notes in Electrical Engineering 596,
https://doi.org/10.1007/978-981-32-9852-1_11

11.2 The Low-Pass Corrector Algorithm

The subject of the passband group delay correction was visited for a long period of time now [1]. It was stated as a nonlinear optimization problem [2] which will here be solved by a variant of the Remez algorithm [3].

Given the transfer function of a low-pass filter and consequently its group delay as a function of frequency $T_{df}(\omega)$, one is to find the transfer function of an all-pass filter whose group delay $T_{dc}(\omega)$ adds to $T_{df}(\omega)$ so that together:

$$T_d(\omega) = T_{df}(\omega) + T_{dc}(\omega) \tag{11.1}$$

are approximating a constant within a prescribed error δ (%).

For the filter's group delay function one may use (10.7) and (10.10) for n even and n odd respectively, while for the corrector similar has to be rewritten:

$$T_{dc}(\omega) = 2 \cdot \sum_{l=1}^{k} \frac{-\alpha_l^2/\alpha_l}{(\omega - \beta_l)^2 + \alpha_l^2} \tag{11.2}$$

for k even. After substituting $\beta_l = |\beta_l|$ it may be rewritten as

$$T_{dc}(\omega) = \sum_{l=1}^{\lfloor k/2 \rfloor} T_l(\omega) \tag{11.3}$$

where

$$T_l(\omega) = \frac{-2 \cdot \alpha_l}{\alpha_l^2 + (\omega - \beta_k)^2} + \frac{-2 \cdot \alpha_l}{\alpha_l^2 + (\omega + \beta_l)^2} \tag{11.4}$$

When k is odd, after denoting the real pole as $z = \alpha_0$, one uses

$$T_{dc}(\omega) = T_0(\omega) + \sum_{l=1}^{\lfloor k/2 \rfloor} T_l(\omega) \tag{11.5}$$

with

$$T_0(\omega) = \frac{-2\alpha_0}{\omega^2 + \alpha_0^2}. \tag{11.6}$$

In the above $\lfloor . \rfloor$ is denoting the flooring function.

Note that in (11.3)–(11.6) the poles of the corrector are considered only. One is not to confuse them with the poles of the filter which, for convenience, in Chap. 10 were denoted by different letters.

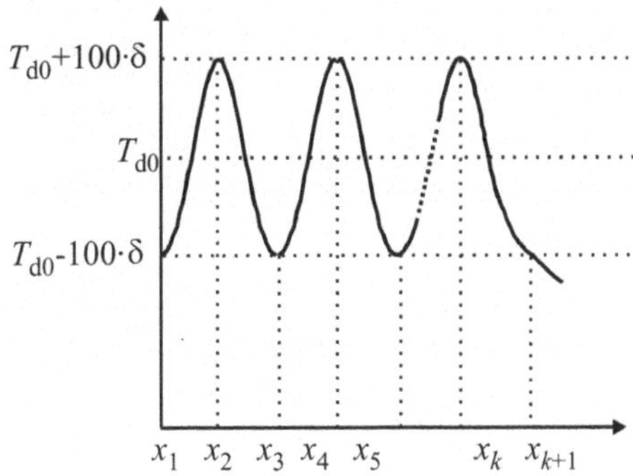

Fig. 11.1 Definition of the low-pass corrector's group delay approximation problem (Case: *k*-even)

In fact, according to Fig. 11.1 (for a corrector of even order), the following system of nonlinear equations is to be solved iteratively:

$$T_{\mathrm{d}}(\mathbf{p}, x_i) = T_{\mathrm{d0}} + 100 \cdot \delta \cdot (-1)^{r+i}, \quad \text{for} \quad i = 1, 2, \ldots, k+1, \tag{11.7}$$

where k is the order of the corrector, $\mathbf{p}$ is a vector of unknown coordinates of the poles of the all-pass corrector (two per a complex pole and one additional-real if k-odd), T_{d0} is the unknown average value of the passband group delay while $r = 1$ for k-odd and $r = 0$ for k-even.

Note since x_i, $i = 1, 2, \ldots, k + 1$ are unknown, in every iteration of correction of $\mathbf{p}$, one has to first recalculate their value. A proper initial solution for $\mathbf{p}$ is necessary which, in general, may be an obstacle to the implementation of the algorithm. If, however, the so called iteration damping is implemented [4], i.e. if at the beginning of the iterative process only fractions of the corrections are added to $\mathbf{p}$, as an initial solution for the complex poles one may simply use

$$p_l = \sigma_l + j\omega_l = -\frac{1}{k+n} \pm j\frac{2 \cdot l - 1}{k+1}, \quad l = 1, 2, \ldots, \lfloor k/2 \rfloor \tag{11.8}$$

while for the real pole (if k-odd) one may use $p_k = -1/(k + n)$, where n is, as earlier, the order of the filter to be corrected. In addition, as initial value, one may use $T_{\mathrm{d0}} = (n + 2 \cdot k) \cdot \pi/2$. Above, $\lfloor . \rfloor$ stands for the flooring function.

To find the location of the abscissas of the extremal points of the group delay, here denoted by x_i, $i = 1, 2, \ldots, k + 1$, one has to solve the following nonlinear equation.

$$\frac{dT_{\mathrm{d}}(\omega)}{\mathrm{d}\omega} = \frac{dT_{\mathrm{df}}(\omega)}{\mathrm{d}\omega} + \frac{dT_{\mathrm{dc}}(\omega)}{\mathrm{d}\omega} = 0. \tag{11.9}$$

For $dT_{\mathrm{df}}(\omega)/\mathrm{d}\omega$, in this expression, one may use (10.18) and (10.19) for n even and n odd, respectively, while for $dT_{\mathrm{dc}}(\omega)/\mathrm{d}\omega$ adjustment is needed in the following form

$$\frac{dT_{\mathrm{dc}}(\omega)}{\mathrm{d}\omega} = \frac{dT_0(\omega)}{\mathrm{d}\omega} + \sum_{l=1}^{\lfloor k/2 \rfloor} \frac{dT_l(\omega)}{\mathrm{d}\omega} = 0. \tag{11.10}$$

One is to substitute in (11.10) the following

$$\frac{dT_0(\omega)}{\mathrm{d}\omega} = \frac{4 \cdot \alpha_0 \cdot \omega}{\left(\omega^2 + \alpha_0^2\right)} \tag{11.11a}$$

and

$$\frac{dT_l(\omega)}{\mathrm{d}\omega} = \frac{4\alpha_l(\omega - \beta_l)}{\left[\alpha_l^2 + (\omega - \beta_l)^2\right]^2} + \frac{4\alpha_l(\omega + \beta_l)}{\left[\alpha_l^2 + (\omega + \beta_l)^2\right]^2}. \tag{11.11b}$$

Of course, for k even, (11.11a) is zero valued.

Note, again, that in (11.10) to (11.11) the poles of the corrector are considered only. One is not to confuse them with the poles of the filter which, for convenience, in Chap. 10 were denoted by different letters.

In our opinion there is no need for a sophisticated procedure to find the zeros of (10.9). If sophisticated, one may search the ω-axis with a coarse step to identify the intervals where the zeros are located. That should enable implementation of the modified secant method as described earlier in this book. Otherwise, one may set a fine step for ω (say $\Delta\omega = 10^{-6}$ rad/s) and look for change of the sign.

To solve (11.7) for the unknown coordinates of the poles of the corrector one needs to implement Newton-Raphson linearization as follows

$$\begin{aligned} &\sum_{l=1}^{\lfloor k/2 \rfloor} \left(\frac{\partial T_{\mathrm{dc}}(\mathbf{p}, x_i)}{\partial \sigma_l} \Delta\sigma_l + \frac{\partial T_{\mathrm{dc}}(\mathbf{p}, x_i)}{\partial \omega_l} \Delta\omega_l \right) \\ &= -T_{\mathrm{d}}(\mathbf{p}, x_i) + T_{\mathrm{d}0} + 100 \cdot \delta \cdot (-1)^{r+i}, \end{aligned} \tag{11.12}$$

for $i = 1, 2, \ldots, k$.

In this expression the number of unknown is $k + 1$, the first k being elements of $\mathbf{p}$ while the last is $T_{\mathrm{d}0}$. On the other side, the number of equation is k. So, alike in the previous chapter, here, there will be no calculation of the increment of $T_{\mathrm{d}0}$. Instead $T_{\mathrm{d}0}$ will be recalculated after every iteration according to

$$T_{\mathrm{d}0} = \frac{1}{k} \sum_{i=1}^{k} T_{\mathrm{d}}(\mathbf{p}, x_i). \tag{11.13}$$

For the partial derivatives one may use

$$\frac{\partial T_{\mathrm{dc}}(\omega)}{\partial \sigma_l} = 2 \cdot \frac{\sigma_l^2 - (\omega - \omega_l)^2}{\left[\sigma_l^2 + (\omega - \omega_l)^2\right]^2} + 2 \cdot \frac{\sigma_l^2 - (\omega + \beta_l)^2}{\left[\sigma_l^2 + (\omega + \beta_l)^2\right]^2} \tag{11.14}$$

$$\frac{\partial T_{\mathrm{dc}}(\omega)}{\partial \omega_l} = \frac{-4 \cdot \sigma_k \cdot (\omega - \omega_l)}{\left[\sigma_l^2 + (\omega - \omega_l)^2\right]^2} + \frac{4 \cdot \sigma_l \cdot (\omega + \omega_l)}{\left[\sigma_l^2 + (\omega + \omega_l)^2\right]^2} \tag{11.15}$$

and

$$\frac{\partial T_0(\omega)}{\partial \alpha_0} = 2 \cdot \frac{(\alpha_0^2 - \omega^2)}{(\omega^2 + \alpha_0^2)^2}. \tag{11.16}$$

This completes the description of the procedure of computing the pole locations of the corrector. In the next some examples will be discussed highlighting the results which may be obtained using the *corrector* program of the *RM* software and, in the same time, exposing the difficulties coming with the synthesis of group delay corrector as such.

We took the LSM paradigm as a prototype whose group delay will be corrected. The reason for that is the fact that in this example the difference between the minimum and the maximum of the passband group delay is relatively small as compared with Chebyshev, Halpern, and Papoulis filters and their derivatives with improved selectivity including the Elliptic filters. This is partially illustrated by Fig. 11.2a. Here four corrected group delay characteristics are depicted with the order of the corrector increased from 0 (no corrector) to 6 while keeping the maximum relative group delay error to be 1% of the average value of the low frequency passband group delay.

From the same figure we may see how is the corrector filling the valley on the top of the group delay and so increasing the average and the peak value at the end of the passband. Yes, by increasing the order of the corrector the band-width in which the group delay is corrected is increased, too. That may be easily seen from Fig. 11.2b. Note, in the $k = 6$ case the peak of the group delay at cut-off artificially contributes to the approximation interval making the solution to cover the whole passband which is even better than planned.

There is a limiting case of k for a given value of the error, however. Namely, further increasing the order of the corrector is not possible since any pole contributes by π to the area of the curve under the group delay. There is no room for additional increase of that area in Fig. 11.2a. Unfortunately, it is very difficult to estimate the maximum value of k for a given relative error in advance. One is supposed to cut and try until satisfied.

The reader probably already noticed (from Fig. 11.2a) that the absolute value of the error is rising with the rise of the order of the corrector. The reason for that is the increase of the overall area under the curve which, in turn, is increased by 2π in every step. This is inevitable when relative error is an approximation criterion

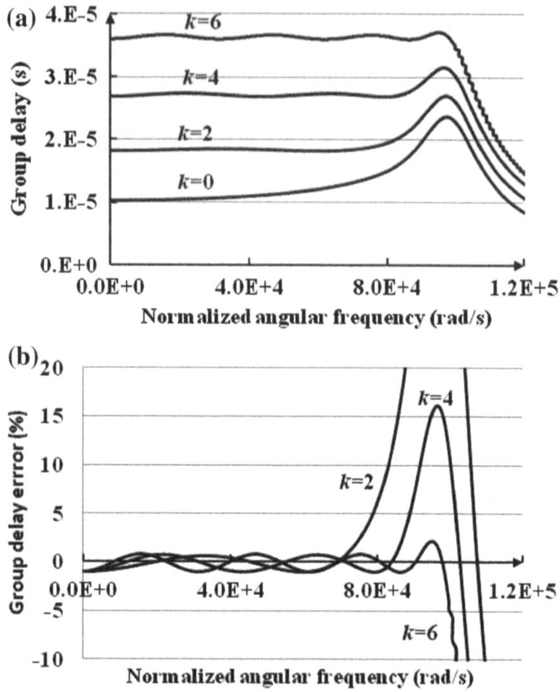

Fig. 11.2 LSM filter of 9th order with corrected group delay. The order of the corrector was varied **a** Overall group delay; **b** relative group delay error

which, unfortunately, is unavoidable. Namely, the designer has no knowledge in advance on the absolute values which will be produced by the given prototype filter (to be corrected) and the given complexity of the corrector for the selected δ. If, as it is usual, the absolute value of the delay error is imposed, one will need to repeat the corrector synthesis procedure with variable δ until satisfied.

That opens the issue of the size of δ and its influence to the approximation procedure and solution. To investigate that, since we did not expect any problems for $k = 2$ and $k = 4$, we repeated the above synthesis of the corrector with $k = 6$ and with increasing value of δ from 1% via 5% to 10%. The results are depicted in Fig. 11.3.

Care should be paid to distinguish the traces in Fig. 11.3a which depicts the group delay of the corrected LSM filter with $n = 9$ and $k = 6$. The lines are interwoven so that it is not easy to find which one is which. To facilitate that Fig. 11.3b depict the relative error for the same solutions. One may first recognize in Fig. 11.3a the 1% curve which is repeated (and properly marked) on both sides. Then, one may see that the increase of δ to 5% leads to a curve which has an average delay larger than the peak value at the end of the pass-band. As a consequence the approximation interval is significantly reduced when compared with the 1% solution.

The most interesting result is related to the case of δ = 10%. Namely, here, the last pole of the corrector is located outside of the passband allowing for an

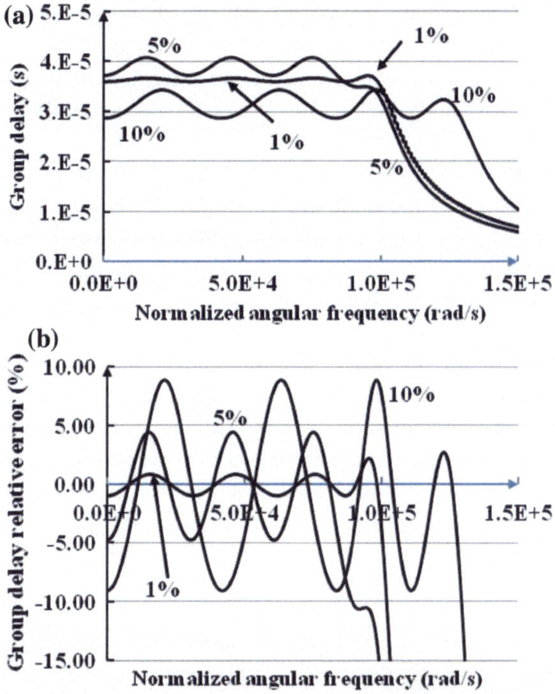

Fig. 11.3 LSM filter of 9th order with corrected group delay with a 6th order corrector. The value of the group delay error was varied. **a** Overall group delay; **b** relative group delay error

equi-ripple curve in which the abscissa of the peak at the cut-off is taking the role of x_k (with reference to Fig. 11.1). In return, the approximation interval is again artificially extended as it is with the 1% case but now far outside the passband. Note, this solution could be obtained only thanks to the high robustness of the *corrector* program.

To further exemplify the problem of a corrector synthesis we produced a function with Chebyshev prototype of 7th order (the same selectivity as the LSM of 9th order discussed above) with 1 dB maximum attenuation in the passband but renormalized to 3 dB at cut-off. The corrector was of 8th order while the approximation error was 1%. The resulting group delay characteristic is depicted in Fig. 11.4.

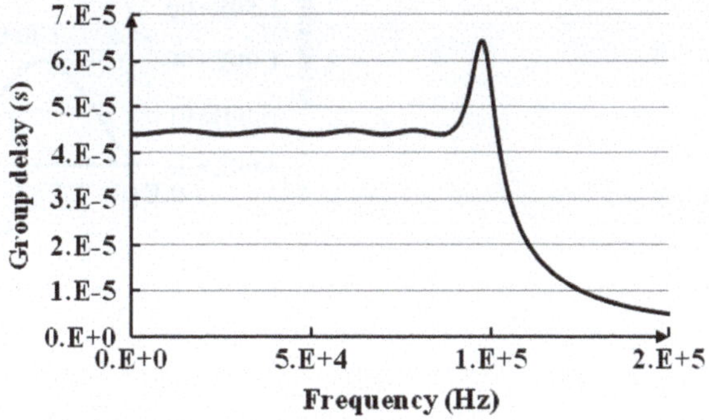

Fig. 11.4 The group delay characteristic of a corrected 7th order Chebyshev filter (1 dB passband) using an 8th order corrector

To cope with the lobes of the passband group delay one needed higher order a corrector than the "monotonic" group delay of the LSM filter (which was of higher order) asks for. In addition, due to extremely high value of the peak at the end of the passband, only 86% of the passband was covered by constant group delay with 1% error.

The LSM plus corrector and the Chebyshev plus corrector solutions described above are of the same order: $n + k = 15$. If we accept that the excess second order cell of the corrected Chebyshev filter realizing the corrector and the excess second order cell of the corrected LSM filter realizing a pair of poles, are of the same complexity (which is not in favor to the LSM), we may conclude the following. The corrected LSM solution while having almost identical stopband performance as the corrected Chebyshev, exhibits much better overall group delay and passband amplitude characteristics. To support the claim in Fig. 11.5 we include the attenuation characteristics of both. This issue was discussed in more details in [5].

To summarize, the synthesis of a corrector is a tricky job. The un-experienced designer is supposed to repeat the design sequence several times with different values of k and δ in order to find the possible and, in the same time, acceptable solution.

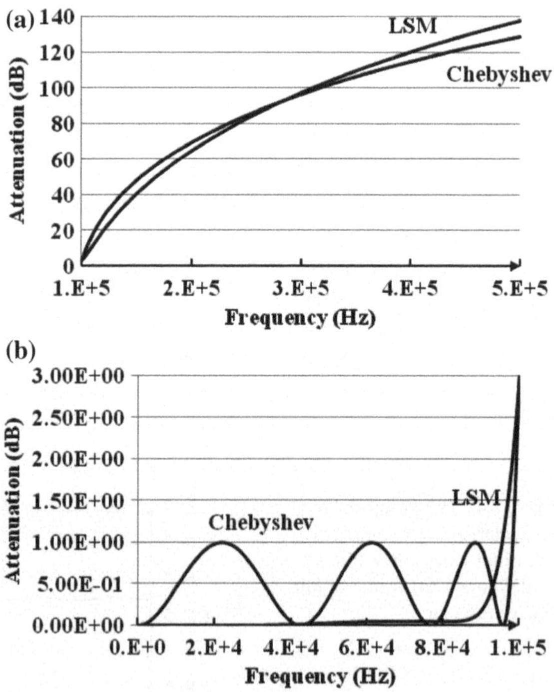

Fig. 11.5 Attenuation characteristics of the Chebyshev (7th order) and LSM (9th order) filters having the group delay corrected, **a** stopband and **b** passband attenuation

11.3 Design Example of a Low-Pass Corrector

The example described here is related to a 9th order LSM filter corrected by a 6th order corrector with relative error of 1%. Since the attenuation characteristic of the LSM filter is already given earlier in this book, and the group delay was discussed in the previous paragraph, we will first here depict, in Fig. 11.6, the time domain responses obtained by the *LP_analysis* program.

There are important differences between the time domain responses of minimum phase circuit (circuits with transmission zeros limited to the left half of the complex frequency plane) and the non-minimum phase ones having transmission zeros in the right half of the s-plane. That may be observed from Fig. 11.6 where both the step and pulse responses (the latter being a derivative of the former) have significant ringing before the arrival of the main signal. This ringing, when looked at in the eye diagram, contributes to reduction to its height (opening) from below.

Apart of that, looking to the step response we recognize that the overshoot and the undershoot are not significantly influenced as compared with the polynomial LSM case. On the other side, the main lobe of the pulse response is noticeably more symmetric which is a property highly valued when reduction of inter-symbol interference is considered. That may be stated as the prime benefit of the introduction of the corrector.

Here are excerpts from the reports produced by the *RM* software. First goes the report produced by the transfer function synthesis program *corrector*.

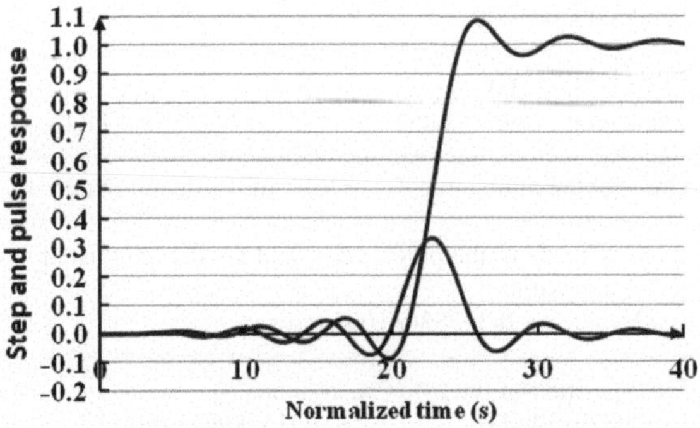

Fig. 11.6 Step and pulse responses of a 7th order LSM filter corrected by a 6th order corrector to produce 1% group delay error in the passband

Start.E

```
+++++++++++++++++++++++++++++++++++++++++++++++++++++++++++++++
Synthesis of a corrector filter.  PROJECT NAME: Cor_LSM_9_6_1
+++++++++++++++++++++++++++++++++++++++++++++++++++++++++++++++
Read in data for corrector synthesis
 Order of the numerator: n=0 Order of the denominator m=9        Order of the corrector
lk=6
------------------------------------------------------------------------------------------
Input data
 Order of the numerator n=0  Order of the denominator m=9
 Order of the corrector lk=6  Required group delay error tepstd=1.000000 %
------------------------------------------------------------------------------------------
Real part of the pole
 -8.171005975e-001  -8.171005975e-001  -1.076807322e-001  -1.076807322e-001
-5.046817384e-001  -5.046817384e-001  -3.092196716e-001  -3.092196716e-001
-9.159576574e-001
 Imaginary part of the pole
 -1.028090737e-001   1.028090737e-001   9.825844638e-001   -9.825844638e-001
5.215009666e-001   -5.215009666e-001   -8.327774086e-001   8.327774086e-001
0.000000000e+000
------------------------------------------------------------------------------------------
 The original function AFTER NORMALIZATION
------------------------------------------------------------------------------------------
 Real part of the pole
 -8.171005975e-001  -8.171005975e-001  -1.076807322e-001  -1.076807322e-001
-5.046817384e-001  -5.046817384e-001  -3.092196716e-001  -3.092196716e-001
-9.159576574e-001
 Imaginary part of the pole
 -1.028090737e-001   1.028090737e-001   9.825844638e-001   -9.825844638e-001
5.215009666e-001   -5.215009666e-001   -8.327774086e-001   8.327774086e-001
0.000000000e+000
+++++++++++++++++++++++++++++++++++++++++++++++++++++++++++++++
 FINAL SOLUTION
 The overall transfer function AFTER OPTIMIZATION
------------------------------------------------------------------------------------------
 Order of the numerator=6   Order of the denominator=15
===============================================================
The results about the poles, zeros and coefficients listed below are valid for frequency nor-
malization with
wc=2*pi*fc=6.28318530e+005 rad/sec
===============================================================
 Final position of the zeros_in columns
2.3690829334e-001        7.8051578902e-001
2.5927579155e-001        4.7291712338e-001
2.6330940206e-001        1.5799384088e-001
2.6330940206e-001        -1.5799384088e-001
2.5927579155e-001        -4.7291712338e-001
2.3690829334e-001        -7.8051578902e-001
Final position of the poles_in columns
```

```
-8.1710059750e-001          -1.0280907370e-001
-8.1710059750e-001          1.0280907370e-001
-1.0768073220e-001          9.8258446380e-001
-1.0768073220e-001          -9.8258446380e-001
-5.0468173840e-001          5.2150096660e-001
-5.0468173840e-001          -5.2150096660e-001
-3.0921967160e-001          -8.3277740860e-001
-3.0921967160e-001          8.3277740860e-001
-9.1595765740e-001          0.0000000000e+000
-2.3690829334e-001          7.8051578902e-001
-2.5927579155e-001          4.7291712338e-001
-2.6330940206e-001          1.5799384088e-001
-2.6330940206e-001          -1.5799384088e-001
-2.5927579155e-001          -4.7291712338e-001
-2.3690829334e-001          -7.8051578902e-001
------------------------------------------------------------------------------------------
Coefficients of the denominator of the transfer function (ascending in s):
 4.603482651e-003   6.675813690e-002   4.753288564e-001   2.180410139e+000
 7.249489406e+000   1.853251112e+001   3.773087928e+001   6.249390069e+001
 8.520316679e+001   9.603809602e+001   8.922552867e+001   6.755738774e+001
 4.070100212e+001 1.863530686e+001 5.912310111e+000 1.000000000e+000

Coefficients of the numerator of the transfer function (ascending in s):
 1.824847927e-002   -1.474431769e-001   5.611268373e-001   -1.209348482e+000
 1.818796953e+000 -1.518986974e+000 1.000000000e+000
------------------------------------------------------------------------------------------
 epstd=1.000e+000 (%)     t0=3.62230e-005 (s)     first    before    last    freq.
 (err=epstd)=6.38945e+004                                                    (Hz)
==========================================================================================
```

The next is part of the report produced by the *LP_Analysis* program.

```
*************************************************************************
 The cut-off frequency defined by 4.00000e+001 dB is equal to 1.49166e+005
*************************************************************************
 delay time (at 0.5)=3.62155e-005      rise time (0.1-0.9)=4.30521e-006
*************************************************************************
 The number of oscillations at the beginning of the step response is=8
 time=3.18310e-009          value of the extremum=-1.60956e-014
 time=6.68451e-008          value of the extremum=4.93182e-015
 time=8.87458e-006          value of the extremum=1.16396e-002
 time=1.32036e-005          value of the extremum=-7.39864e-003
 time=1.76019e-005          value of the extremum=2.62159e-002
 time=2.21495e-005          value of the extremum=-2.70272e-002
 time=2.67795e-005          value of the extremum=5.49823e-002
 time=3.14822e-005          value of the extremum=-8.72682e-002
*************************************************************************
 The number of oscillations at the end of the step response is=64
 time=4.11109e-005          value=1.08651e+000
```

```
time=4.59784e-005          value=9.65299e-001
time=5.08885e-005          value=1.02960e+000
time=5.58472e-005          value=9.89155e-001
time=6.08288e-005          value=1.01609e+000
time=6.59859e-005          value=9.96774e-001
time=7.04666e-005          value=1.00624e+000
time=7.57286e-005          value=9.95506e-001
*******************************************************************
```

End.E

11.4 The Band-Pass Corrector Algorithm

The necessity for constant group delay in broadband selective filters comes from the fact that they carry collections of channels located next to each other on the frequency axis. Since they form a single package to be communicated, it is expected all of them to arrive at the target receiver's input simultaneously or, in other words, without mutual delay. If we suppose that the transmission medium has flat amplitude and phase characteristic (does not contribute to any kind of distortions either linear or nonlinear) then the filter at the transmitting side should be the one which has to have constant group delay characteristic in order not to contribute to linear phase distortions.

To create a corrector in the case of a band-pass selective prototype, slight modifications are to be introduced in the procedure just described for low-pass cases. According to Fig. 11.7 the following nonlinear system of equations is to be solved iteratively

$$T_{\rm d}(\mathbf{p}, x_i) = T_{\rm d0} + 100 \cdot \delta \cdot (-1)^i, \text{for } i = 1,\ 2,\ \ldots,\ k, \tag{11.17}$$

Since there will always be a spare minimum either at low or at high frequencies inside the passband, alternative definition is possible in which the x-sequence starts with a maximum and ends with minimum of the overall group delay. Due to the strong arithmetical asymmetry of the amplitude characteristics observed in band-pass filters produced by frequency transformation (having geometrical symmetry) the group delay of such filters is also strongly arithmetical asymmetric as

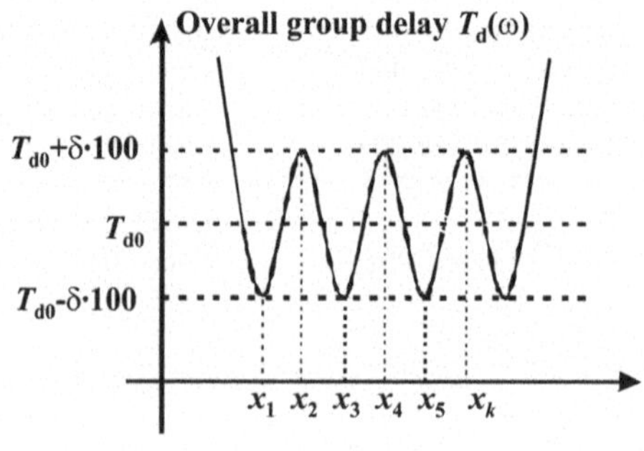

Fig. 11.7 Definition of the band-pass corrector's group delay approximation problem (k-even, always) with $T_{\rm d0}$ calculated in every iteration by (11.13)

shown in Fig. 4.7. That means that the above alternative will not lead to the same final solution as the original one depicted in Fig. 11.7. Since, as is the case with the low-pass filters just explained, the creation of the corrector is a tricky matter, using this alternative may become a gateway for a designer stacked by divergences or unacceptable convergences. Of course, the price paid will be an almost completely new program for bandpass corrector design.

As for the mathematical formulation, one should notice that there are no odd order solutions so that one may use the complete formulation from Paragraph 2 with parts related to odd order filters and odd order correctors excluded. That stands for the expression for the derivative of the overall group delay with respect to ω, and the derivatives of the corrector's group delay with respect to the real and imaginary parts of the poles. What must be different are the initial solutions which may be chosen to be

$$\sigma_l = -\frac{1}{2 \cdot n \cdot (\omega_{\max 2} - \omega_{\max 1})}, \quad i = 1, 2, \ldots, k/2 \tag{11.18a}$$

$$\omega_l = \omega_{\max 1} + 1.2 \cdot \frac{\omega_{\max 2} - \omega_{\max 1}}{k} \cdot i, \quad i = 1, 2, \ldots, k/2, \tag{11.18b}$$

and $\sigma_{k/2+i} = \sigma_i$ and $\omega_{k/2+i} = -\omega_i$, for i = 1, 2, ..., k/2. Of course, since k is an even number there is no necessity for the flooring function used earlier.

In (11.18) $\omega_{\max 1}$ is the normalized angular frequency of the first maximum of the group delay of the prototype selective filter. $\omega_{\max 2}$ is the normalized angular frequency of the last maximum being on the opposite side of the passband. n stands for the order of the low-pass selective prototype filter used to be transformed into a band-pass one.

As an illustration of the influence of the order of the corrector to the final approximation of constant group delay, a 14th order LSM filter with improved selectivity exhibiting 60 dB attenuation in the stopband will be used. It was obtained by LP-to-BP transformation with a relative bandwidth of 40%. The group delay error required was 1% as it was for the low-pass case above. Four solutions are depicted in Fig. 11.8a, for k = 0, k = 2, k = 4, k = 6, and k = 8. To expose the shape of the group delay in the passband Fig. 11.8b depicts the same solutions with a stretched ω-axis.

To expose the influence of the group delay error and the limitations which are coming from that side, we created Fig. 11.9 where everything is the same as in Fig. 11.8 but the group delay error is 5%. Note the absence of the 8th order corrector. Even the curve representing the group delay for k = 6 is a small miracle since it catches the last chance to include the last peak of the original group delay as part of the overall solution.

By observing the group delay characteristics of Fig. 11.9 one may recognize the influence of the choice of the first abscissa in the statement of the solution of Fig. 11.7. A choice where x_1 would start with a maximum would transpose the incomplete ringing of the group delay which is now at the upper part of the passband into the lower part. Since the difference among the peaks of the uncorrected group delay is significant one should expect really different solution. This option is not elaborated within the *corrector_bp* program which was used for creation of Figs. 11.8 and 11.9.

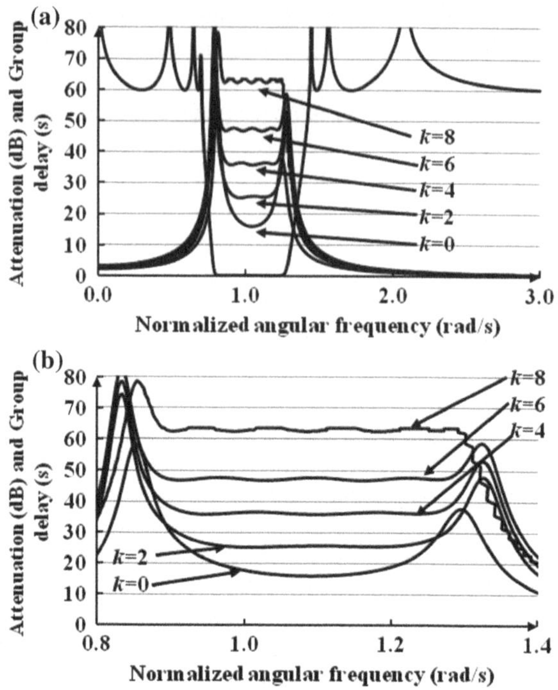

Fig. 11.8 Frequency characteristics of a 14th order LSM filter with improved selectivity whose group delay characteristic was corrected with a corrector of rising complexity. Relative group delay error is 1% for all solutions, **a** attenuation and group delay **b** group delay highlighted

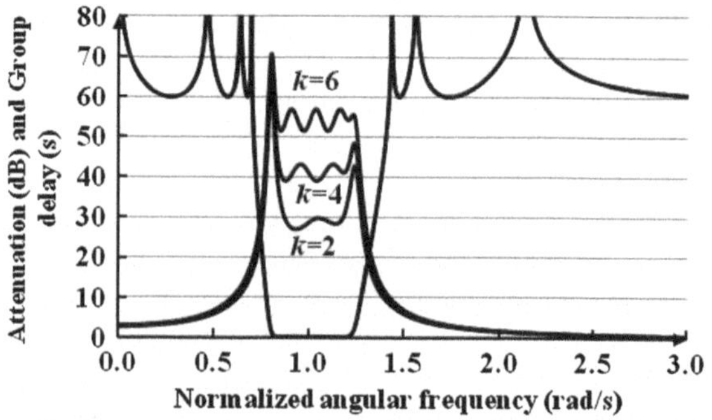

Fig. 11.9 Frequency characteristics of a 14th order LSM filter with improved selectivity whose group delay was corrected with a corrector of rising complexity. Relative group delay error is 5% for all solutions

11.5 Design Example of a Band-Pass Corrector

Here is an excerpt of the report created by the *corrector_bp* program of the *RM* software for filter design. The example is one among the ones whose characteristics are depicted in Fig. 11.8. It is the $k = 8$ case. To reduce the length of the text the normalized numerical values of the zeros and poles are rearranged into a table.

Start.E

++

Synthesis of a corrector of a BANDPASS selective filter.

PROJECT NAME: cor_bp_lsm_14_8_60_1

++

Order of the numerator: n=13 Order of the denominator m=14 Order of the corrector lk=8

==

INPUT DATA

Required group delay error=1.00000e+000

Real part of the zero

0.000000000e+000 -0.000000000e+000 0.000000000e+000 0.000000000e+000
8.200387628e-009 -8.200387628e-009 -8.200387628e-009 8.200387628e-009
0.000000000e+000 -0.000000000e+000 0.000000000e+000 0.000000000e+000
0.000000000e+000

Imaginary part of the zero

1.433678590e+000 -6.975064054e-001 -1.433678590e+000 6.975064054e-001
1.560422865e+000 -6.408519268e-001 -1.560422865e+000 6.408519268e-001
2.137240548e+000 -4.678930507e-001 -2.137240548e+000 4.678930507e-001
0.000000000e+000

Real part of the pole

-4.068082046e-002 -2.618671244e-002 -2.618671244e-002 -4.068082046e-002
-3.084482587e-001 -3.084482587e-001 -1.354522616e-001 -9.271389173e-002
-9.271389173e-002 -1.354522616e-001 -2.616955131e-001 -2.107388101e-001
-2.107388101e-001 -2.616955131e-001

Imaginary part of the pole

1.245727100e+000 -8.018888750e-001 8.018888750e-001 -1.245727100e+000
9.512411218e-001 -9.512411218e-001 1.201092500e+000 -8.221196065e-001
8.221196065e-001 -1.201092500e+000 1.083196998e+000 -8.722795581e-001
8.722795581e-001 -1.083196998e+000

==

FINAL SOLUTION

The overall transfer function AFTER OPTIMIZATION

Order of the numerator=21 Order of the denominator=22

Zeros		Poles	
Real part	Imaginary part	Real part	Imaginary part
7.9163270830e-002	8.8323977779e-001	-4.0680820463e-002	1.2457271003e+000
8.3812239515e-002	9.8312260662e-001	-2.6186712439e-002	-8.0188887500e-001
8.3322386989e-002	1.0833244330e+000	-2.6186712439e-002	8.0188887500e-001
7.7376865987e-002	1.1818800018e+000	-4.0680820463e-002	-1.2457271003e+000
7.7376865987e-002	-1.1818800018e+000	-3.0844825866e-001	9.5124112176e-001
8.3322386989e-002	-1.0833244330e+000	-3.0844825866e-001	-9.5124112176e-001
8.3812239515e-002	-9.8312260662e-001	-1.3545226162e-001	1.2010925003e+000
7.9163270830e-002	-8.8323977779e-001	-9.2713891726e-002	-8.2211960650e-001
0.0000000000e+000	1.4336785902e+000	-9.2713891726e-002	8.2211960650e-001
0.0000000000e+000	-6.9750640544e-001	-1.3545226162e-001	-1.2010925003e+000
0.0000000000e+000	-1.4336785902e+000	-2.6169551305e-001	1.0831969982e+000
0.0000000000e+000	6.9750640544e-001	-2.1073881008e-001	-8.7227955812e-001
0.0000000000e+000	1.5604228655e+000	-2.1073881008e-001	8.7227955812e-001
0.0000000000e+000	-6.4085192682e-001	-2.6169551305e-001	-1.0831969982e+000
0.0000000000e+000	-1.5604228655e+000	-7.9163270830e-002	8.8323977779e-001
0.0000000000e+000	6.4085192682e-001	-8.3812239515e-002	9.8312260662e-001
0.0000000000e+000	2.1372405479e+000	-8.3322386989e-002	1.0833244330e+000
-0.0000000000e+000	-4.6789305068e-001	-7.7376865987e-002	1.1818800018e+000
0.0000000000e+000	-2.1372405479e+000	-7.7376865987e-002	-1.1818800018e+000
0.0000000000e+000	4.6789305068e-001	-8.3322386989e-002	-1.0833244330e+000
0.0000000000e+000	0.0000000000e+000	-8.3812239515e-002	-9.8312260662e-001
		-7.9163270830e-002	-8.8323977779e-001

End.E

11.6 Developer's Corner

There are several aspects related to the design of group delay correctors.

From the software developer's point of view one is forced to first create routines for solution of systems of linear equations, finding the zeros of the derivatives of the group delay, and elaborating the damping procedure including limitation of the maximum number of iterations. Since really difficult situations may arise (from the convergence point of view) one is supposed to build in an effective monitoring of the convergence process by periodically printing the value of the modulus of the corrective vector ($|\Delta \mathbf{x}|$) along with the number of iterations.

The designer's task is not much easier even if a good program is to be used. Finding the right selective prototype is the first task. Namely, as we already demonstrated, the final complexity of the system is mainly defined by the complexity of the corrector. That means that one is to be careful with the selective prototype in order not to produce a very complex final solution. Chebyshev type passband attenuation as shown in Fig. 7.4 may become a nightmare in finding the right value of the order of the corrector for a given error. The final result may be a really complex corrector since it has to fill all the valleys in the original group delay. So, low ripple Chebyshev type or LSM solutions would lead to a best compromise in satisfying the selectivity and group delay requirements by a system of minimum complexity. Furthermore, one is not to forget that, as shown in Chap. 8, a transmission zero at the ω-axis contributes much more to the selectivity than an increase of the order of the filter while, in the same time, reducing the lobes in the passband attenuation and group delay responses.

The group delay error, as mentioned in the case of low-pass corrector, influences the final value of the average passband delay and the absolute value of the delay error. If the latter was specified, a tedious task is in front of the designer to find the relative error which is adequate to the required absolute. Both the order of the corrector and the value of the relative error are to be varied in order to get satisfactory result.

References

1. Gregorian R, Temes GC (1978) Design techniques for digital and analog all-pass circuits. IEEE Trans Circuits Syst CAS 25(12):981–988
2. Skwirzinski J, Zdunek J (1964) Design of networks with prescribed delay and amplitude characteristics. Marconi Rev Fourth Quarter:162–187
3. Remez EY (1934) Sur la détermination des polynômes d'approximation de degré donnée. Comm Soc Math Kharkov 10(41):41–63
4. Litovski V, Zwolinski M (1997) VLSI circuit simulation and optimization. Chapman and Hall, London
5. Andrejević Stošović M, Topisirović D, Litovski V (2019) Frequency and time domain comparison of selective polynomial filters with corrected phase characteristics. Int J Electron 106(5):770–778

Chapter 12
Direct Bandpass Synthesis of Linear and Parabolic Phase All-Pass Filters

In this chapter the subject of synthesis of all-pass filters approximating the group delay in the maximally flat sense around a given non-zero frequency is introduced. Requirements asking for constant and linearly rising or falling group delay are accepted. The corresponding phase would be linear or parabolic. Iterative procedure is described in full detail allowing realization of a computer program performing the synthesis. Examples are given for all three situations. The influence of the change of the slope to the approximation interval is studied.

12.1 Introduction

All-pass networks behaving as delay lines at a given carrier frequency are often of interest. These exhibit constant group delay (a zero slope line) in a range of fre quencies arround the central one [1–3]. Their application in contemporrary software defined radio is indispesible [4]. Others, exhibiting a group delay characteristic approximating a line with finite slope arround the carrier, are used for pulse shaping [5–12]. Namely, a frequency modulated signal input to such a filter will be transformed in the sense that the modulation intensity will be enhanced or diminished due to the (parabolic) phase characteristic of the filter. As can be seen from the list of references, the scientific and industrial interest on the subjec of band-pass all-pass filters spans for more than half a century.

There are different approaches in the design of all-pass filters. Some go for equi-ripple approximation [13], other for time domain optimization [8] and some [2, 10] go for maximally flat approximation.

To get the feeling as to how the all-pass effect is created Fig. 12.1 depicts the pole-zero pattern of an 18th order allpass filter approximating constant group delay around the central frequency which is here normalized to unity.

Two vectors are depicted in the figure representing jω-z_1 and jω-p_1, z_1 and p_1 being a zero and a pole of the overall transfer function. The resulting contribution of

V. Litovski, *Electronic Filters*, Lecture Notes in Electrical Engineering 596,
https://doi.org/10.1007/978-981-32-9852-1_12

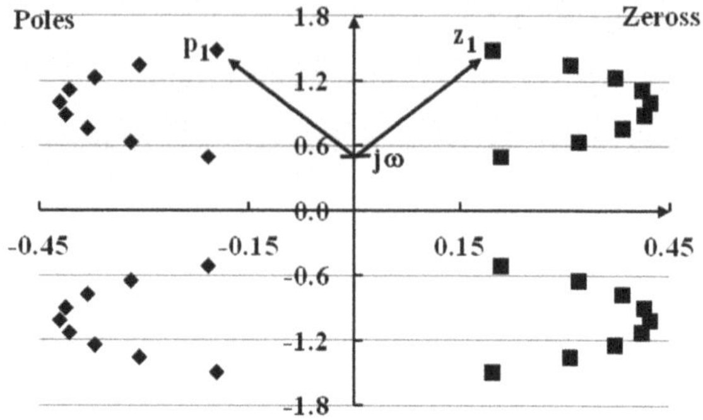

Fig. 12.1 Poles and zeros of an all-pass band-pass system

this pole-zero pair to the transfer function is (jω-z_1)/(jω-p_1). Since, according to Fig. 12.1, the moduli of the numerator and the denominator of this expression are equal, no matter the value of ω, the resulting function will always have unity gain. That stands for all pole-zero pairs of the all-pass system. The phase angle, however, always doubles and changes with the change of the frequency.

In this chapter the concept based on maximally flat approximation will be adopted and elaborated. Complete procedure will be given for synthesis of band-pass all-pass filters approximating both constant and linearly changing group delay. The procedure is based on an original development [10] for the *n*th derivatives of the group delay with respect to the angular frequency and consequently with respect to the coordinates of the transmission zeros.

The program *B0010* performs synthesis of such all-pass filters with arbitrary parameters.

12.2 The Algorithm

The group delay of an *n*th order allpass network whose RHP transmission *zeros* in the first quadrant are $z_k = \sigma_k + j\omega_k$, k = 1,2,…, n/2, may be calculated as

$$\tau_d(\omega) = \sum_{k=1}^{n/2} T_k(\omega) \tag{12.1}$$

where

$$T_k(\omega) = r(\omega) + q(\omega) = \frac{2 \cdot \sigma_k}{\sigma_k^2 + (\omega - \omega_k)^2} + \frac{2 \cdot \sigma_k}{\sigma_k^2 + (\omega + \omega_k)^2}. \tag{12.2}$$

To get a maximally flat approximation of the group delay around the central frequency ω_0, with $\tau_d(\omega_0) = t_0$ and $\mathbf{d}\tau_d/\mathbf{d}\omega_{|(\omega=\omega_0)} = s$ (or $\mathbf{d}\tau_d/\mathbf{d}\omega_{|(\omega=\omega_0)} = -s$), where t_0 and s are positive constants, one has to solve the following system of nonlinear equation with respect to the coordinates of z_k

$$F_1(\boldsymbol{\sigma}, \boldsymbol{\omega}) = \tau_d(\omega_0) - t_0 = 0 \tag{12.3a}$$

$$F_2(\boldsymbol{\sigma}, \boldsymbol{\omega}) = \mathbf{d}\tau_d/\mathbf{d}\omega_{|(\omega=\omega_0)} - s = 0 \tag{12.3b}$$

$$F_i(\boldsymbol{\sigma}, \boldsymbol{\omega}) = \mathbf{d}^i\tau_d/(\mathbf{d}\omega)^i_{|(\omega=\omega_0)} = 0 \quad \text{for } i = 3, \ldots, n. \tag{12.3c}$$

where $\boldsymbol{\sigma}$ and $\boldsymbol{\omega}$ are vectors of the unknown coordinates of the zeros.

In (12.3b) the parameter s defines the slope at the central frequency of the passband of the final group delay characteristic. If $s = 0$ one goes for constant group delay which is the most frequent case. Constant group delay means linear phase around the central frequency.

If $s \neq 0$ one requires a linearly rising ($s > 0$) or linearly decreasing ($s < 0$) group delay. That, in both cases means, a parabolic shape of the phase characteristic around the central frequency.

After linearization the following system of linear equations arises

$$\sum_{j=1}^{n/2}\left\{\frac{\partial F_i(\boldsymbol{\sigma}, \boldsymbol{\omega})}{\partial \sigma_j}\Delta\sigma_j + \frac{\partial F_i(\boldsymbol{\sigma}, \boldsymbol{\omega})}{\partial \omega_j}\Delta\omega_j\right\} = -F_i(\boldsymbol{\sigma}, \boldsymbol{\omega}) \quad \text{for } i = 1, \ldots, n. \tag{12.4}$$

It should be solved iteratively.

A strongly damped iterative process was implemented in order for the solution to converge for slopes of any (positive and negative) value and for filters of any order.

We will address the issue of the derivatives with respect to ω, now. To find them we will process the addends of the group delay $r(\omega)$ and $q(\omega)$. So, according to (12.2) we have

$$r(\omega) = \frac{2 \cdot \sigma_k}{\sigma_k^2 + (\omega - \omega_k)^2}. \tag{12.5}$$

It may be rewritten in the following form

$$r(\omega) \cdot \left[\sigma_k^2 + \left(\omega - \omega_k\right)^2\right] - 2 \cdot \sigma_k = 0 \tag{12.6}$$

which allows creation of a recurrent expression for calculating the derivatives. Namely, the first five derivatives may be obtained by solving the following set of expressions in succession.

$$\frac{dr(\omega)}{d\omega}\cdot\left[\sigma_k^2+\left(\omega-\omega_k\right)^2\right]+2\cdot(\omega-\omega_k)\cdot r(\omega)=0 \tag{1.7a}$$

$$\frac{d^2r(\omega)}{d\omega^2}\cdot\left[\sigma_k^2+\left(\omega-\omega_k\right)^2\right]+4\cdot\frac{dr(\omega)}{d\omega}\cdot(\omega-\omega_k)+2\cdot r(\omega)=0 \tag{12.7b}$$

$$\frac{d^3r(\omega)}{d\omega^3}\cdot\left[\sigma_k^2+\left(\omega-\omega_k\right)^2\right]+6\cdot\frac{d^2r(\omega)}{d\omega^2}\cdot(\omega-\omega_k)+6\cdot\frac{dr(\omega)}{d\omega}=0 \tag{12.7c}$$

$$\frac{d^4r(\omega)}{d\omega^4}\cdot\left[\sigma_k^2+\left(\omega-\omega_k\right)^2\right]+8\cdot\frac{d^3r(\omega)}{d\omega^3}\cdot(\omega-\omega_k)+12\cdot\frac{d^2r(\omega)}{d\omega^2}=0 \tag{12.7d}$$

and

$$\frac{d^5r(\omega)}{d\omega^5}\cdot\left[\sigma_k^2+\left(\omega-\omega_k\right)^2\right]+10\cdot\frac{d^4r(\omega)}{d\omega^4}\cdot(\omega-\omega_k)+20\cdot\frac{d^3r(\omega)}{d\omega^3}=0. \tag{12.7e}$$

For example, the first derivative is

$$\frac{dr(\omega)}{d\omega}=\frac{-2\cdot(\omega-\omega_k)\cdot r(\omega)}{\left[\sigma_k^2+\left(\omega-\omega_k\right)^2\right]}. \tag{12.8}$$

The sequence (1.7a) may be generalized into the form

$$\begin{aligned}&\frac{d^jr(\omega)}{d\omega^j}\cdot\left[\sigma_k^2+\left(\omega-\omega_k\right)^2\right]\\&+2\cdot j\cdot\frac{d^{j-1}r(\omega)}{d\omega^{j-1}}\cdot(\omega-\omega_k)+j\cdot(j-1)\cdot\frac{d^{j-2}r(\omega)}{d\omega^{j-2}}=0,\end{aligned} \tag{12.9}$$

which allows for calculation of the derivative of order j.

Similar series of expression may written for $q(\omega)$ which would lead to

$$\begin{aligned}&\frac{d^jq(\omega)}{d\omega^j}\cdot\left[\sigma_k^2+\left(\omega+\omega_k\right)^2\right]\\&+2\cdot j\cdot\frac{d^{j-1}q(\omega)}{d\omega^{j-1}}\cdot(\omega+\omega_k)+j\cdot(j-1)\cdot\frac{d^{j-2}q(\omega)}{d\omega^{j-2}}=0.\end{aligned} \tag{12.10}$$

Now, having this result, finding the partial derivatives with respect to the parameters (σ_l and ω_l), is a trivial task. The recurrent expressions for the partial derivatives with respect to ω_l are as follows.

$$\left[\sigma_k^2 + \left(\omega - \omega_l\right)^2\right] \frac{\partial\left(\frac{\mathrm{d}^j r(\omega)}{\mathrm{d}\omega^j}\right)}{\partial \omega_l} - 2 \cdot \frac{\mathrm{d}^j r(\omega)}{\mathrm{d}\omega^j} (\omega - \omega_l)$$
$$+ 2 \cdot j \cdot \frac{\partial\left(\frac{\mathrm{d}^{j-1} r(\omega)}{\mathrm{d}\omega^{j-1}}\right)}{\partial \omega_l} \cdot (\omega - \omega_l) - 2 \cdot j \cdot \frac{\mathrm{d}^{j-1} r(\omega)}{\mathrm{d}\omega^{j-1}} \tag{12.11a}$$
$$+ j \cdot (j-1) \cdot \frac{\partial\left(\frac{\mathrm{d}^{j-2} r(\omega)}{\mathrm{d}\omega^{j-2}}\right)}{\partial \omega_l} = 0$$

with

$$\frac{\partial\left(\frac{\mathrm{d}r(\omega)}{\mathrm{d}\omega}\right)}{\partial \omega_l} \cdot \left[\sigma_l^2 + \left(\omega - \omega_l\right)^2\right]$$
$$- 2 \cdot (\omega - \omega_l) \cdot \frac{\mathrm{d}r(\omega)}{\mathrm{d}\omega} - 2 \cdot r(\omega) + (\omega - \omega_l) \cdot \frac{\partial r(\omega)}{\partial \omega_l} = 0 \tag{12.11b}$$

and

$$\frac{\partial r(\omega)}{\partial \omega_l} \cdot \left[\sigma_l^2 + \left(\omega - \omega_l\right)^2\right] - 2 \cdot (\omega - \omega_l) \cdot r(\omega) \cdot = 0. \tag{12.11c}$$

$$\left[\sigma_l^2 + \left(\omega + \omega_l\right)^2\right] \frac{\partial\left(\frac{\mathrm{d}^j q(\omega)}{\mathrm{d}\omega^j}\right)}{\partial \omega_l} + 2 \cdot \frac{\mathrm{d}^j q(\omega)}{\mathrm{d}\omega^j} (\omega + \omega_l)$$
$$+ 2 \cdot j \cdot \frac{\partial\left(\frac{\mathrm{d}^{j-1} q(\omega)}{\mathrm{d}\omega^{j-1}}\right)}{\partial \omega_l} \cdot (\omega + \omega_l) - 2 \cdot j \cdot \frac{\mathrm{d}^{j-1} q(\omega)}{\mathrm{d}\omega^{j-1}} \tag{12.12a}$$
$$+ j \cdot (j-1) \cdot \frac{\partial\left(\frac{\mathrm{d}^{j-2} q(\omega)}{\mathrm{d}\omega^{j-2}}\right)}{\partial \omega_l} = 0$$

with

$$\frac{\partial\left(\frac{\mathrm{d}q(\omega)}{\mathrm{d}\omega}\right)}{\partial \omega_l} \cdot \left[\sigma_l^2 + \left(\omega + \omega_l\right)^2\right]$$
$$+ 2 \cdot (\omega + \omega_l) \cdot \frac{\mathrm{d}q(\omega)}{\mathrm{d}\omega} + 2 \cdot r(\omega) + (\omega + \omega_k) \cdot \frac{\partial q(\omega)}{\partial \omega_l} = 0 \tag{12.12b}$$

and

$$\frac{\partial q(\omega)}{\partial \omega_l} \cdot \left[\sigma_k^2 + \left(\omega + \omega_k\right)^2\right] + 2 \cdot (\omega + \omega_k) \cdot q(\omega) \cdot = 0. \tag{12.12c}$$

Similar set of expressions may be produced for the derivatives with respect to σ_l. Partial derivatives of $r(\omega)$ and its derivatives with respect to ω are

$$\begin{aligned} &\frac{\partial\left(\frac{\mathrm{d}^j r(\omega)}{\mathrm{d}\omega^j}\right)}{\partial \sigma_l} \cdot \left[\sigma_l^2 + \left(\omega - \omega_l\right)^2\right] + 2 \cdot \sigma_l \cdot \frac{\mathrm{d}^j r(\omega)}{\mathrm{d}\omega^j} \\ &+ 2 \cdot j \cdot \frac{\partial\left(\frac{\mathrm{d}^{j-1} r(\omega)}{\mathrm{d}\omega^{j-1}}\right)}{\partial \sigma_l} \cdot (\omega - \omega_l) \\ &+ j \cdot (j-1) \cdot \frac{\partial\left(\frac{\mathrm{d}^{j-2} r(\omega)}{\mathrm{d}\omega^{j-2}}\right)}{\partial \sigma_l} = 0 \end{aligned} \tag{12.13a}$$

with

$$\begin{aligned} &\frac{\partial\left(\frac{\mathrm{d}r(\omega)}{\mathrm{d}\omega}\right)}{\partial \sigma_l} \cdot \left[\sigma_l^2 + \left(\omega - \omega_l\right)^2\right] \\ &+ 2 \cdot \sigma_l \cdot \frac{\mathrm{d}r(\omega)}{\mathrm{d}\omega} + 2 \cdot (\omega - \omega_k) \cdot \frac{\partial r(\omega)}{\partial \sigma_l} = 0 \end{aligned} \tag{12.13b}$$

and

$$\frac{\partial r(\omega)}{\partial \sigma_l} \cdot \left[\sigma_l^2 + \left(\omega - \omega_l\right)^2\right] + 2 \cdot \sigma_l \cdot r(\omega) - 2 = 0 \tag{12.13c}$$

Partial derivatives of $q(\omega)$ and its derivatives with respect to ω are

$$\begin{aligned} &\frac{\partial\left(\frac{\mathrm{d}^j q(\omega)}{\mathrm{d}\omega^j}\right)}{\partial \sigma_l} \cdot \left[\sigma_l^2 + \left(\omega + \omega_l\right)^2\right] + 2 \cdot \sigma_l \cdot \frac{\mathrm{d}^j q(\omega)}{\mathrm{d}\omega^j} \\ &+ 2 \cdot j \cdot \frac{\partial\left(\frac{\mathrm{d}^{j-1} q(\omega)}{\mathrm{d}\omega^{j-1}}\right)}{\partial \sigma_l} \cdot (\omega + \omega_k) \\ &+ j \cdot (j-1) \cdot \frac{\partial\left(\frac{\mathrm{d}^{j-2} q(\omega)}{\mathrm{d}\omega^{j-2}}\right)}{\partial \sigma_l} = 0 \end{aligned} \tag{12.14a}$$

with

$$\frac{\partial\left(\frac{dq(\omega)}{d\omega}\right)}{\partial\sigma_l}\cdot\left[\sigma_l^2+\left(\omega-\omega_l\right)^2\right]$$
$$+2\cdot\sigma_l\cdot\frac{dq(\omega)}{d\omega}+2\cdot(\omega-\omega_l)\cdot\frac{\partial q(\omega)}{\partial\sigma_l}=0 \tag{12.14b}$$

and

$$\frac{\partial q(\omega)}{\partial\sigma_l}\cdot\left[\sigma_l^2+\left(\omega-\omega_l\right)^2\right]+2\cdot\sigma_l\cdot q(\omega)-2=0. \tag{12.14c}$$

Here ends the developments related to the formulation of the linear system of Eq. (12.4).

To start, however, one needs initial solutions. We would recommend the following normalized coordinates [10]

$$\sigma_i=\frac{\pi\cdot n}{t_0}\cdot\frac{n+6}{2t_0} \tag{12.15a}$$

$$\pm\omega_i=1+\frac{\pi\cdot n}{2t_0}\cdot\left(\frac{2i-1}{n}-0.5\right)\text{ for } i=1,2,\ldots,n/2, \tag{12.15b}$$

with $t_0=2\cdot\pi\cdot T_0\cdot\Omega_0$. Here T_0 is the requested un-normalized group delay and Ω_0 is the requested un-normalized central frequency of the filter. Note the scaling factor $\pi\cdot n/t_0$ represents width (in angular frequency) since it is a quotient of the area under the curve and its height. Again, the real parts are taken to be equal since, most frequently, constant group delay is sought.

Having in mind the fact that the order of the filter, the band-width of approximation, the slope of the group delay (positive and negative), and the value of the nominal group delay are bound by the area under the group delay curve, on should pay attention in the choice of T_0 and n when starting the design. Large T_0 and small n will lead to extremely narrow approximation interval. On the contrary, small T_0 and large n, while asking for broad approximation interval, may happen to be a serious obstacle to convergence since half of the area has to be below Ω_0. When this happens the program *b0010* of the *RM* software will stop and suggest the nominal group delay to be increased or the proposed order of the filter decreased.

All these bring us again to the necessity of a strong damping of the iteration process since the same set of initial conditions are to provide for convergence for vide intervals of design parameters and types of approximation.

To illustrate the results which may be obtained by implementation of the procedure described above, we will consider an all-pass filter wit T_0 = 50 μs.

Approximation of a constant group delay is illustrated first. The results are depicted in Fig. 12.2 where two solutions (n = 10 and n = 18) are presented.

Figure 12.2a represents the group delay characteristics. The broadening of the approximation interval with the rise of the order of the filter is clearly noticeable. On the other side, in Fig. 12.2b the phase characteristics are depicted. Here from it is easy to recognize the interval of linearity but not as clear as in the case of the group delay.

It is obvious from Fig. 12.2a that further rise of the order of the filter will soon reach its limiting value since not much of area is left above the $n = 18$ curve at low frequencies.

Next, we will demonstrate the influence of the order of the filter to the approximation interval in the case of linearly rising group delay. The results are depicted in Fig. 12.3. The group delay characteristics for $n = 6$, $n = 10$ and $n = 16$ are depicted in Fig. 12.3.a. Steady broadening of the approximation interval with the rise of the order of the filter may be noticed. That is clearly represented in Fig. 12.3b where the relative deviation of the group delay from a straight line imposed as a target is depicted. Note, if 5% of an error was set as a margin, the 16th order solution covers an approximation interval which makes 90% of the central frequency of the filter. Finally, Fig. 12.3c is depicting the phase characteristics of these three solutions. Here, unfortunately, it is very difficult to recognize the parabolic shape of the phase in the passband which, as usual, imposes the group delay as a merit of success.

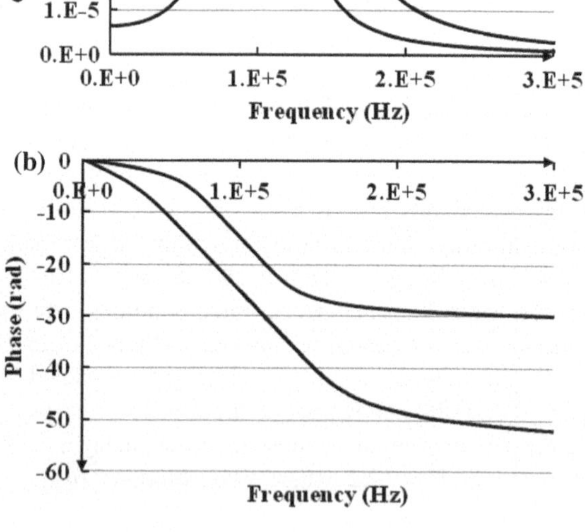

Fig. 12.2 All-pass filters approximating constant group delay around the central frequency in maximally flat sense. **a** Group delay and **b** phase characteristics

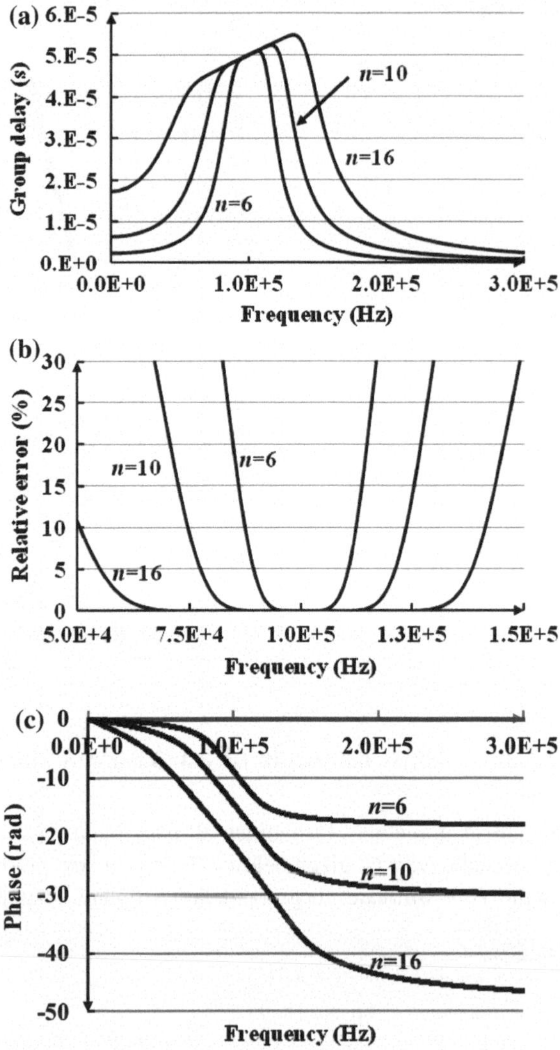

Fig. 12.3 All-pass filters approximating linearly rising group delay around the central frequency in maximally flat sense. **a** Group delay, **b** relative approximation error and **c** phase characteristics

Finally, Fig. 12.4 depicts the synthesis results for a decreasing group delay of the all-pass filter. The order of the filter was kept to be $n = 10$ while the required slope of the group delay was varied from −25% via −15% to −5%. Figure 12.4a depicts the group delay characteristics while Fig. 12.4b depicts the absolute deviation from the corresponding strait line. Of course, with the rise of the modulus of the slope the approximation interval decreases.

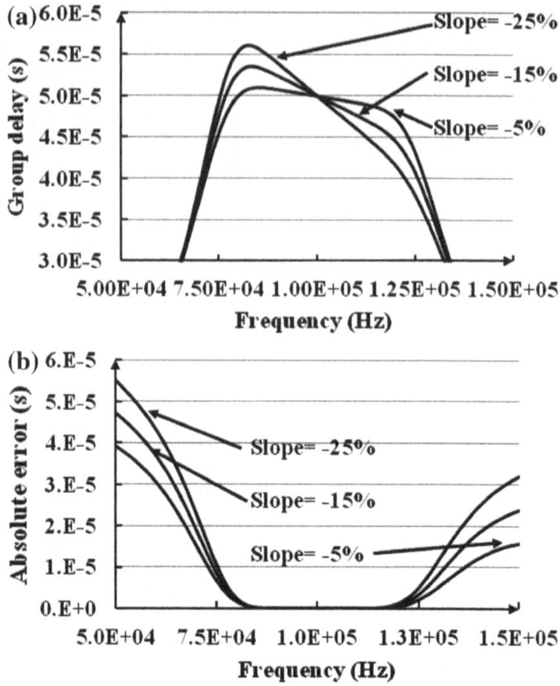

Fig. 12.4 All-pass filters of 10th order approximating linearly decreasing group delay around the central frequency in maximally flat sense. **a** Group delay, **b** absolute approximation error

12.3 Design Example of an All-Pass Band-Pass Filter

As an example the 18th order all-pass band-pass filter approximating constant group delay will be given below. The report was produced by the *B0010* program of the *RM* software. Its group delay and phase responses are depicted in Fig. 12.2.

Start.E

==

Project name: bp_ap_18_0

==

Read in synthesis data:

Order of the filter n=18

Central angular frequency W0=6.28319e+005

Nominal value of the delay T0=5.00000e-005

Slope of the characteristics B=0.00000e+000 %

++

ALLPASS AMPLITUDE BANDPASS GROUP DELAY FILTER

Maximally flat group delay with controllable slope

SOLUTION

Order of the filter_n=18

==

The results about the poles, zeros and coefficients listed below are valid for frequency normalization with

wc=2*pi*fc=6.28318530e+005

==

Final position of the zeros :

zs[1]=3.196274315e-001 zw[1]=6.433747107e-001
zs[2]=2.092498889e-001 zw[2]=5.034317134e-001
zs[3]=3.814830929e-001 zw[3]=7.689960822e-001
zs[4]=4.135607799e-001 zw[4]=8.884053715e-001
zs[5]=4.220290968e-001 zw[5]=1.004836348e+000
zs[6]=4.085981999e-001 zw[6]=1.120240854e+000
zs[7]=3.719495701e-001 zw[7]=1.236393138e+000
zs[8]=1.962690375e-001 zw[8]=1.485216279e+000
zs[9]=3.066621481e-001 zw[9]=1.355887585e+000
zs[10]=3.196274315e-001 zw[10]=-6.433747107e-001
zs[11]=2.092498889e-001 zw[11]=-5.034317134e-001
zs[12]=3.814830929e-001 zw[12]=-7.689960822e-001
zs[13]=4.135607799e-001 zw[13]=-8.884053715e-001
zs[14]=4.220290968e-001 zw[14]=-1.004836348e+000
zs[15]=4.085981999e-001 zw[15]=-1.120240854e+000
zs[16]=3.719495701e-001 zw[16]=-1.236393138e+000
zs[17]=1.962690375e-001 zw[17]=-1.485216279e+000
zs[18]=3.066621481e-001 zw[18]=-1.355887585e+000

Final position of the poles:

s[1]=-3.196274315e-001 w[1]=6.433747107e-001
s[2]=-2.092498889e-001 w[2]=5.034317134e-001
s[3]=-3.814830929e-001 w[3]=7.689960822e-001
s[4]=-4.135607799e-001 w[4]=8.884053715e-001
s[5]=-4.220290968e-001 w[5]=1.004836348e+000
s[6]=-4.085981999e-001 w[6]=1.120240854e+000
s[7]=-3.719495701e-001 w[7]=1.236393138e+000
s[8]=-1.962690375e-001 w[8]=1.485216279e+000
s[9]=-3.066621481e-001 w[9]=1.355887585e+000
s[10]=-3.196274315e-001 w[10]=-6.433747107e-001
s[11]=-2.092498889e-001 w[11]=-5.034317134e-001
s[12]=-3.814830929e-001 w[12]=-7.689960822e-001
s[13]=-4.135607799e-001 w[13]=-8.884053715e-001
s[14]=-4.220290968e-001 w[14]=-1.004836348e+000
s[15]=-4.085981999e-001 w[15]=-1.120240854e+000

```
s[16]=-3.719495701e-001   w[16]=-1.236393138e+000
s[17]=-1.962690375e-001   w[17]=-1.485216279e+000
s[18]=-3.066621481e-001   w[18]=-1.355887585e+000
=====================================================================
Coefficients of the numerator of the transfer function:

Ascending exponent of frequency (s)
 1.32551964e+000 -8.97007189e+000 4.03725957e+001 -1.24908827e+002
 3.07010105e+002 -6.04069533e+002 9.99483785e+002
-1.39042344e+003 1.66739594e+003 -1.71345800e+003 1.53241714e+003
-1.17738207e+003 7.85936887e+002 -4.44347525e+002 2.15050375e+002
-8.42140879e+001 2.71619991e+001 -6.05885849e+000 1.00000000e+000
Descending exponent of frequency (s)
 1.00000000e+000 -6.05885849e+000 2.71619991e+001 -8.42140879e+001
 2.15050375e+002 -4.44347525e+002 7.85936887e+002
-1.17738207e+003 1.53241714e+003 -1.71345800e+003 1.66739594e+003
-1.39042344e+003 9.99483785e+002 -6.04069533e+002 3.07010105e+002
-1.24908827e+002 4.03725957e+001 -8.97007189e+000 1.32551964e+000

Coefficients of the denominator of the transfer function:

Ascending exponent of frequency (s)
 1.32551964e+000 8.97007189e+000 4.03725957e+001 1.24908827e+002
3.07010105e+002 6.04069533e+002 9.99483785e+002 1.39042344e+003
1.66739594e+003 1.71345800e+003 1.53241714e+003 1.17738207e+003
7.85936887e+002 4.44347525e+002 2.15050375e+002 8.42140879e+001
2.71619991e+001 6.05885849e+000 1.00000000e+000
Descending exponent of frequency (s)
 1.00000000e+000 6.05885849e+000 2.71619991e+001 8.42140879e+001
2.15050375e+002 4.44347525e+002 7.85936887e+002 1.17738207e+003
1.53241714e+003 1.71345800e+003 1.66739594e+003 1.39042344e+003
9.99483785e+002 6.04069533e+002 3.07010105e+002 1.24908827e+002
4.03725957e+001 8.97007189e+000 1.32551964e+000
------------------------------------------------------------
Normalized group delay(w0)=3.14144e+001
=====================================================================
```

End.E

12.4 Developer's Corner

Speaking from the software development point of view, one has to carefully examine the expressions for the derivatives either with respect to the frequency or with respect to the zero's coordinates. One is to exploit the recurrence formulas given which will make the program readable and facilitate the debagging.

Extreme care must be paid to the monitoring of the iterative process and control of convergence. That would include extensive damping of the iterations. Of course, these actions will lead to spending excessive computer time but any solution is better than no solution.

Speaking from the designer's point of view, considering convergence and, of course, the shape of the resulting function, there is an important difference between the positive and the negative slope cases. Namely, when small positive slopes in conjunction with high order filters are sought, it becomes much more difficult to achieved convergence since large amount of the area under the curve is to be pushed below the central frequency. That may be seen from Fig. 12.3 where the 16th order filter seems to be the limiting case for the given value of the nominal delay and the slope.

Finally, the procedure described above is by no means limited to linear and parabolic phase characteristic. Namely, by choosing the derivatives higher than first to be nonzero, one may go for higher order polynomial approximation of the phase.

References

1. Crane R (1968) All-pass network synthesis. IEEE Trans Circuit Theory 15(4):474–478
2. Halpern PH (1971) Solution of flat time delay at finite frequencies. IEEE Trans Circuit Theory CT-18(2):247–246
3. Tonne JL (2017) Allpass networks in a speech chain. Updated 4 January 2017. http://wwwtonnesoftware.com/appnotes/allpass/Allpass.pdf
4. Gift SJG (2000) Applications of all-pass filters in the design of multiphase sinusoidal systems. Microelectr J 31(1):9–13
5. Klauder JR, Price AC, Darlington S, Albersheim WJ (1960) The theory and design of chirp radars. Bell Syst Tech Journal 39(4):745–808
6. O'Meara TR (1960) Linear slope delay filter for compression. Proc IRE 48(11):1916–1918
7. Fowle EN, Carey DR, Vander Schuur RE, Yost RC (1963) A pulse compression system employing a linear FM Gaussian signal. Proc IEEE 51(2):312–314
8. Lazovich SM (1972) Transfer-function synthesis of allpass filters for pulse compression. Electron Lett 8(3):77–79
9. Barton DK (ed) (1975) Radars, vol 3. Pulse Compression, Artech House, Boston
10. Litovski V, Lazović S (1979) A new method of synthesis of filter with linearly decreasing group delay. (In Serbian) Proceedings of the XXIII conference ETAN, Maribor: I.169–I.176
11. Li L, Coon M, McLinden M (2013) Radar range sidelobe reduction using adaptive pulse compression techniques. NASA Tech Brief GSC-16458-1

12. Soltanalian M (2014) Signal design for active sensing and communications. Uppsala Dissertations from the Faculty of Science and Technology, Elanders, Sverige AB
13. Gregorian R, Temes GC (1978) Design techniques for digital and analog all-pass circuits. IEEE Trans. on Circuits and Systems CAS-25(12):981–988

Chapter 13
Direct Bandpass Synthesis of Linear and Parabolic Phase Selective Filters

A procedure is proposed and verified (including the computer program used for implementation) which allows for simultaneous fulfillment of requirements for the amplitude and group delay characteristic of band-pass filters. The final result is reminiscent to the one described in Chap. 10 and the previous chapter in the sense that, in an iterative loop, the group delay characteristic is synthesized (in the maximally flat sense) first and transmission zeros on the ω-axis are subsequently added to boost the selectivity. The nominal group delay is adjusted as a parameter within the optimization loop until proper passband-width (of the amplitude characteristic) is obtained. All three cases (constant, linearly rising and linearly falling group delay) are encompassed. Detailed description of the algorithm is given accompanied by a comprehensive set of examples.

13.1 Introduction

Direct synthesis of band-pass filters is a special challenge. It is so even when requirements on the amplitude characteristics are imposed only [1]. More challenging is, however, synthesis of band-pass filters with constant group delay in the passband [2]. Additional challenge is to have a group delay approximating a line with a given slope. In such cases additional limiting the frequency spectrum of the transmitted pulses may be achieved [3].

The solution offered below may be considered as a combination of use of band-pass group delay approximating function (in a maximally flat sense around the central frequency) and a band-pass amplitude corrector which is not affecting the group delay. These two approximants are alternatively implemented until the

V. Litovski, *Electronic Filters*, Lecture Notes in Electrical Engineering 596,
https://doi.org/10.1007/978-981-32-9852-1_13

desired pass-band width is achieved with the group delay at the central frequency being the variable parameter enabling adjustment. Namely, by varying the nominal group delay of the now band pass filter approximating constant group delay one may change the approximation width due to the fixed area under the group delay curve. The approximation width is directly related to the passband width. Of course, the passband width is affected by the amplitude corrector too. So a value of the nominal group delay exists which allows the prescribed passband width to be obtained. That is in fact the stopping criterion needed.

One may, of course, approximate linearly changing group delay (with positive or negative slope) while keeping the rest of the requirements the same.

The transfer function used for this class of filters may be stated as

$$H(s) = H_0 \cdot s^l \frac{\prod_{k=1}^{m/2} \left(s^2 - z_k^2\right)}{\prod_{k=1}^{n} \left(s - p_k\right)} \tag{13.1}$$

where

$$H_0 = \frac{1}{(j\omega_0)^l} \frac{\prod_{k=1}^{n} \left(j \cdot \omega_0 - p_k\right)}{\prod_{k=1}^{m/2} \left(-\omega_0^2 - z_k^2\right)}. \tag{13.2}$$

In the above expression $m + l$ is the order of the numerator, n is the order of the filter, z_k is the coordinate of the transmission zero at the imaginary axis, and $p_k = \sigma_k + j \cdot \omega_k$ is the kth complex pole of the filter. s is the normalized complex angular frequency. It is assumed here that ω_0 is the central frequency of the passband.

No matter how usual this function looks like, due to the specific way of the filter synthesis implemented below, it needs to be explained shortly. Namely, the denominator of the function is defining both the group delay and the attenuation in the passband. Without the transmission zeros contained in the numerator one would have a low-pass function approximating constant group delay around the central frequency. The numerator's transmission zeros are supposed to shape the attenuation characteristic into a band-pass one. There should be three "categories" of them: the ones in the upper stopband, the ones in the lower stopband and some in the origin.

13.2 The Algorithm

In this paragraph the algorithm implemented within the *bptdam* program of the *RM* software will be described.

The group delay may be calculated from

$$\tau_{d}(\omega) = \sum_{k=1}^{n/2} T_{k}(\omega) \tag{13.3a}$$

where

$$T_{k}(\omega) = \frac{-\sigma_{k}}{\sigma_{k}^{2} + (\omega - \omega_{k})^{2}} + \frac{-\sigma_{k}}{\sigma_{k}^{2} + (\omega + \omega_{k})^{2}}. \tag{13.3b}$$

Note the similarity with the expression (12.2) from the previous chapter. The difference is in the fact that we are doing here with the poles hence the negative sign. Furthermore, there are no complex zeros hence the missing 2 in the numerators.

To get a maximally flat approximation of the group delay around the central frequency ω_0, with $\tau_d(\omega_0) = t_0$ and $\mathrm{d}\tau_d/\mathrm{d}\omega_{|(\omega=\omega_0)} = b$, where t_0 and b are positive constants, one has to solve the following system of nonlinear equation with respect to the coordinates of p_k

$$F_{1}(\boldsymbol{\sigma}, \boldsymbol{\omega}) = \tau_{d}(\omega_{0}) - t_{0} = 0 \tag{13.4a}$$

$$F_{2}(\boldsymbol{\sigma}, \boldsymbol{\omega}) = \mathrm{d}\tau_{d}/\mathrm{d}\omega_{|(\omega=\omega_{0})} - b = 0 \tag{13.4b}$$

$$F_{i}(\boldsymbol{\sigma}, \boldsymbol{\omega}) = \mathrm{d}^{i}\tau_{d}/(\mathrm{d}\omega \overset{i}{\omega} |(\omega = \omega_{0}) = 0 \quad \text{for } i = 2, \ldots, n-1. \tag{13.4c}$$

where $\boldsymbol{\sigma}$ and $\boldsymbol{\omega}$ are vectors of the unknown coordinates of the poles $p_k = \sigma_k + j\omega_k$.

After linearization the following system of linear equations arises

$$F_{i} + \sum_{j=1}^{n/2} \left\{ \frac{\partial F_{i}}{\partial \sigma_{j}} \mathrm{d}\sigma_{j} + \frac{\partial F_{i}}{\partial \omega_{j}} \mathrm{d}\omega_{j} \right\} = 0 \text{ for } i = 0, \ldots, n-1. \tag{13.5}$$

It should be solved iteratively. To get the proper derivatives needed to solve this system one may use all theory developed in the previous chapter. Adjustments are to be made, however, related to the facts that there are no complex zeros and there are poles to be manipulated as mentioned above.

In order to produce a bandpass characteristic the numerator of the transfer function is considered an amplitude corrector. In that way only the coordinates of the transmission zeros will be considered variable (unknown). With reference to Fig. 13.1, the following system of nonlinear equations is to be solved iteratively

$$H(s) \cdot H(-s)_{|s=j \cdot y_{i}} = H_{\max}^{2} \text{ for } i = 1, \ldots, m/2. \tag{13.6}$$

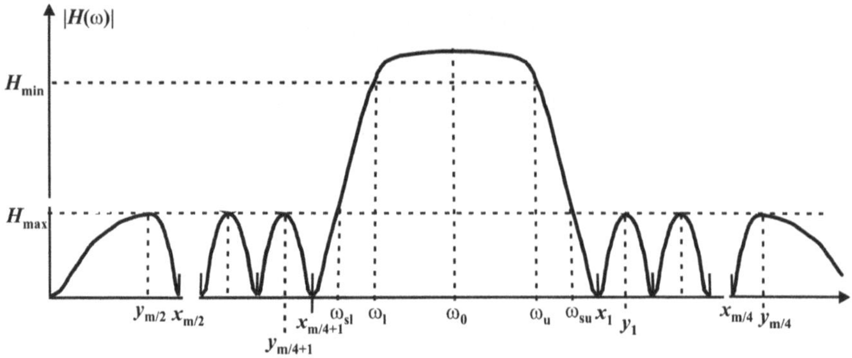

Fig. 13.1 Definition of the parameters of the amplitude characteristic of a band-pass filter

with respect to the unknown coordinates of the attenuation poles $\mathbf{p}$. The frequencies y_i are the locations of the maxima of the stopband gain which are to be calculated in every step of correction of the attenuation poles $\mathbf{p}$.

To get symmetry around ω_0 the transfer function is supposed to have the transmission zeros at the imaginary axis distributed symmetrical (in number) around the passband. So, for the overall number of finite transmission zeros, as depicted in Fig. 13.1 the number of zeros located at the positive half of the ω-axis should be a multiple of 2, and the overall number of finite transmission zeros a multiple of 4. Furthermore, to get symmetry one has to have attenuation poles in the origin. We choose an even number of them to be the best solution. So, it may rewritten in the form

$$H(s) = H_0 \cdot \frac{s^{2\cdot l} \cdot \prod_{k=1}^{m/2-l}\left(s^2 - z_k^2\right)}{\prod_{k=1}^{n}(s - p_k)} \tag{13.7a}$$

with

$$H_0 = \frac{\prod_{k=1}^{n}(j \cdot \omega_0 - p_k)}{(\mathrm{j} \cdot \omega_0)^{2\cdot l} \cdot \prod_{k=1}^{m/2-l}\left(-\omega_0^2 - z_k^2\right)}. \tag{13.7b}$$

Keeping l even, one uses (13.1) for programming.

In that way in order to get a transfer function exhibiting band-pass approximation of the group delay in maximally flat sense and a selective amplitude characteristic with attenuation poles at finite frequencies, given $a_{\max}$, b, B_0, n, m, and l, one has to search for the values of p_k, k = 1,2, …, n, and z_k, k = 1, …, $m/2$-l. B_0 stands for the required 3 dB pass-band width.

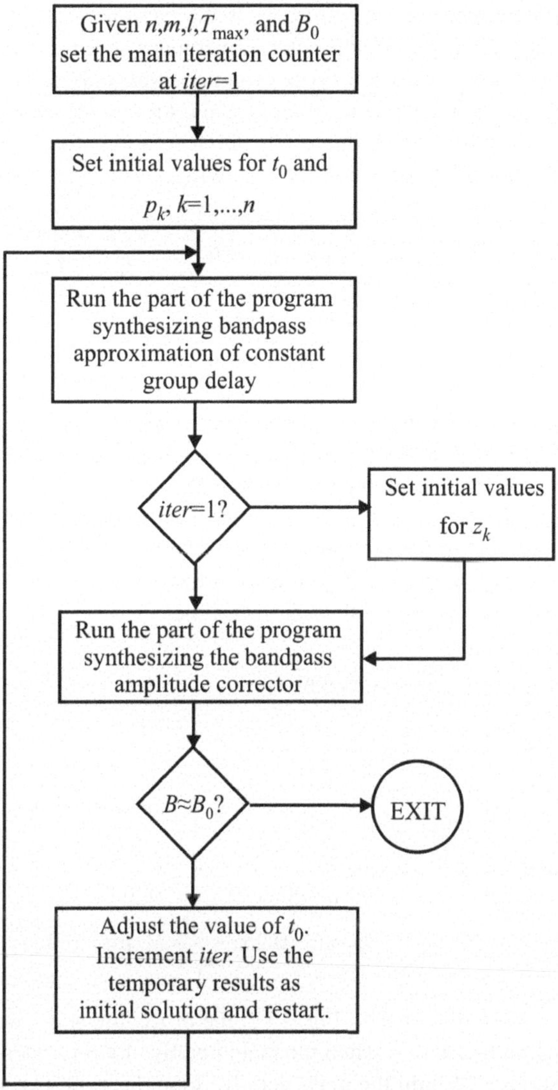

Fig. 13.2 Algorithm implemented within the *bptdam* program

The program finds these values in an iterative process containing two internal iterative loops nested into the main one as depicted in Fig. 13.2.

To start the process one needs an initial value for the nominal group delay. To that end we choose the normalized value

$$T_0 = \pi \cdot n, \tag{13.8}$$

which stems from the area under the curve and is based on the assumption that the approximation width will be equal to 1 rad/s. This choice is not crucial for the convergence as such since T_0 is the variable used to adjust the passband width. It may, however, influence the convergence speed if chosen to far from the final value. Of course, what we propose is accompanied by strong iteration damping which means convergence is assured while the speed is under control of the damping.

Further, to start one needs initial solutions for the poles of the transfer function. We would recommend the following normalized coordinates

$$\sigma_i = -\frac{\pi \cdot n}{T_0} \cdot \frac{n+6}{T_0} \tag{13.9a}$$

$$\pm\omega_i = 1 + \frac{\pi \cdot n}{T_0} \cdot \left(\frac{2i-1}{n} - 0.5\right) \text{for } i = 1, 2, \ldots, n/2, \tag{13.9b}$$

These formulas are slight modification of (12.15) in the sense that the scaling factor now $\pi \cdot n/T_0$ since the area is reduced by factor of 2.

Finally, speaking on initial solutions, one needs a guess for the zeros of the transfer function. For this purpose we propose first the upper and lower edges of the stopband of the current solution to be found. These are denoted in Fig. 13.1 as ω_{su} and ω_{sl}, respectively. We also introduce $B_s = (\omega_{su} - \omega_{sl})$ and $z_i = \mathrm{j}\ x_i$. With that one may use the following recurrence

$$x_i = \sqrt{a_i} \text{ and } x_{m/4+i} = \sqrt{b_i} \text{ for } i = 1, 2, \ldots, m/4, \tag{13.10a}$$

with

$$a_i = (1 + \alpha_i)^2 \text{ and } b_i = (1 - \alpha_i)^2, \tag{13.10b}$$

and

$$\alpha_i = 0.25 \cdot \frac{i \cdot B_s}{m/4}, \quad \text{for } i = 1, 2, \ldots, m/4. \tag{13.10c}$$

With reference to Fig. 13.2, the procedure starts with setting the initial values as explained above. Then the main iterative loop is entered and the iteration counter is set to 1. Within the main iterative loop there are two more iterative loops. The first performs iterative search for the poles of the transfer function under the restriction of maximally flat approximation as stated by (13.4a). In other words in this loop (13.5) is solved.

The second loop is after the coordinates of the transmission zeros which is in fact solving (13.6) iteratively. That done the passband width B is evaluated by solving numerically the following

$$H(s) \cdot H(-s)_{|s=\mathrm{j}\cdot\omega_u} = H^2_{\min} \tag{13.11a}$$

and

$$H(s) \cdot H(-s)_{|s=j \cdot \omega_l} = H^2_{\min}. \tag{13.11a}$$

Now B is calculated as

$$B = \omega_u - \omega_l \tag{13.12}$$

and compared with its prescribed value B_0. If satisfactory the process is leaving the main loop and the solution is reached. In the opposite T_0 is changed.

When not satisfactory, one should change T_0. It should be decreased in order for B to increase and the opposite. The increment, ΔT_0, of course, should be reduced gradually.

To avoid complicated search for T_0, a fixed increment may be adopted say $\Delta T_0 = 10^{-5} \cdot T_0$, which will be added to T_0 in order to decrease or subtracted in order to increase B. If so, depending on the mutual values of the initial and final T_0, the search will produce monotonic rise or decrease of T_0. The stopping criterion, in this case, will be simply the necessity of change of the sign of ΔT_0.

The procedure described above was implemented in the *bptdam* program of the *RM* software. In the sequel we will describe several examples of implementation of *bptdam*. Since this procedure and the results obtained are unique and new, some more attention will be paid to the properties of the obtained functions.

Figures 13.3, 13.4, 13.5, 13.6 are representing the properties of a transfer function of a 14th order filter with 8 finite transmission zeros at the ω-axis and two zeros at the origin leaving two zeros for the infinity. This structure was selected in order to enable symmetry per se. Stopband attenuation $a_{\min}$ = 40 dB was required.

Figure 13.3 depicts the attenuation characteristic. The whole characteristic is depicted in Fig. 13.3a while Fig. 13.3b depicts the passband response only. Slight asymmetry may be observed since the upper edge of the passband is smaller than the required 110 kHz. That however is incomparably less damaging than in the case of selective filters obtained by frequency transformations as could be seen in Fig. 11.8 and Fig. 11.9. From the design example given in the next paragraph one may find that the edges of the stopband are approximately f_{sl} = 72.6 kHz and f_{su} = 127 kHz which means that the upper transition region is f_{su}–f_u = 127–110 = 17 kHz which is approximately 1.7 times larger than the upper half of the passband. Thinking in halves, this would correspond to a low-pass filter of 7th order with 4 transmission zeros at the ω-axis.

That is to be compared with the attenuation characteristic of the Thomson filter with improved selectivity depicted in Fig. 10.6b (trace T) which is of similar selectivity but of an order higher by two. We think that this improvement comes from the accommodation of the group delay approximation in favor to the selectivity.

Figure 13.4 depicts for the first time the amplitude characteristic of this kind filters. To illustrate, Fig. 13.4b depicts the stopband attenuation of the filter.

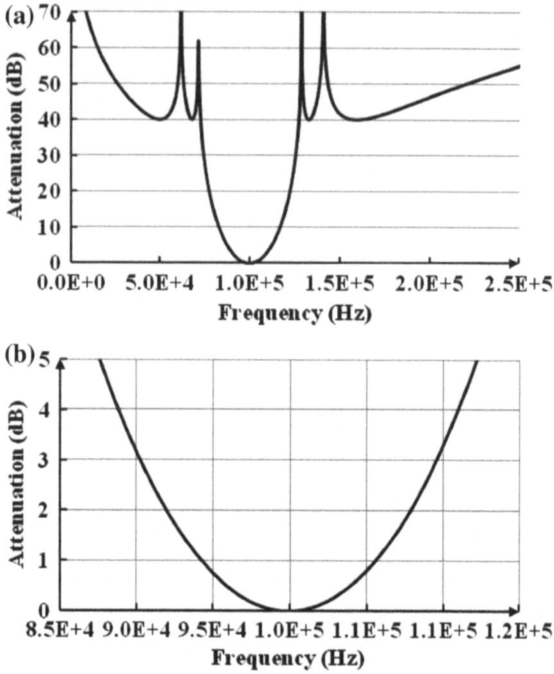

Fig. 13.3 Attenuation characteristic of the 14th order filter simultaneously approximating constant group delay and selective attenuation, **a** Overall attenuation and **b** passband attenuation

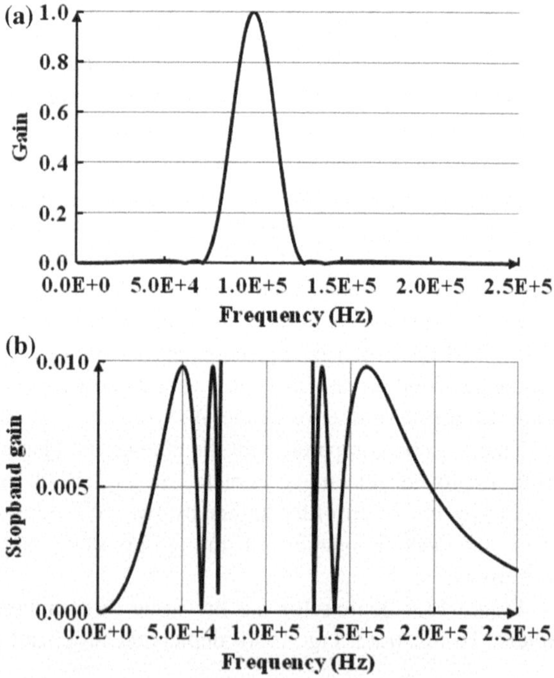

Fig. 13.4 Amplitude characteristic (Gain) of the 14th order filter simultaneously approximating constant group delay and selective attenuation, **a** Overall gain and **b** stopband gain

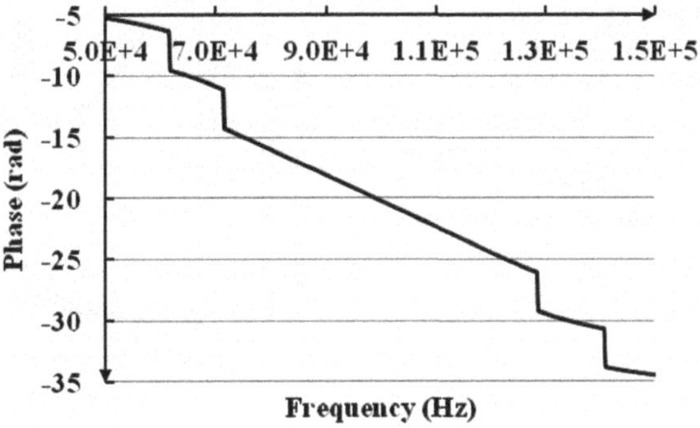

Fig. 13.5 Passband phase characteristic of the 14th order filter simultaneously approximating constant group delay and selective attenuation

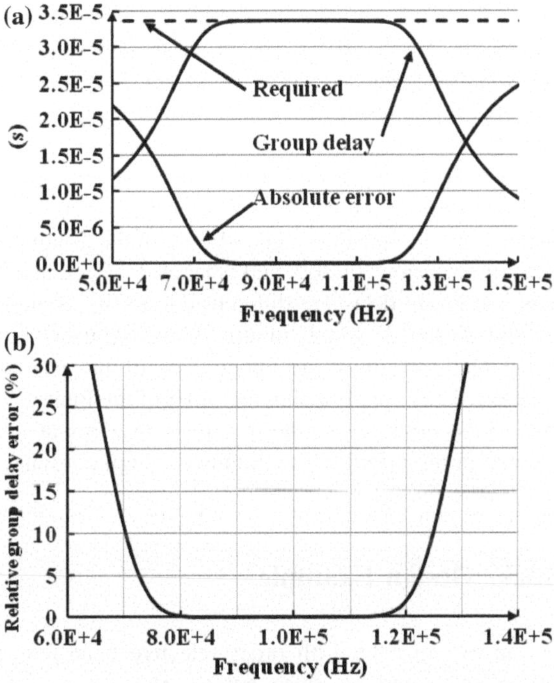

Fig. 13.6 **a** Passband group delay and absolute error characteristics of the 14th order filter simultaneously approximating constant group delay and selective attenuation and **b** Passband relative error characteristics

Next, the passband phase characteristic is depicted in Fig. 13.5. Apart of the extreme good linearity which covers the whole passband and beyond, there is not much to discuss.

Finally, Fig. 13.6 depicts the group delay and the approximation errors. Figure 13.6a depicts the group delay characteristic and the absolute deviation from the constant here presented by a dashed line. Figure 13.6b depicts the relative deviation of the group delay. As can be seen even in this case the approximation interval is much broader than necessary. It is also slightly asymmetric which, being far outside of the passband, has not much of importance.

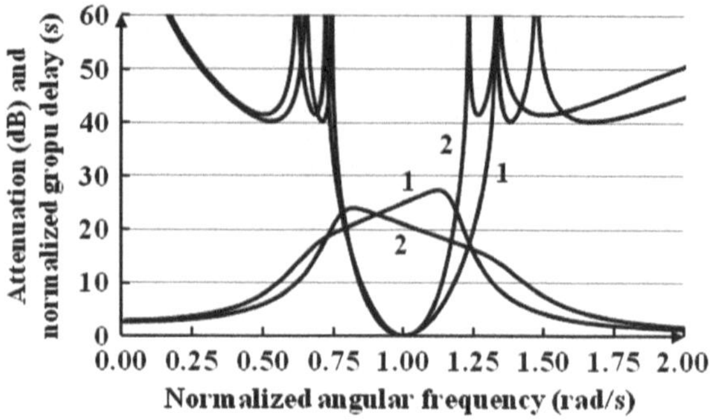

Fig. 13.7 Group delay and attenuation characteristics of the 14th order filter simultaneously approximating linearly changing group delay and selective attenuation. (1) positive slope 20% and (2) negative slope 20%

To further investigate the behaviour of the design procedure proposed and the capabilities of the *bptdam* program two examples were elaborated. Both are approximating linearly changing group delay having, however, opposite signs of the slopes. The resulting attenuation and group delay characteristics are depicted in Fig. 13.7.

As earlier, it is 14th order filter function with 8 transmission zeros at the ω-axis and two zeros at the origin exhibiting $a_{\min} = 40$ dB in the stopband. The first one, labeled "1", has positive slope of 20% of the group delay characteristic. The second has negative slope of 20% and is labeled "2". As can be seen the final values of the nominal group delay are different due to the asymmetry of the amplitude characteristic of case "1" which, in turn, is a consequence of the "gathering" the poles with small real parts near the upper edge of the passband. This asymmetry can be "corrected" by simply rising the order of the filter by 2 (or more) which would give rise to the influence of the transmission zeros at infinity as opposed to the ones located in the origin whose number is kept constant.

13.3 Design Example

As an example the 14th order selective band-pass filter approximating constant group delay will be given below. The report was produced by the *bptdam* program of the *RM* software. Its attenuation, amplitude phase and group delay responses are depicted in Figs. 13.3, 13.4, 13.5 and 13.6. To reduce the length of the report the pole-zero coordinates were rearranged into a table.

Start.E

BPTDAM, Program for band-pass synthesis
of constant group delay in maximally flat sense
and an amplitude corrector using finite transmission zeros
Project name: BP_SIM_14_8_2_20p

==

Read in synthesis data:
Order of the filter n=14
Number of zeros at the origin k0=2
Number of zeros at the imaginary axis m0=8
Central angular frequency fc=1.00000e+005 Hz
Initial (evaluated internally) value of the delay T0=7.00000e-005
Minimum stopband attenuation=4.00000e+001 Required relative pass-band width=2.0000e-001. Required slope of the group delay=0.00000e+000

==

AMPLITUDE (SELECTIVITY) CORRECTION OF A
CONSTANT GROUP DELAY BANDPASS FILTER

SOLUTION
Order of the filter_n=14
Number of zeros in the origin_k0=2
Number of zeros at the imaginary axis_n0=8
Order of the numerator_m=10

==

The results about the poles, zeros and coefficients listed below are valid for frequency normalization with wc=2*pi*fc=6.28318530e+005

==

Zeros		Poles	
Real part	Imag. part	Real part	Imag. part
0.000000000e+000	0.000000000e+000	-1.300627052e-001	7.141266708e-001
0.000000000e+000	0.000000000e+000	-1.966151002e-001	8.064308820e-001
0.000000000e+000	1.284856433e+000	-2.294116708e-001	8.923336045e-001
0.000000000e+000	-1.284856433e+000	-2.393232414e-001	9.761194435e-001
0.000000000e+000	1.406737429e+000	-2.287358256e-001	1.059705097e+000
0.000000000e+000	-1.406737429e+000	-1.954218884e-001	1.144987168e+000
0.000000000e+000	7.137798208e-001	-1.288047125e-001	1.236174310e+000
0.000000000e+000	-7.137798208e-001	-1.300627052e-001	-7.141266708e-001
0.000000000e+000	6.164873291e-001	-1.966151002e-001	-8.064308820e-001
0.000000000e+000	-6.164873291e-001	-2.294116708e-001	-8.923336045e-001
		-2.393232414e-001	-9.761194435e-001
		-2.287358256e-001	-1.059705097e+000
		-1.954218884e-001	-1.144987168e+000
		-1.288047125e-001	-1.236174310e+000

==

Coefficients of the numerator of the transfer function:
Ascending exponent of frequency (s)

```
 0.00000000e+000  0.00000000e+000  3.26495378e+000  0.00000000e+000
1.86266956e+001   0.00000000e+000  3.45261708e+001  0.00000000e+000
2.33257979e+001   0.00000000e+000  5.16136894e+000
Descending exponent of frequency (s)
 5.16136894e+000  0.00000000e+000  2.33257979e+001  0.00000000e+000
3.45261708e+001   0.00000000e+000  1.86266956e+001  0.00000000e+000
3.26495378e+000   0.00000000e+000  0.00000000e+000
==================================================================
 Coefficients of the denominator of the transfer function:
Ascending exponent of frequency (s)
 7.87217438e+001  2.30221361e+002  8.94152075e+002  1.69838640e+003
3.58082548e+003   4.83784654e+003  6.92448976e+003  6.85215700e+003
7.12885191e+003   5.11034424e+003  3.92345925e+003  1.90728636e+003
1.05676186e+003   2.78435569e+002  1.03248555e+002
Descending exponent of frequency (s)
 1.03248555e+002  2.78435569e+002  1.05676186e+003  1.90728636e+003
3.92345925e+003   5.11034424e+003  7.12885191e+003  6.85215700e+003
6.92448976e+003   4.83784654e+003  3.58082548e+003  1.69838640e+003
8.94152075e+002   2.30221361e+002  7.87217438e+001
==================================================================
DENORMALIZED VALUES
Final value of the nominal delay=3.36329e-005 (s)
Value of (group delay) slope=0.00000e+000 %
Final relative width of the pass-band=2.02531e+001 %
Lower edge of the pass-band fl=9.22250e+004 Hz     Upper edge of the pass-band
fh=1.12478e+005 Hz
Final width of the pass-band=2.02531e+004 Hz
Edges of the stopband
f_low=7.26070e+004 Hz     f_high=1.27145e+005 Hz
 Bandwidth below A_min=4.00000e+001 dB      is BW=5.45383e+004 Hz
 Relative bandwidth below A_min=4.00000e+001 dB is BW=5.45383e+001 %
==================================================================
```

End.E

13.4 Developer's Corner

Since the synthesis algorithm proposed above relies on two procedures already described in the previous chapters, there is not much to advice when concerns software development. The main issue is the choice of the procedure of advancement with ΔT_0. It is our advice to proceed with the simplified procedure proposed above especially having in mind that when T_0 is changed for small amount the new positions of the transmission zeros will be obtained in a single or in at most two iterations what would compensate for the increase of the overall number of iterations in the main loop (having sophisticated incrementing of T_0 is applied).

Design of filters with maximally flat group delay and a selective amplitude characteristic, however, must always be viewed in the context of group delay corrected selective band-pass filters discussed in Chap. 11. Namely, trade off is done here which gives advantage to the group delay as opposed to the case of use phase corrector where we already had highly selective solution. Which one is better is up to the application at hand.

In addition, even if we choose a solution where the group delay is the main factor to decide, there are doubts as to what should be the order of the filter, and how to distribute the zeros to get proper selectivity. In these considerations the *final* value of the nominal (central) group delay will play a decisive role since in most cases it may even be the design requirement. If so, having in mind that a_{min} is affecting the final value of the nominal group delay (via the width of the passband) cut and try should be performed until the right values of n, m, and l are found.

References

1. Watanabe H (1961) Approximation theory for filter-networks. IRE Trans. Circuit Theory 8 (3):341–356
2. Wellekens CJ (1974) Simultaneous approximation of the attenuation and group delay of reciprocal lossless networks. In: Proceedings of IEEE symposium on circuits & systems, ISCAS-1974, San Francisco, pp 274–277
3. Li L, Coon M, McLinden M (2013) Radar range sidelobe reduction using adaptive pulse compression techniques. NASA Tech Brief GSC-16458-1

Part III
Circuit and System Synthesis

Chapter 14
Passive RLC Cascade Circuit Synthesis

Abstract The theory of synthesis of LC filters dates from the very beginning of filter circuit implementations. The theory of synthesis of LC filters dates from the very beginning of filter circuit implementations. Here, however, following the overall concept of the book, only cascade synthesis will be described. It creates a cascade of cells realizing one transmission zero each. To that end cells realizing zeros: in the origin, at infinity, at a frequency on the ω-axis, in the left half plane, and in the right-half plane are encompassed. Depending of the numerical values of the transmission zeros, alternative circuits realizing the same transmission zero will be necessary. That is most frequently related to the necessity to avoid implementation of transformers or coupled inductances. In order to simplify the synthesis use of circuit transformations in place of frequency transformations is exemplified for non-low-pass filters. The order of extraction of the transmission zeros is discussed and advice is given. Three examples are elaborated realizing a low-pass filter having transmission zeros at the ω-axis and a complex pair, a band-pass, and an all-pass filter. The synthesis needs solution of a polynomial of order twice the order of filter. In addition, polynomial subtraction and division are undertaken frequently. To cope with numerical problems of that kind use of extended precision arithmetic is advised.

14.1 Introduction

Passive lossless LC circuits are the first filters to be synthesized of all. That technology dominated the trade for a very long time and it had permanent influence to the development of synthesis methods including even the terminology. So, the reader noticed the attenuation is almost the most frequently used term in this book while not all contemporary technologies introduce attenuation in the passband while filtering the input signal. The synthesis methods, as it usually happens, went through many phases [1] and many solutions were offered with the cascaded (or its simplified version-ladder) taking finally over.

V. Litovski, *Electronic Filters*, Lecture Notes in Electrical Engineering 596,
https://doi.org/10.1007/978-981-32-9852-1_14

Cascaded connection of four terminal circuits means a connection in which the output pair of terminals of the preceding section (cell) is driving the input terminals of the succeeding one. Figure 14.1 depicts one such a structure.

The passive synthesis concept goes for creating a cascaded connection of cells each one of them creating one transmission zero of the transfer function. Note, some of the cells may realize a pair of zeros which means $k \leq n$, where n is the order of the filter.

There should be cells realizing any kind of possible position (any type) of the transmission zeros including the ones at the origin and at the infinity, the ones located on the ω-axis, and ending with the complex zeros (usually located in the right half of the complex frequency plane).

Realization of passive LC cascade filter based on their transfer function in the presence of transmission zeros laying out of the imaginary axis needs a somewhat specific procedure. In fact one needs to use non-minimum phase circuits of a rather large complexity. There are descriptions of these procedures in the literature e.g. [2]. Here, however, the method described by Scanlan and Rhodes [3] will be adopted since it is convenient for implementation in automatic circuit design suit. It is known as the *cascade synthesis* (CS).

The main problems introduced by cascade synthesis are related to the "order of extraction" of transmission zeros which may influence the realizability of the circuit (may produce negative values of the circuit elements) and the numerical error which is accumulated during the extraction in the case of large circuits.

As it is well known by specific choices of the order of extraction we may produce several networks having the same transfer function. Are the networks so obtained equally desirable? We will discuss those issues later on.

For the passive cascade synthesis of filters within the *RM* software the *cascade* program is in charge and will be used below for creation synthesis examples.

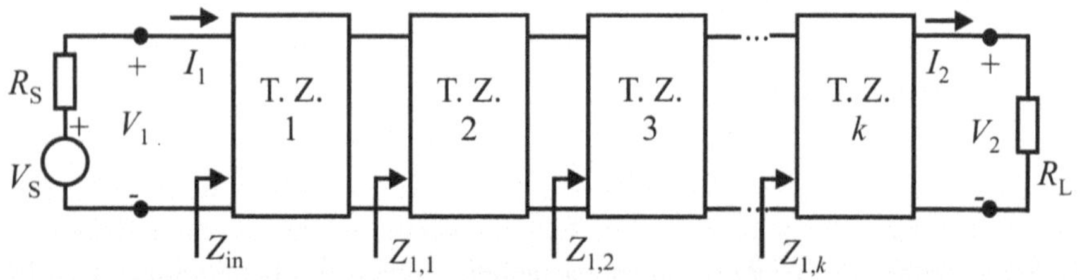

Fig. 14.1 A cascaded structure. A cell denoted as T. Z. i, $i = 1,2, \ldots, k$ is realizing one out of the complete set of transmission zeros of the filter

14.2 Finding the Input Impedance

It is usual to start the realization by generating the input impedance of the doubly terminated passive network as depicted in Fig. 14.2. It is obtained by the Darlington's procedure [1, 4] using the following.

The input impedance is expressed as

$$Z_{\text{in}} = R_{\text{in}} + \text{j} \cdot X_{\text{in}}, \tag{14.1}$$

where R_{in} is its real and X_{in} is its imaginary part. The input current is given by

$$I_1 = \frac{V_S}{R_S + Z_{\text{in}}}. \tag{14.2}$$

Now, the power delivered to the circuit is

$$P_{\text{in}} = R_{\text{in}} \cdot |I_1(\text{j}\omega)|^2. \tag{14.3}$$

Its maximum value is reached only when impedance matching

$$R_s = Z_{\text{in}} \tag{14.4}$$

is present. In such a case $V_1 = V_s/2$. So the value of the maximum power is

$$P_{\text{in,max}} = R_{\text{in}} \cdot |I_1(\text{j}\omega)|^2_{|\text{max}} = R_S \cdot \left(\frac{|V_S(\text{j}\omega)|}{2} \frac{1}{R_S}\right)^2 = \frac{|V_S(\text{j}\omega)|^2}{4 \cdot R_S} \tag{14.5}$$

On the other end of the circuit, for the power delivered to the load we have

$$P_{\text{L}} = \frac{|V_2(\text{j}\omega)|^2}{R_L}. \tag{14.6}$$

Since the network is lossless, under the condition (14.4), one may write

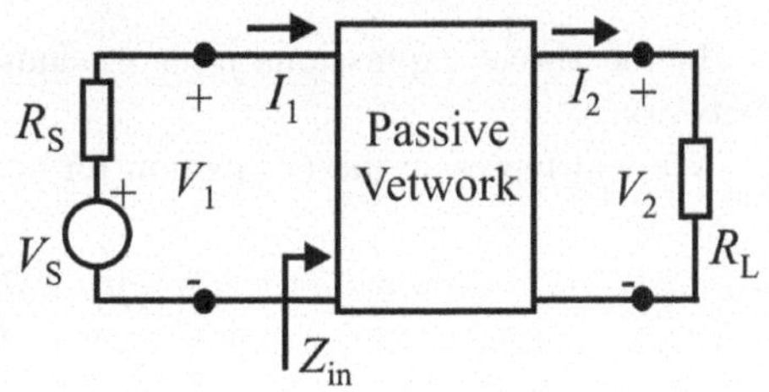

Fig. 14.2 A doubly terminated passive network

$$|H(\mathrm{j}\omega)|^2 = \frac{P_\mathrm{L}}{P_\mathrm{in,max}} = \frac{4 \cdot R_S}{R_\mathrm{L}} \cdot \left|\frac{V_2(\mathrm{j}\omega)}{V_S(\mathrm{j}\omega)}\right|^2 = \frac{4 \cdot R_S}{R_\mathrm{L}} \cdot |G(\mathrm{j}\omega)|^2 \tag{14.7a}$$

where $|G(\mathrm{j}\omega)| = |V_2(\mathrm{j}\omega)|/|V_S(\mathrm{j}\omega)|$. One has to recognize here that due to the specific relations at the input of the circuit, instead of the normal gain which would be $|V_2(\mathrm{j}\omega)|/|V_1(\mathrm{j}\omega)|$, we use $|V_2(\mathrm{j}\omega)|/|V_S(\mathrm{j}\omega)|$. So, in the case of passive cascaded circuits what was in the previous chapters named gain is in fact $|V_2(\mathrm{j}\omega)|/|V_S(\mathrm{j}\omega)|$.

Taking roots of both sides of (14.7a) leads to

$$|H(\mathrm{j}\omega)| = 2\sqrt{\frac{R_S}{R_\mathrm{L}}} \cdot |G(\mathrm{j}\omega)|. \tag{14.7b}$$

Now, combining (14.2), (14.3) and (14.6) and having in mind that the circuit is lossless we get

$$\left|\frac{V_2(\mathrm{j}\omega)}{V_S(\mathrm{j}\omega)}\right|^2 = |H(\mathrm{j}\omega)|^2 = \frac{R_\mathrm{L} R_\mathrm{in}}{|R_S + Z_\mathrm{in}|^2}. \tag{14.8}$$

We will introduce now the reflection coefficient as a measure of the reflected power which is

$$P_\mathrm{R} = P_\mathrm{in,max} - P_\mathrm{L}. \tag{14.9a}$$

The reflection coefficient is now

$$\rho^2 = (P_\mathrm{in,max} - P_\mathrm{L})/P_\mathrm{in,max} \tag{14.9b}$$

which, after substitution of $s = \mathrm{j}\omega$, may be transformed into

$$\rho(s) \cdot \rho(-s) = 1 - \frac{4 \cdot R_S}{R_\mathrm{L}} G(s) \cdot G(-s) = 1 - \frac{4 \cdot R_S}{R_\mathrm{L}} \cdot |G(j\omega)|^2 \tag{14.10}$$

and, for the needs of the subsequent developments, may be expressed by

$$\rho(s) \cdot \rho(-s)_{|s=\mathrm{j}\omega} = |G(j\omega)|^2 f_\mathrm{n}(\omega^2). \tag{14.11}$$

In the above expressions $f_\mathrm{n}(\omega^2)$ stands for the characteristic function of the network.

We will represent the G function for convenience in the following form

$$G(s) = \frac{N(s)}{D(s)} \tag{14.12}$$

where $N(s)$ and $D(s)$ are the numerator and the denominator polynomials, respectively.

Having in mind that: $G(0) = 2 \cdot R_L/(R_L + R_S)$ we may write $R_L = R_S/[2/G(0) - 1]$. So, by adopting $R_S = 1$, one gets $R_L = 1/[2/G(0) - 1]$. All further computation will be related to a load resistance value normalized by the source resistance value. That may be substituted in (14.10) to get

$$\rho(s) \cdot \rho(-s)_{|s=j\omega} = 1 - 4 \cdot \left(\frac{2}{G(0)} - 1\right) \cdot |G(j\omega)|^2 = \frac{R(j\omega) \cdot R(-j\omega)}{D(j\omega) \cdot D(-j\omega)} \quad (14.13)$$

where

$$R(j\omega) \cdot R(-j\omega) = D(j\omega) \cdot D(-j\omega) - 4 \cdot \left(\frac{2}{G(0)} - 1\right) \cdot N(j\omega) \cdot N(-j\omega). \quad (14.14)$$

By combining (14.8) and (14.10) one gets

$$\rho(s) \cdot \rho(-s)_{|s=j\omega} = \frac{|R_S - Z_{in}|^2}{|R_S + Z_{in}|^2} \quad (14.15)$$

and after solving for Z_{in} one gets

$$Z_{in} = \frac{R_S - \rho(s)}{R_S + \rho(s)} \quad \text{or} \quad Z_{in} = \frac{R_S + \rho(s)}{R_S - \rho(s)}. \quad (14.16)$$

Now, to find the input impedance and to proceed towards synthesis one is first to find the reflection coefficient from (14.13). That means to find its poles and zeros.

Since all the poles of ρ(s) must be located in the left half plane, the denominator of the reflection coefficient is equal to the denominator of the transfer function $D(s)$. On the other side, there are no restrictions on the position of its zeros [5]. Nevertheless, if all the zeros of the reflection coefficient are chosen to be in the left half-plane, minimal sensitivity of the network with respect to the changes of the value of R_S will be obtained. On the other hand, if all zeros are selected to be in the right half-plane minimal selectivity with respect to the load resistance (R_L) will be obtained [6]. Accordingly, one is not capable to choose in advance the best position of the reflection zeros and one is to accommodate to the specific problem under consideration.

There is one more problem which is mostly noticeable when realizing narrow band band-pass filters. Since the numerator of the squared reflection coefficient is equal to the numerator of the characteristic function, the order of that polynomial is twice the order of the transfer function while all imaginary parts of its zeros lie in the (narrow) pass-band. That brings numerical difficulties in finding the roots of that polynomial. These are heavier in cases of still narrower pass-band. Further difficulties are encountered in the above case if the reflection zeros are not simple. These difficulties may be mitigated if while transforming the low-pass function into a band-pass one, one calculates the reflection zeros of the prototype-filter first and then transforms them, too.

The reflection zeros of the prototype low-pass filter are either known from the transfer function approximation process or obtained by solving a polynomial of twice lower order.

In the case when the band-pass filter is directly synthesized (no transformation but the approximant is found directly in the band-pass form) it is convenient to use the transformation

$$p = \frac{1}{B}\left(s^2 + \omega_0^2\right) \tag{14.17}$$

which (if $B << \omega_0$) maps a circle of a radius of $B/2$ centered at $s = j\cdot\omega_0$ into a circle with a radius of ω_0 centered in the origin. The polynomial obtained by the mapping is easier to solve since it is of halved order. Practical experience claims that one gains at least one decimal digit in accuracy if one uses this trick.

14.3 Unified Theory of Cascade Synthesis

The realization process we are implementing here is based on successive extraction four-terminal cells which in turn realize the selected transmission zero [3]. It exploits the property of the network that a transmission zero of a loss-less passive circuit is, in the same time, zero of the real part of its input impedance. The procedure is named "unified" since the same algorithm is implemented no matter where the transmission zero is located on the s-plane.

In the next we will consider all special cases with the exception of the trivial ones when the input impedance has a zero or a pole at the location of the transmission zero. Of course, discovering the existence of this kind of zeros is not a trivial job. The classical procedure for extraction of that kind of transmission zeros may be realized in a specific way as described in [7].

The expression for the input impedance may be represented as follows

$$Z(s) = \frac{m_1(s) + n_1(s)}{m_2(s) + n_2(s)} \tag{14.18}$$

where m_1(s) and m_2(s) are the even, and n_1(s) and n_2(s) are the odd parts of the numerator and denominator polynomials, respectively. Note, the order of the numerator and the denominator must not differ by more than one. $Z(s)$ should be realized in the form of a LC network terminated by a resistor. The transmission zeros are zeros of the expression

$$Z(s) + Z(-s) = \frac{m_1(s) \cdot m_2(s) - n_1(s) \cdot n_2(s)}{m_2^2(s) + n_2^2(s)} \tag{14.19}$$

Now, the network of Fig. 14.1 is described using the transmission matrix by the following system of equations

$$\begin{bmatrix} V_1 \\ I_1 \end{bmatrix} = M \cdot \begin{bmatrix} V_2 \\ I_2 \end{bmatrix} = \begin{bmatrix} A & B \\ C & D \end{bmatrix} \cdot \begin{bmatrix} V_2 \\ I_2 \end{bmatrix}. \tag{14.20}$$

If a particular transmission zero s_0 is to be extracted from $Z(s)$ using a cell whose transmission matrix is

$$M' = \frac{1}{f(s)} \begin{bmatrix} A' & B' \\ C' & D' \end{bmatrix} \tag{14.21}$$

where $f(s)$ is a polynomial containing only the zeroes to be extracted, the input impedance of the residual network will be (as depicted in Fig. 14.3)

$$Z_1(s) = \frac{D' \cdot Z(s) - B'}{A' - C' \cdot Z(s)} \tag{14.22}$$

Before proceeding one is to find the conditions for $Z_1(s)$ not to contain the extracted transmission zero.

If one writes the doubled real part of the input impedance (from Eq. 14.19) in the form

$$Z(s) + Z(-s) = f^i(s) \cdot R(s) \tag{14.23}$$

where $i = 1$ or $i = 2$ depending on the order of the zero to be extracted,

$$A' \cdot D' - B' \cdot C' = f^2(s), \tag{14.24}$$

and $R(s)$ is not containing the zero s_0, for the doubled real part of the remaining impedance one may write [8]

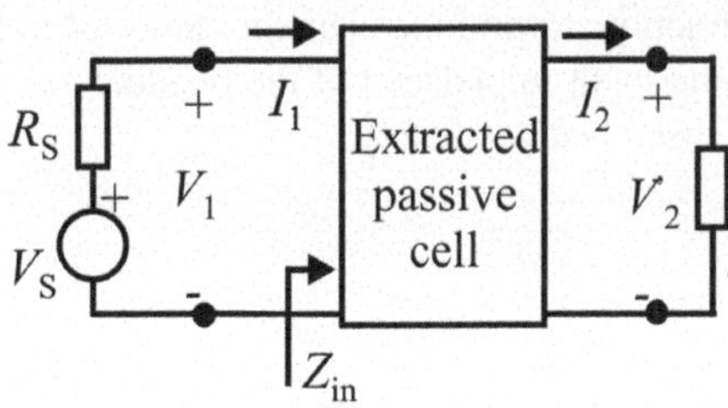

Fig. 14.3 Extracting a cell whose transmission matrix is M'

$$Z_1(s) + Z_1(-s) = \frac{f^{2+i}(s) \cdot R(s)}{[A' - C' \cdot Z(s)] \cdot [A' - C' \cdot Z(s) + C' \cdot f^i(s) \cdot R(s)]} \tag{14.25}$$

Since $Z_1(s) + Z_1(-s)$ must not have $f(s)$ as a factor, $f^{2+i}(s)$ must be a factor of the denominator, which requires that $A' - C' \cdot Z(s)$ should have a factor $f^2(s)$. This fact is used to calculate the element values of the C, D, and Brune networks (described later on) using the appropriate $f(s)$. To develop the above expression the fact that *A*' and *D*' are even, and *C*' and *B*' are odd function of *s*, was utilized. It is easy now to conclude that the even part of the remaining impedance will have a zero at s_0 if

$$\Phi(s) = A' - C' \cdot Z(s) \tag{14.26}$$

has a double zero at $s = s_0$. That gives the conditions for evaluating the elements of the matrix (14.21), and then, the circuit elements of the cell realizing the proper matrix:

$$\frac{d\Phi(s)}{ds}_{|s=s_0} = 0 \quad \text{and} \quad \frac{d^2\Phi(s)}{ds^2}_{|s=s_0} = 0. \tag{14.27}$$

Note, the following is valid [8]

$$Z(s) = \frac{m_1(s) + n_1(s)}{m_2(s) + n_2(s)}, \tag{14.28}$$

$$Z(s_0) = \frac{m_1(s)}{n_2(s)}_{|s=s_0}, \tag{14.29}$$

and

$$\frac{dZ(s)}{ds}_{|s=s_0} = \frac{d}{ds}\left(\frac{m_1(s)}{n_2(s)}\right)_{|s=s_0} \tag{14.30}$$

provided that s_0 is second-order zero of $Z(s) + Z(-s)$.

In the next the cell extraction procedure will be developed in two parts. The zeros at the imaginary axis of the complex frequency plane will be considered first and followed by the complex zeros. Two goals are to be accomplished in every extraction: to find the element values of the extracted cell and to enable calculation of the input impedance of the residual circuit. Every type of transmission zero will be discussed separately.

14.3.1 Zeros at the Imaginary Axis

In this section cell extraction will be developed for the transmission zeros laying on the imaginary axis.

14.3.1.1 Zero at Infinity

Zero at infinity is created by a parallel capacitor or by a parallel inductor.

Parallel capacitor (Fig. 14.4a) is extracted when the order of the denominator of *Z*(*s*) is higher than the order of the numerator. In the case of parallel capacitor one has

$$C = lim_{s\to\infty}\left(\frac{1}{s\cdot Z(s)}\right) \tag{14.31a}$$

and

$$\frac{1}{Z_1(s)} = \frac{1}{Z(s)} - s\cdot C. \tag{14.31b}$$

Note the subtraction in (14.31b).

In the case when the order of the numerator is larger than the order of the denominator of Z(s) one extracts an inductor (Fig. 14.3b). The corresponding value is given by

$$L = \begin{matrix} lim \\ s\to\infty \end{matrix}\left(\frac{Z(s)}{s}\right) \tag{14.32a}$$

while the residual impedance is obtained from

$$Z_1(s) = Z(s) - s\cdot L. \tag{14.32b}$$

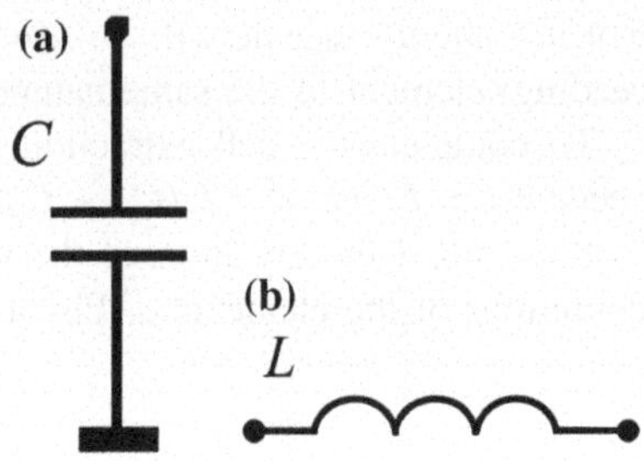

Fig. 14.4 Realization of a zero at infinity, **a** parallel capacitor and **b** series inductor

14.3.1.2 Zero at a Finite Real Frequency

The transfer matrix of a cell that realizes a zero at a real frequency (imaginary axis of the complex frequency plane) has the following form:

$$\frac{1}{1+s^2/\omega_0^2}\cdot\begin{bmatrix}1+a\cdot s^2 & b\cdot s\\ c\cdot s & 1+d\cdot s^2\end{bmatrix} \tag{14.33}$$

where $s_0 = j\omega_0$. For extraction of the zero at $s^2 = -\omega_0^2$, we require that $1+a\cdot s^2 - c\cdot s\cdot Z(s)$ to have a factor $(1+s^2/\omega_0^2)^2$ which leads to

$$1-a\cdot\omega_0^2+c\cdot\omega_0\cdot X(\omega_0)=0, \tag{14.34a}$$

and

$$2\cdot a\cdot\omega_0 - c\cdot X(\omega_0) - c\cdot\omega_0\cdot X'(\omega_0)=0 \tag{14.34b}$$

since real frequency transmission zeros are always of even order. In (14.34) we use $Z(j\omega_0) = j{\cdot}X(\omega_0)$ and $Z'(j\omega_0) = j{\cdot}X'(\omega_0)$.

Using (14.34) we come to the following expressions for the values of the constants a and c:

$$a=\frac{\omega_0\cdot X'(\omega_0)+X(\omega_0)}{\omega_0^2\cdot[\omega_0\cdot X'(\omega_0)-X(\omega_0)]} \tag{14.35a}$$

$$c=\frac{2}{\omega_0\cdot[\omega_0\cdot X'(\omega_0)-X(\omega_0)]}. \tag{14.35b}$$

Based on (14.35) one may develop the following expressions for the element values of the B-section (Brune's) as depicted in Fig. 14.5a, b, c.

$$C=c \quad L_P=a/c \quad M=\frac{1}{\omega_0^2\cdot c}=\sqrt{L_P\cdot L_S}. \tag{14.36}$$

One should note that the value of one of the elements C_p or C_s of Fig. 14.5b and of the elements L_p or L_S of Fig. 14.5c will be negative but that is nothing to be worried about since it will be absorbed by the neighboring (preceding or succeeding) element of the same nature.

To complete the cell extraction i.e. to complete (14.22) the following is still needed: $b = L$, and $d = L{\cdot}(C_m + C_p)$.

By a small modification of the transfer matrix in (14.33) one may obtain the definitions of the elements of Fig. 14.6a, b. Namely, one may write the following

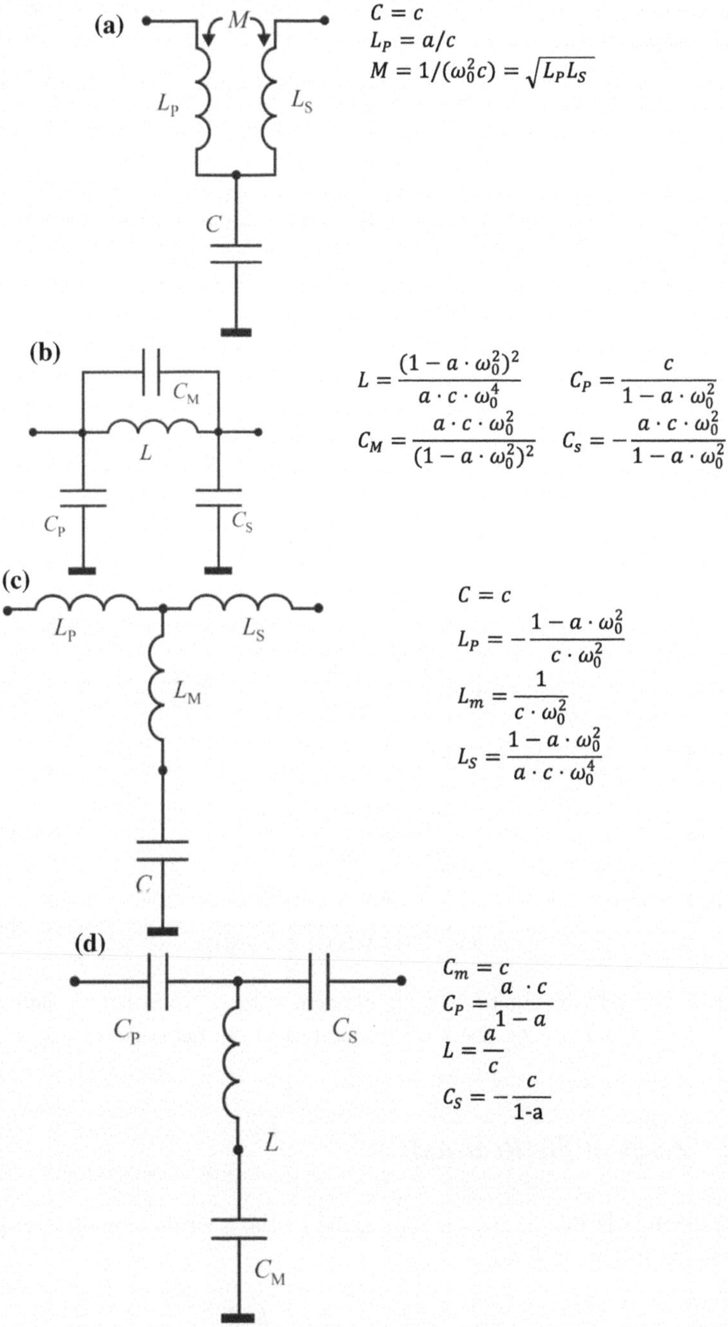

Fig. 14.5 Brune's cell (section) implementations with and without a transformer, **a** variant with coupled inductances, **b** PI-cell, **c** low-pass T-cell and **d** high-pass T-cell

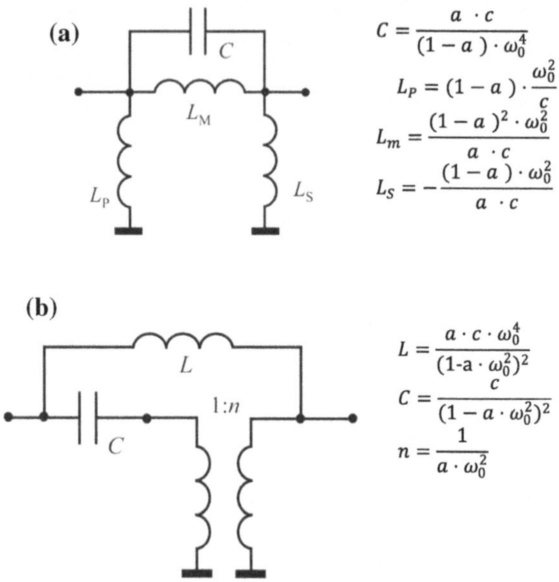

Fig. 14.6 Brune's cell (section) implementations **a** without and **b** with a transformer (extension of Fig. 14.5)

$$\frac{1}{1+s^2/\omega_0^2} \cdot \begin{bmatrix} a' + s^2/\omega_0^2 & b \cdot s \\ c \cdot s & d' + s^2/\omega_0^2 \end{bmatrix} \tag{14.37}$$

Combining with (14.34) one gets

$$a' = \frac{\omega_0 \cdot X'(\omega_0) - X(\omega_0)}{\omega_0 \cdot X'(\omega_0) + X(\omega_0)} \tag{14.38a}$$

$$c' = \frac{2}{\omega_0 \cdot [\omega_0 \cdot X'(\omega_0) + X(\omega_0)]} \tag{14.38b}$$

which allows for the evaluation of the element values. Note that C_S and L_S are always negative and will be absorbed by the rest of the network.

14.3.2 Zeros at the Real Axis

The transfer matrix that realizes a zero at the real axis of the complex frequency plane has the following form

$$\frac{1}{1+s^2/\sigma_0^2} \cdot \begin{bmatrix} 1 + a \cdot s^2 & b \cdot s \\ c \cdot s & 1 + d \cdot s^2 \end{bmatrix} \tag{14.39}$$

Combining with

$$1 - a \cdot \sigma_0^2 + c \cdot \sigma_0 \cdot X(\omega_0) = 0, \tag{14.40a}$$

and

$$2 \cdot a \cdot \sigma_0 - c \cdot X(\sigma_0) - c \cdot \sigma_0 \cdot X'(\sigma_0) = 0, \tag{14.40b}$$

one gets

$$a = -\sigma_0^2 \frac{\sigma_0 \cdot X'(\sigma_0) + X(\sigma_0)}{\sigma_0 \cdot X'(\sigma_0) - X(\sigma_0)} \tag{14.41a}$$

and

$$c = -\sigma_0 \frac{2}{\sigma_0 \cdot X'(\sigma_0) - X(\sigma_0)]}. \tag{14.41b}$$

This transfer matrix may be realized by the cell depicted in Fig. 14.5a with the following element values

$$\begin{aligned} L_P &= a/\left(c \cdot \sigma_0^2\right) \\ M &= -\tfrac{1}{c} = -\sqrt{L_P \cdot L_S}\,. \\ C &= c/\sigma_0^2 \end{aligned} \tag{14.42a}$$

Since the mutual inductance of this cell is always negative we may conclude that one cannot avoid the mutual inductance or a transformer (shown in Fig. 14.5g).

Following element values are offered for the solution using a transformer:

$$\begin{aligned} L &= \frac{-a \cdot c \cdot \sigma_0^4}{(1 - a \cdot \sigma_0^2)^2} \\ C &= \frac{c}{(1 - a \cdot \sigma_0^2)^2}\,. \\ n &= \frac{1}{a \cdot \sigma_0^2}\,. \end{aligned} \tag{14.42b}$$

To complete the extraction of the cell by finding $Z_1(s)$ one needs the following: $b = (L_P + L_S - 2 \cdot M)$ and $d = L_S \cdot C$.

14.3.3 Complex Zeros

The transfer matrix of a cell that realizes a complex transmission zero is presented by the expression

$$\begin{aligned}&\frac{1}{s^4+2\cdot\left(\omega_0^2-\sigma_0^2\right)\cdot s^2+|s_0|^4}\\&\times\begin{bmatrix}a_2\cdot s^4+a_1\cdot s^2+|s_0|^4 & s\cdot(b_1+b_2\cdot s^2)\\ s\cdot(c_1+c_2\cdot s^2) & d_2\cdot s^4+d_1\cdot s^2+|s_0|^4\end{bmatrix}\end{aligned} \tag{14.43}$$

where $|s_0|^2 = \omega_0^2 + \sigma_0^2$.

Following the previous method, we require

$$a_2\cdot s^4+a_1\cdot s^2+|s_0|^4-s\cdot\left(c_1+c_2s^2\right)Z(s)_{|s=s_0}=0 \tag{14.44}$$

and

$$\begin{aligned}&4\cdot a_2\cdot s^3+2\cdot a_1\cdot s+|s_0|^4\\&-\left[\left(c_1+3\cdot c_2s^2\right)Z(s)\right.\\&\left.+s\cdot\left(c_1+c_2s^2\right)Z'(s)\right]_{|s=s_0}=0\end{aligned} \tag{14.45}$$

where $s_0 = \pm\sigma_0 \pm j\omega_0$.

Evaluating (14.44) and (14.45) at $s_0 = \sigma_0 + j\omega_0$ gives

$$a_1=2\cdot\omega_n^2+4\cdot\frac{\left[I_2\left\{X\cdot\sigma_0\cdot J_2+R\cdot\omega_0\cdot\left(3\sigma_0^2-\omega_0^2\right)\right\}+I_1\omega_m^2J_1\right]}{I_2\cdot I_3-I_1\cdot I_4} \tag{14.46a}$$

$$a_2=1+\frac{4\cdot\{I_1(X\cdot\sigma_0+R\cdot\omega_0)-I_2\cdot J_1\}}{I_2\cdot I_3-I_1\cdot I_4} \tag{14.46b}$$

$$c_2=\frac{8\cdot\sigma_0\cdot\omega_0\cdot\omega_m\cdot I_2}{I_2\cdot I_3-I_1\cdot I_4} \tag{14.46c}$$

$$c_3=\frac{8\cdot\sigma_0\cdot\omega_0\cdot I_2}{I_2\cdot I_3-I_1\cdot I_4} \tag{14.46d}$$

where

$$I_1=X\cdot\sigma_0^3-R'\cdot\sigma_0\cdot\omega_0\cdot\omega_m+R\cdot\omega_0^3 \tag{14.47a}$$

$$\begin{aligned}I_2=&-X\cdot\sigma_0^3-R'\cdot\sigma_0\cdot\omega_0\cdot\omega_n\\&+R\cdot\omega_0^3-2\cdot X'\cdot\sigma_0^2\cdot\omega_0^2\end{aligned} \tag{14.47b}$$

$$I_3=X'\cdot\omega_m+3\cdot R\cdot\omega_0-3\cdot\sigma_0\cdot X \tag{14.47c}$$

$$I_4 = X' \cdot \omega_n - 2 \cdot R' \cdot \sigma_0 \cdot \omega_0 + R \cdot \omega_0 + \sigma_0 \cdot X \tag{14.47d}$$

$$Z(\sigma_0 + j \cdot \omega_0) = R + j \cdot X \tag{14.47e}$$

$$Z'(\sigma_0 + j \cdot \omega_0) = R' + j \cdot X' \tag{14.47f}$$

$$\omega_m^2 = \sigma_0^2 + \omega_0^2 \tag{14.47g}$$

$$\omega_n^2 = -\sigma_0^2 + \omega_0^2 \tag{14.47h}$$

$$J_1 = (R \cdot \omega_0 - X \cdot \sigma_0) \tag{14.47i}$$

$$J_2 = 3\omega_0^2 - \sigma_0^2 \tag{14.47j}$$

This transfer matrix may be realized by the D-section with the element values expressed by the proper variant of the section in Fig. 14.7a, b, c.

The appropriate versions of the D-cell having one transformer or even no transformer are depicted in Fig. 14.7b, c. One of the capacitances C_1 or C_2 (Fig. 14.7c) is always negative but will be absorbed by the rest of the circuit. The condition for existence of a cell realizing complex transmission zero without a transformer is

$$\omega_0^2 > \sigma_0^2 \tag{14.48}$$

$$b_1/b_2 \geq \frac{|s_0|^4}{2 \cdot \left(\omega_0^2 - \sigma_0^2\right)}. \tag{14.49}$$

The values of the parameters b and d may be easily calculated from the expressions used for a and c parameters, respectively, if instead of $Z(s_0)$, one puts $Y(s_0) = G + j \cdot B$ and, instead of $Z'(s_0)$, one puts $Y'(s_0) = G' + j \cdot B'$.

14.3.4 All-Pass Lattice

We distinguish here a first order cell and a second order cell. In both cases $Z(s) \equiv 1$ and $Z'(s) \equiv 0$.

Based on that, the expressions given for a transmission zero at real axis (first order all-pass cell, Brune cell of Fig. 14.6) changes into

$$a = \frac{1}{\sigma_0^2} = d \tag{14.50a}$$

(a)

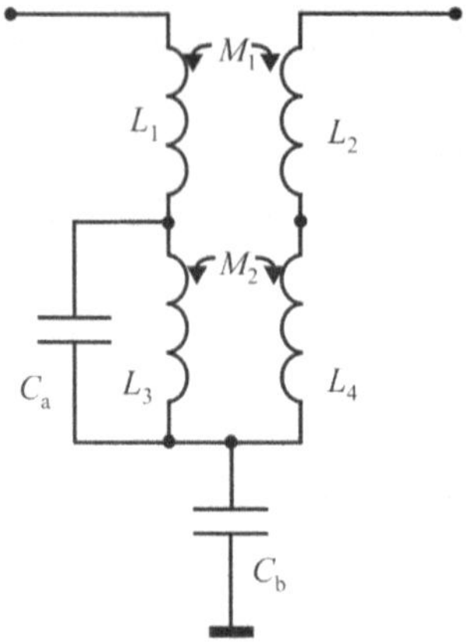

$$M_1 = 1/c_2 \qquad L_4 = M_2^2 / L_3$$

$$L_1 = a_2 / c_2 \qquad C_b = c_1 / \omega_m^4$$

$$L_2 = M_1^2 / L_1 \qquad C_a = c_2 / (c_1 L_3)$$

$$M_2 = -\frac{1}{c_1^2 c_2}\left\{c_1^2 - 2\cdot\omega_n^2\cdot c_1\cdot c_2 + c_2^2\cdot\omega_m^4\right\}$$

$$L3 = \frac{1}{c_1^2\cdot c_2}\left(a_1\cdot c_1\cdot c_2 - a_2\cdot c_1^2 - \omega_m^4\cdot c_2^2\right)$$

(b)

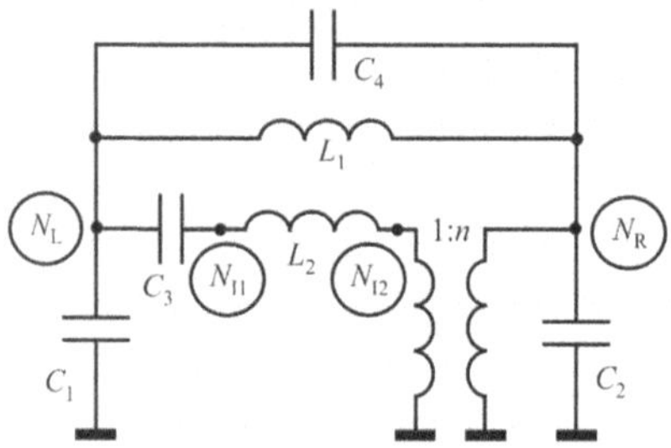

$$C_1 = \frac{d_2 - 1}{b_2} \qquad C_2 = -\frac{d_2 - 1}{b_2 d_2}$$

$$C_4 = \frac{1}{b_2}$$

$$L_1 = \frac{b_1}{\omega_m^4} \qquad L_2\cdot C_3 = b_2 / b_1$$

$$C_3 = \frac{b_2}{b_1^2}\left(\frac{b_1}{b_2}\cdot d_1 - \frac{b_1\cdot d_2}{b_2^2} - \omega_m^4\right)$$

$$n = \frac{\omega_m^4 - \frac{b_1}{b_2}d_1 + \frac{b_1^2}{b_2^2}d_2}{\omega_m^4 - 2\frac{b_1}{b_2}\omega_n^2 + \frac{b_1^2}{b_2^2}}$$

(c)

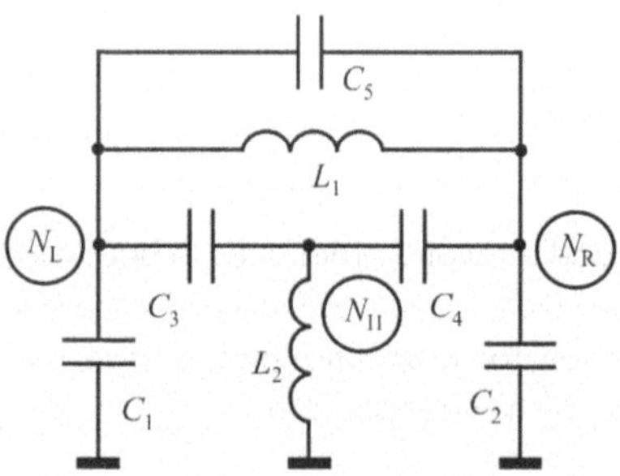

$$L_1 = |s_0|^4 / b_1 \qquad C_1 = \frac{1 - a_2}{a_2 b_2}$$

$$C_2 = -\frac{1 - a_2}{b_2}$$

$$C_3 = \frac{d_1 - 2\cdot\omega_n^2 - \frac{b_1(1 - a_2)}{a_2\cdot b_2}}{b_1}$$

$$C_4 = \frac{a_1 - 2\cdot\omega_n^2 - b_1\cdot(a_2 - 1)/b_2}{b_1}$$

$$C_5 = \left[2\cdot\omega_n^2 - \frac{b_2}{b_1}\cdot\omega_m^4\right] / b_1$$

$$L_2 = \frac{b_2}{a_1 + d_1 - 4\omega_n^2 - \frac{b_1(a_2 - 1)^2}{a_2\cdot b_2}}$$

Fig. 14.7 D-section and alternatives, **a** variant with two pairs of coupled inductances, **b** variant with one transformer, and **c** variant without transformer (Conditions:$\omega_0^2 > \sigma_0^2$ and $\frac{b_1}{b_2} \geq \frac{\omega_m^4}{2\cdot\omega_n^2}$)

and

$$c = \frac{2}{\sigma_0} = b. \tag{14.50b}$$

As for the second order all-pass cell, since we are speaking of a cell with complex transmission zeros, one uses the D-section with

$$\begin{aligned} a_2 &= d_2 = 1 \\ a_1 &= d_1 = 2 \cdot \left(\omega_0^2 + 3 \cdot \sigma_0^2\right) \\ c_1 &= b_1 = 4 \cdot \sigma_0 \cdot |s_0|^2 \\ c_2 &= b_2 = 4 \cdot \sigma_0 \end{aligned} \tag{14.51}$$

Now, for the circuit of Fig. 14.6b one may use

$$L_1 = b_1/\omega_m^4 \tag{14.52a}$$

$$C_1 = (d_2 - 1)/b_2 \tag{14.52b}$$

$$C_2 = (1 - d_2)/(b_2 \cdot d_2) \tag{14.52c}$$

$$C_3 = \frac{b_2 \cdot \left(b_1 \cdot d_1/b_2 - b_1^2 d_2/b_2^2 - \omega_m^4\right)}{b_1^2}. \tag{14.52d}$$

$$C_4 = 1/b_2 \tag{14.52e}$$

$$L_2 = b_2/(b_1 \cdot C3) \tag{14.52f}$$

$$n = \frac{\omega_m^4 - \frac{b_1}{b_2} d_1 + \frac{b_1^2}{b_2^2} d_2}{\omega_m^4 - 2\frac{b_1}{b_2}\omega_n^2 + \frac{b_1^2}{b_2^2}} \tag{14.52g}$$

For the circuit of Fig. 14.6c one may use

$$L_1 = \omega_m^2/(4 \cdot \sigma_0) \tag{14.53a}$$

$$C_1 = C_2 = 0 \tag{14.53b}$$

$$C_3 = C_4 = (2 \cdot \sigma_0)/\omega_m^2 \tag{14.53c}$$

$$C_5 = \frac{\omega_0^2 - 3 \cdot \sigma_0^2}{4\sigma_0 \cdot \omega_m^2} \tag{14.53d}$$

$$L_2 = 1/(4 \cdot \sigma_0). \tag{14.53e}$$

14.4 On the Order of Extraction of the Transmission Zeros

It was already mentioned above that the order of extraction of the transmission zeros has important influence to the properties of the resulting circuit. It was shown in [9] that the extraction order influences the dispersion of the element values, the sensitivity of the network to the variation of the element values as well as the possibility of realization of a transformerless network. It was shown that for low-pass filter with zeros exclusively on the imaginary axis an order of extraction exists that minimizes the sensitivity and the dispersion while in the same time providing for a transformer-less network. In the next we will here shortly discuss that results and recommendations.

Let a filter has m pairs of zeroes on the imaginary axis. If these are ordered by size so that the smallest (by absolute value) becomes enumerated first while the largest becomes mth, then the optimal order of extraction is when they are ordered alternatively to both ends of the new series, starting with the largest from the previous series, as follows:

$$\omega_m, \omega_{m-2}, \ldots, \omega_1, \ldots, \omega_{m-3}, \omega_{m-1})$$

or

$$\omega_{m-1}, \omega_{m-3}, \ldots, \omega_1, \ldots, \omega_{m-2}, \omega_m)$$

This result was theoretically proven in [9]. It is our experience that it works very well [10].

In the case of complex transmission zeros, in order to obtain a network without a transformer the conditions:

$$\omega_0^2 > \sigma_0^2. \tag{14.54}$$

and

$$\frac{b_1}{b_2} \geq \frac{\omega_m^4}{2 \cdot \omega_n^2} \tag{14.55}$$

are to be satisfied.

The order of extraction, in this case, may influence the fulfillment of (14.55) only. Nevertheless, if the condition (14.54) is satisfied one may expect to get a transformer-less network if the complex zeros are ordered according to the size of the imaginary part. In cases when (14.54) is not satisfied it is recommendable to put the cells realizing complex transmission zeros at the end of the chain. In that way two goals may be reached: (1) reduction of the cumulative numerical error and (2) use the transformer contained in the cell for some additional purposes.

In the case of band-pass filters there are no theoretical recommendations as to which order of extraction is the best. By experience, however, one extracts first the zeros from the lower part of the stop-band in the form of the cells from Fig. 14.4d and then the ones from the top part of the stop-band in the form of the form of Fig. 14.4b.

Under some specific conditions [11] one may interchange the zeros from the lower and the upper stop-band. Finally, in the case of band-pass filters with complex transmission zeros the situation becomes more complex while it was observed that in this case the condition (14.54) is usually satisfied (especially for narrow-band filters).

14.5 Circuit Transformations

Having a low-pass transfer function one may go for frequency transformation in order to get a transfer function behaving like a band-pass, band-stop or high-pass. In the first two cases the resulting function will become of order twice as large as the prototype. Finding the element values for these functions, in the case of higher order filters, becomes a serious problem from the numerical point of view as well as for the influence of the order of extraction to the realizability point of view.

To avoid these problems while having the prototype low-pass circuit already realized, one may directly transform the circuit elements.

In the next these transformations will be given.

14.5.1 LP_BP Transformation

Putting into the expression for the impedance of an inductor

$$Z_{\mathrm{L}} = \mathrm{j}\,\omega\, L \tag{14.56}$$

the transformed variable according to (4.1b), one gets

$$Z_{\mathrm{BP\_L}} = \mathrm{j}L \cdot \frac{\omega_0}{B_{\mathrm{r}}}\left(\frac{\Omega}{\omega_c} - \frac{\omega_{\mathrm{c}}}{\Omega}\right). \tag{14.57}$$

This is an expression representing series connection of a capacitor $C_{\mathrm{BP\_s}}$ and an inductor $L_{\mathrm{BP\_s}}$. So an inductor is substituted by a series LC circuit whose elements have the following values

$$L_{\mathrm{BP\_S}} = \frac{\omega_0 L_{\mathrm{LP}}}{\omega_c B_r} \tag{14.58a}$$

$$C_{\mathrm{BP\_S}} = \frac{B_r}{\omega_0 \omega_c L_{\mathrm{LP}}}. \tag{14.58b}$$

The indices LP and BP stand for low-pass and band-pass, respectively, while the extension S stands for series. The resulting transformation is depicted in Fig. 14.8.

Following the same procedure we come to a conclusion that a capacitor is substituted by a parallel LC connection, as depicted in Fig. 14.9, the elements of which have the following values

$$C_{\mathrm{BP\_P}} = \frac{\omega_0 C_{\mathrm{LP}}}{\omega_c B_r} \tag{14.59a}$$

$$L_{\mathrm{BP\_P}} = \frac{B_r}{\omega_c \omega_0 C_{\mathrm{LP}}}. \tag{14.59a}$$

Here again the indices LP and BP stand for low-pass and band-pass, respectively, while the extension P stands for parallel.

14.5.2 LP_HP Transformation

Putting into the expression for the impedance of an inductor the transformed variable according to (4.25b), one gets

$$jL\omega = -jL\omega_0 \cdot \omega_c/\Omega = \frac{1}{j\Omega/(L \cdot \omega_0 \cdot \omega_c)} \tag{4.60}$$

Accordingly, a series inductor, as shown in Fig. 14.10, is substituted by a parallel capacitor of a value

$$C_{\mathrm{HP}} = 1/(\omega_0 \omega_c L_{\mathrm{LP}}) \tag{4.61}$$

A series capacitor is substituted by a parallel inductor according to Fig. 14.11. Its value is given by

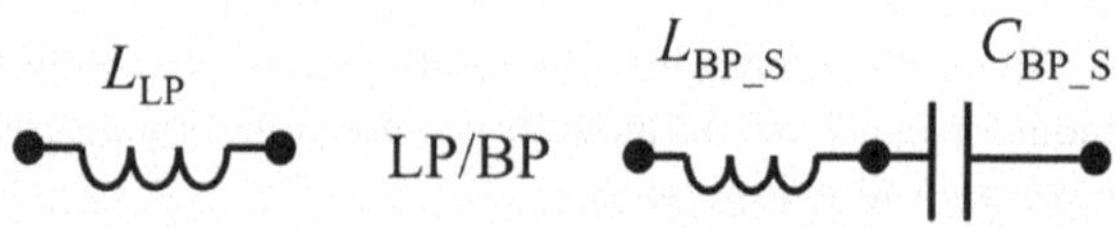

Fig. 14.8 Low-pass to band-pass transformation of an inductor

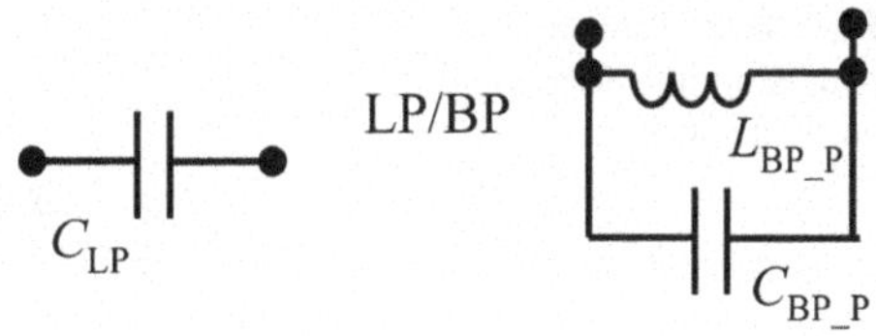

Fig. 14.9 Low-pass to band-pass transformation of a capacitor

Fig. 14.10 Low-pass to high-pass transformation of a series inductor

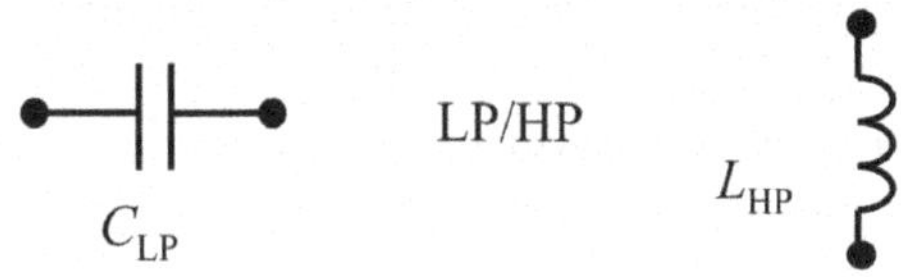

Fig. 14.11 Low-pass to high-pass transformation of a series capacitor

$$L_{HP} = \frac{1}{\omega_0 \omega_c C_{LP}} \tag{4.62}$$

14.5.3 LP_BS Transformation

Putting into the expression for the impedance of an inductor (14.56) the transformed variable according to (4.16a) with $s = \mathrm{j}\omega$ and $z = \mathrm{j}\Omega$, one gets

$$Z_{BS\_L} = -\mathrm{j}\omega_0 L \cdot \frac{B_r \cdot \omega_c \cdot \Omega}{\Omega^2 - \omega_c^2}. \tag{14.63}$$

This is an expression representing parallel connection of a capacitor C_{BS_P} and an inductor L_{BS_P}. So an inductor is substituted by a parallel LC circuit, as depicted in Fig. 14.12, whose elements have the following values

$$C_{BS\_P} = \frac{1}{L_{LP} \cdot B_r \cdot \omega_0 \cdot \omega_c} \tag{14.64a}$$

$$L_{BS\_P} = \frac{L_{LP} \cdot B_r \cdot \omega_0}{\omega_c} \tag{14.64b}$$

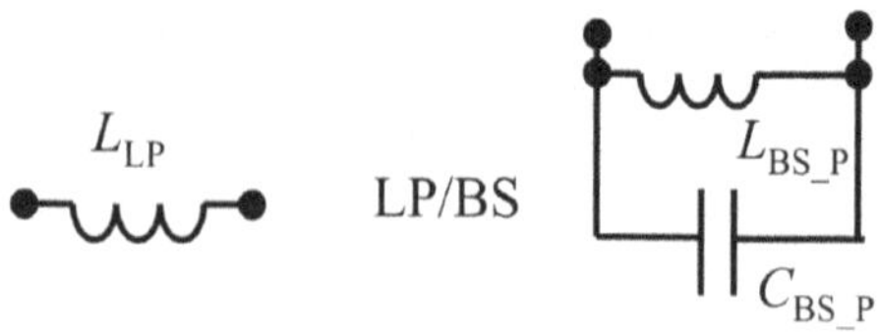

Fig. 14.12 Low-pass to band-stop transformation of an inductor

Fig. 14.13 Low-pass to band-stop transformation of a capacitor

Here the indices LP and BS stand for low-pass and band-stop, respectively, while the extension P stands for parallel.

A capacitor is substituted by a series LC circuit, as depicted in Fig. 14.13, whose elements are calculated as follows

$$L_{BS\_S} = \frac{1}{C_{LP} \cdot B_r \cdot \omega_0 \cdot \omega_c} \tag{14.65a}$$

$$C_{BS\_S} = \frac{C_{LP} \cdot B_r \cdot \omega_0}{\omega_c}. \tag{14.65b}$$

Here the indices LP and BS stand for low-pass and band-stop, respectively, while the extension S stands for series.

14.6 Design Examples

In this paragraph several examples will be given in order to exemplify different design situations. That will cover first a low-pass filter having transmission zeros both: on the ω-axis and complex ones. Then a solution will be demonstrated which realizes a group delay corrected band-pass filter obtained from a selective low-pass prototype and then corrected, as described in Chap. 11. All examples are produced by proper programs of the $\mathcal{RM}$ software.

14.6.1 Example 1

Here a low-pass filter taken from the literature [12] will be considered. Its transfer function was obtained by simultaneous implementation of passband amplitude, stopband amplitude and passband group delay requirements which had as a result a transfer function with zeros being located both: on the ω-axis and in the right-half of the complex frequency plane.

To get the schematic in a form of cascaded network the *cascade_LP* program was used. Below one of the reports of the *cascade_LP* program was copied (with negligible change in the formatting). The reported results cover all phases of the synthesis process which was done with normalized variables. That practice was adopted to avoid very large numbers when high order polynomials are to be manipulated with high values of the frequency variable.

Since, according to the theory given above, it may happen that some cells are not realizable without transformer, the report listed below contains complete set of variants which may be used to substitute each other if necessary.

Then, results of the SPICE simulation is given in Fig. 14.14. Note, the *cascade_LP* program creates a SPICE file automatically. To avoid misunderstanding, the variant of the cell implemented in SPICE was explicitly marked in the report. It is intended to be used for frequency domain simulation. Since, however, it is a text file, after minor changes, the user may create a file for time domain simulation, too.

Next is the report.

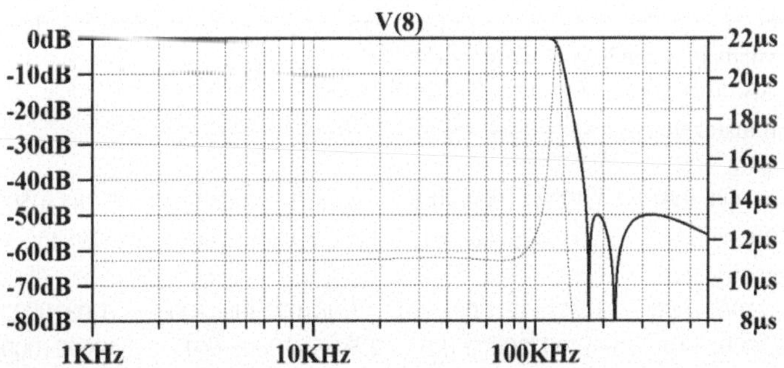

Fig. 14.14 Gain and group delay of the low-pass filter of Example 1

```
Start.E

Program: CASCADE_LP
        SYNTHESIS OF LOW-PASS PASIVELC FILTERS
        Project name: Complex_LP_8_6_50_syn
+++++++++++++++++++++++++++++++++++++++++++++++++++++++++++++++++++
Input data on the transfer function
------------------------------------------------------
Order of the numerator n=6;    Order of the denominator m=8

You entered the transmission zeros as follows (n=6):
sigma[1]=0.000000000e+000    omega[1]=1.729901800e+000
sigma[2]=0.000000000e+000    omega[2]=-1.729901800e+000
sigma[3]=0.000000000e+000    omega[3]=2.257123100e+000
sigma[4]=0.000000000e+000    omega[4]=-2.257123100e+000
sigma[5]=7.455204000e-001    omega[5]=4.335474000e-001
sigma[6]=7.455204000e-001    omega[6]=-4.335474000e-001

You entered the poles of the transfer function as follows (m=8):
sigma[1]=-1.123441000e-001    omega[1]=1.251432100e+000
sigma[2]=-1.123441000e-001    omega[2]=-1.251432100e+000
sigma[3]=-3.785105000e-001    omega[3]=1.081092300e+000
sigma[4]=-3.785105000e-001    omega[4]=-1.081092300e+000
sigma[5]=-5.903942000e-001    omega[5]=6.480072000e-001
sigma[6]=-5.903942000e-001    omega[6]=-6.480072000e-001
sigma[7]=-6.519845000e-001    omega[7]=2.596921000e-001
sigma[8]=-6.519845000e-001    omega[8]=-2.596921000e-001
+++++++++++++++++++++++++++++++++++++++++++++++++++++++++++++++++++
All polynomials in ascending order of s or w**2

The input transfer function ----------------------------------------------
Poles-real part:
   -1.123441000e-001   -1.123441000e-001   -3.785105000e-001   -3.785105000e-001
  -5.903942000e-001   -5.903942000e-001   -6.519845000e-001   -6.519845000e-001
Poles-imaginary part:
   1.251432100e+000   -1.251432100e+000    1.081092300e+000   -1.081092300e+000
   6.480072000e-001   -6.480072000e-001    2.596921000e-001   -2.596921000e-001
+++++++++++++++++++++++++++++++++++++++++++++++++++++++++++++++++++
Ordered transmission zeros
Zeros-real part:
   9.999999990e+009    7.455204000e-001    0.000000000e+000    0.000000000e+000
9.999999990e+008
```

Zeros-imaginary part:
9.999999990e+009 4.335474000e-001 2.257123100e+000 1.729901800e+000
9.999999990e+008
++
Zeros of there flection coefficient
Real part
-5.241728031e-002 -5.241728031e-002 -1.487778877e-001 -1.487778877e-001
-1.164294396e-001 -1.164294396e-001 0.000000000e+000 0.000000000e+000
Imaginary part
1.120629266e+000 -1.120629266e+000 5.400947469e-001 -5.400947469e-001
8.852014701e-001 -8.852014701e-001 0.000000000e+000 0.000000000e+000
++
Coefficients of there flection coefficient (numerator):
-0.000000000e+000 0.000000000e+000 3.148550774e-001 4.167225621e-001
1.768127755e+000 1.101562355e+000 2.494426468e+000 6.352492153e-001
1.000000000e+000
Coefficients of there flection coefficient (denominator):
7.839774840e-001 3.844124590e+000 9.319123209e+000 1.453832366e+001
1.627904128e+001 1.342844969e+001 8.300851029e+000 3.466466600e+000
1.000000000e+000
++
INPUT IMPEDANCE Variant Zin=RS*(1-RO)/(1+RO)
Numerator:
-7.839774840e-001 -3.844124590e+000 -9.004268131e+000
-1.412160110e+001 -1.451091352e+001 -1.232688733e+001 -5.806424562e+000
-2.831217385e+000
Denominator:
-7.839774840e-001 -3.844124590e+000 -9.633978286e+000
-1.495504623e+001 -1.804716903e+001 -1.453001204e+001 -1.079527750e+001
-4.101715815e+000 -2.000000000e+000
++
EXTRACTION OF THE CELLS
++
k=1 (th) ZERO AT INFINITY: parallel capacitance C=1.1242863e-009
Node_UP=1 Node_Down=0

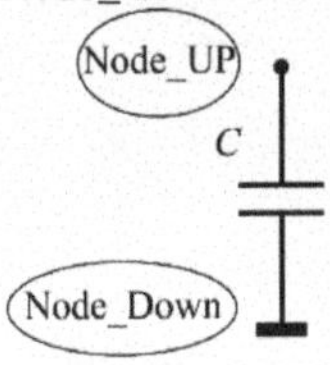

k=2 (th) A PAIR OF COMPLEX ZEROS s=7.4552040e-001 w=4.3354740e-001

Variant with two pairs of coupled inductances
L1=2.1855930e-003 L2=3.0757002e-005 CA=1.6290579e-001 L3=1.4812050e-009
L4=3.9371043e-003 M12=5.4038165e-004 M34=-9.1647078e-001
CB=7.6711677e-009

Variant with one transformer, Implemented in the SPICE description
L1=7.5324584e-003 C4=2.8846422e-010 C3=1.5299972e-009 L2=3.4889189e-003
N=-1.0592958e+000 C1=-2.5424427e-010 C2=2.1432055e-009
Node_i=1 Node_o=5

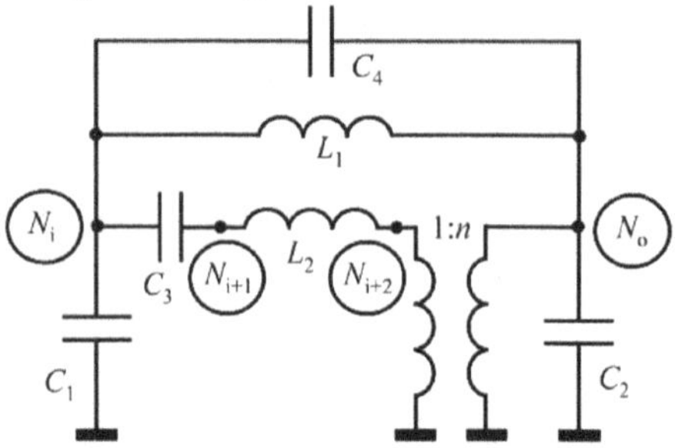

Variant with no transformers
L1=3.3628192e-004 C5=-1.1558889e-009 C1=2.9743503e-009 C3=2.8078561e-009
C4=-2.5424427e-010 C2=2.1432055e-009 L2=9.2318326e-004

----

k=3 (th) ZERO AT THE IMAGINARY AXIS=2.2571231e+000

Variant with coupled inductances (Brunecell):
C=1.9055334e-009 Lp=2.5223343e-003 Ls=2.6991298e-005 M=1.6394307e-001

Variant with a transformer
C=2.5367873e-011 L=2.0274786e-003 n=1.0344525e-001

Variant with a resonant circuit (tank=L:Cm) in the series branch,PI_cell
Implemented in the SPICE description
L=2.0274786e-003
Cp=-2.1986207e-010 Cm=2.4522994e-010 Cs=2.1253955e-009
Node_i=5 Node_o=6

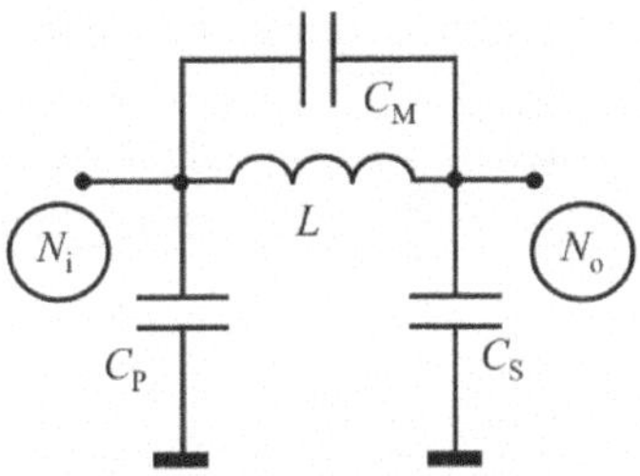

k=4 (th) ZERO AT THE IMAGINARY AXIS=1.7299018e+000

Variant with coupled inductances (Brune cell):
C=2.9356964e-010 Lp=5.1319009e-003 Ls=1.6199223e-003 M=1.8116158e+000

Variant with a transformer
C=4.8266918e-010 L=9.8527113e-004 n=5.6183393e-001

Variant with a resonant circuit (tank=L:Cm) in the series branch, PI_cell
Implemented in the SPICE description
L=9.8527113e-004
Cp=-3.7642664e-010 Cm=8.5909582e-010 Cs=6.6999628e-010
Node_i=6 Node_o=7

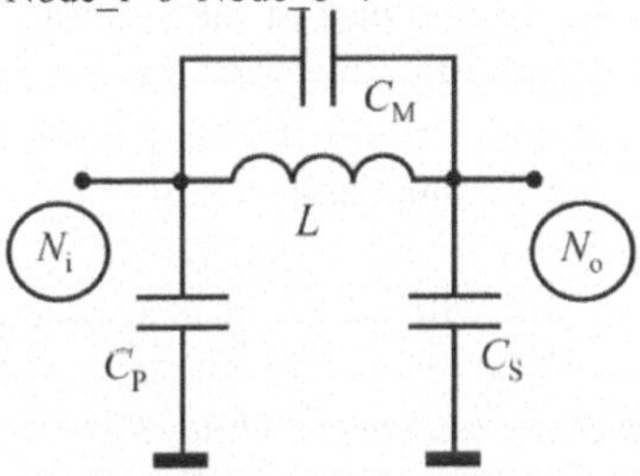

k=5 (th) Zero at infinity: series inductance L=4.4934960e-004
Node_L=7 Node_R=8

------------------ Residual load impedance --------------------

Residual is resistor (Implemented in the SPICE description) and R=1.0000008e+003

Node_end=8

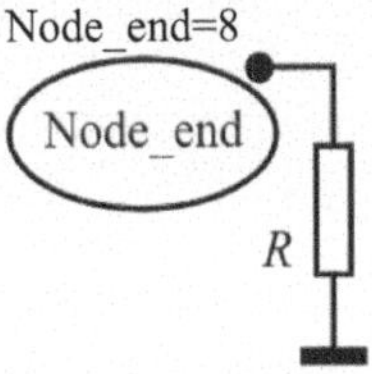

Here ends the synthesis process

End.E

When considering the result depicted in Fig. 14.14 one has to have in mind that in case of $R_S = R_L$, (14.7b) becomes $|T(j\omega)| = 2 \cdot |G(j\omega)|$. This means that if $|G(j\omega)|$ (or what is the same, the output voltage as in Fig. 14.14) is depicted; it would create a reduction of the gain for 6 dB. To avoid that the input signal within the SPICE description of the circuit was set to be 2 V.

14.6.2 Example 2

This example is about synthesis of a cascaded passive LC circuit realizing a 10th order band-pass filter obtained from modified elliptic prototype to which a second order group delay corrector was added. The passband attenuation was 0.1 dB with renormalization of the prototype to 3 dB. The stopband attenuation was 40 dB while the group delay error was 2%.

The $\mathcal{RM}$ software works in the following way. Based on the input data the prototype transfer function is synthesized first and transformed to obtain the bandpass transfer function. To that function a corrector function is created as described in Chap. 11. The circuit synthesis process goes in slightly different way. First the circuit realizing the low-pass prototype is synthesized. Then the lp-to-bp transformation is performed on the circuit level as described in Paragraph 14.5. This action is performed by the program. In that way the selective part of the system is created. The corrector circuit is synthesized separately by the program. Two separate reports are created containing the schematics of both circuits. For verification purposes two SPICE files are created, too. These are automatically merged by a specially written program. The load resistor of the first cascade was deleted.

Below, in order to save space, the synthesis reports will be given in reduced form. In addition, two results of SPICE simulation will be given. The one depicted in Fig. 14.15a represents the whole gain and group delay responses while the other depicted on Fig. 14.15b represents the passband responses only.

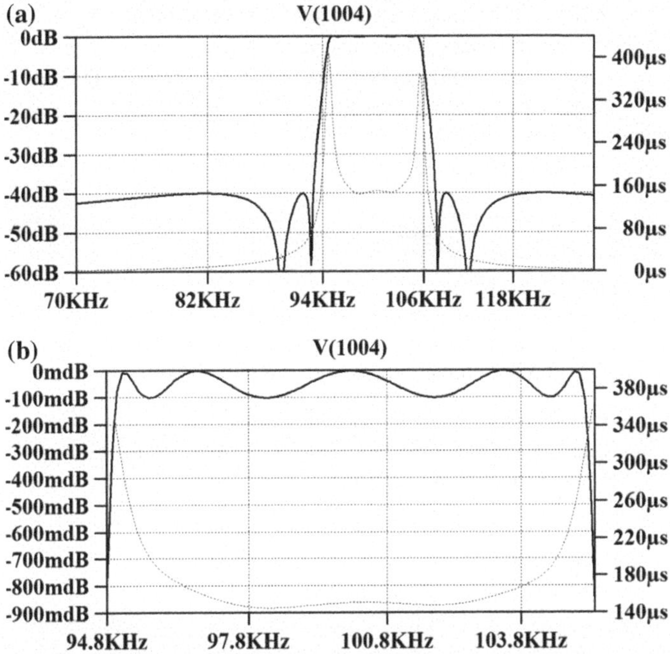

Fig. 14.15 Gain (in dB) and group delay of the low-pass filter of Example 2. **a** The complete responses and **b** enlarged passband responses

Start.E

Program: CASCADE_TR

SYNTHESIS OF TRANSFORMED PASIVE LC FILTERS

Project name: ellip_LP_5_2_01_40_syn

==

No_of_elements of the low_pass=9
Type of transformation=1
Fcut-off of low_pass=6.28319e+005 ,
Central/cut_off of transformed=6.28319e+005
Relative bandwidth=1.02632e-001

Cell No.=1

ZERO AT the imaginary axis_C=1.6074985e-009
Parallel resonant circuit connected to ground
Lmp1=1.6172325770e-004 Cmp1=1.5662741605e-008
Node_UP=1 Node_Down=0

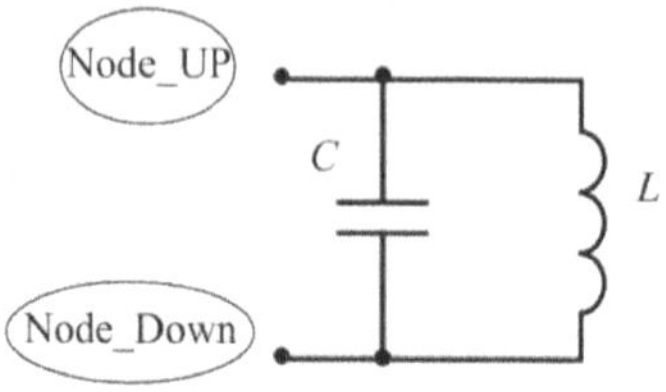

ZERO AT the imaginary axis_C=2.8333839e-010

Parallel resonant circuit in a series branch
Lms1=9.1752443345e-004 Cms1=2.7607216816e-009
Node_LEFT=1 Node_RIGTH=3

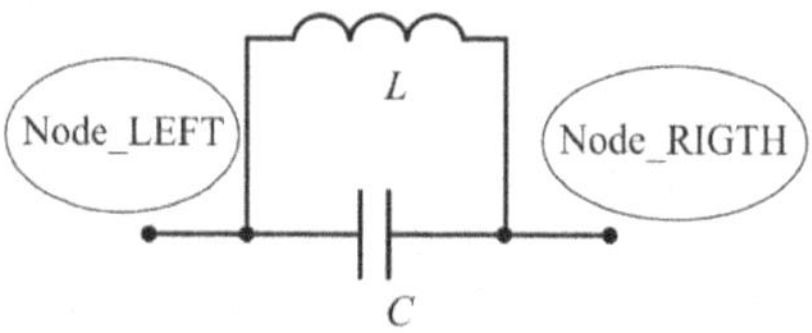

ZERO AT the imaginary axis_L=1.8938700e-003
Series resonant circuit in a series branch
Cs1=1.3726913374e-010 Ls1=1.8453016551e-002
Node_LEFT=1 Node_RIGTH=3

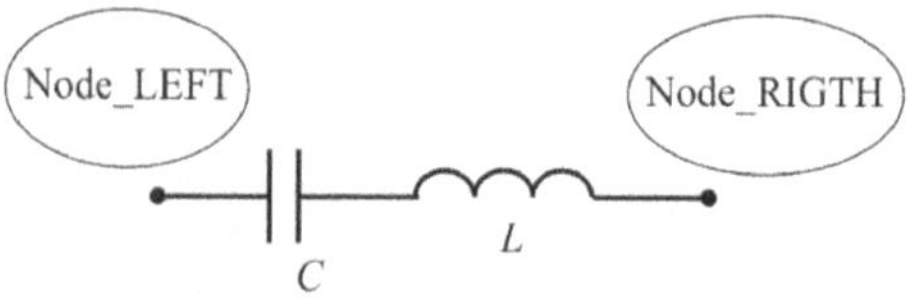

Cell No. =2

ZERO AT the imaginary axis_C=2.5198993e-009
Parallel resonant circuit connected to ground
Lmp2=1.0316677990e-004 Cmp2=2.4552763975e-008
Node_UP=3 Node_Down=0

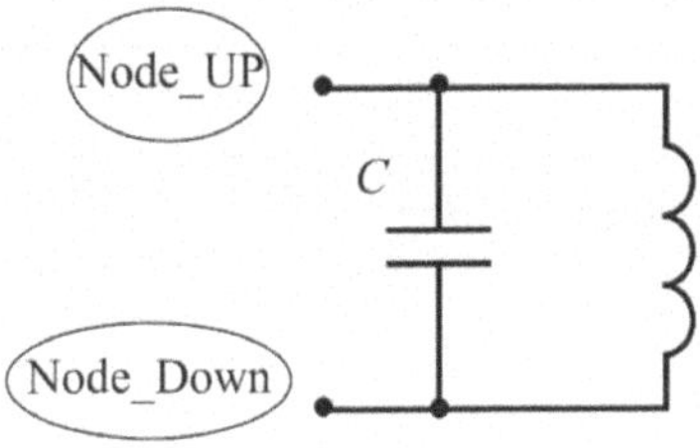

ZERO AT the imaginary axis_C=8.4620274e-010
Parallel resonant circuit in a series branch
Lms2=3.0721939316e-004 Cms2=8.2450185543e-009
Node_LEFT=3 Node_RIGTH=5

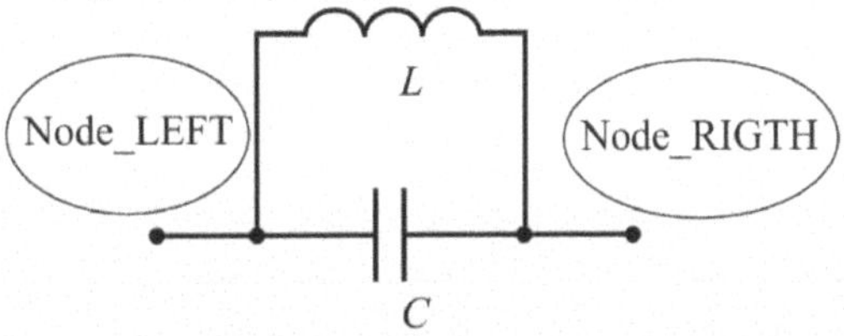

ZERO AT the imaginary axis_L=1.3869696e-003
Series resonant circuit in a series branch
Cs2=1.8743734586e-010 Ls2=1.3514006962e-002
Node_ LEFT =3 Node_ RIGTH =5

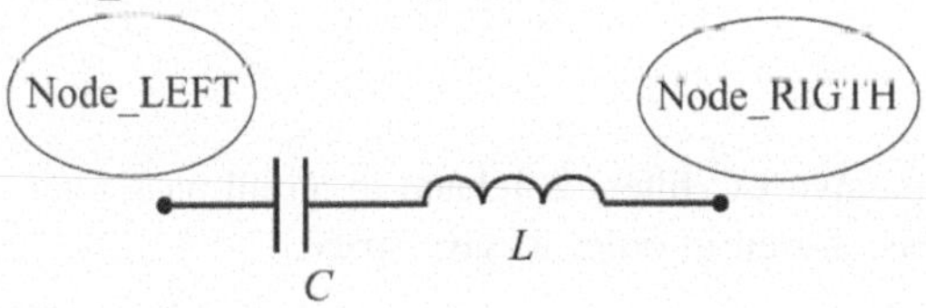

Cell No.=3

ZERO AT INFINITY_C=1.2136028e-009
Parallel resonant circuit in the parallel branch connected from node 5 to ground
C013= 1.1824798835e-008 L013= 2.1421333523e-004
Node_UP=5 Node_Down=0

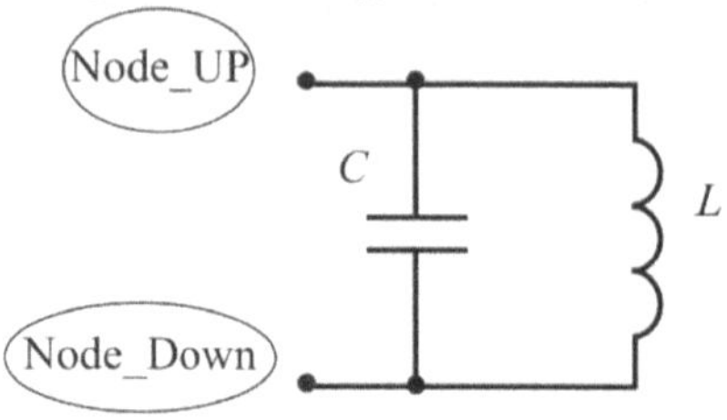

Load resistor
R=1.0000000000e+003
Node_end=5

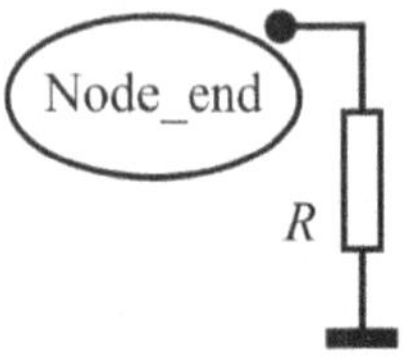

==

Here ends the transformed BAND-PASS synthesis

==

End.E

14.6.3 Example 3

The last example of synthesis of passive LC filters is related to an all-pass filter. A simple second order function was chosen in order to save space.

Start.E

CASCADE_AP, Program for all-pass synthesis of passive LC filters

----

Project name: ellip_LP_5_2_01_40_syn

++

Input data on the transfer function

Order of the numerator n=2; Order of the denominator m=2
The input transfer function

Poles-real part:

-4.739296228e-002 -4.739296228e-002

Poles-imaginary part:

1.003910923e+000 -1.003910923e+000

++

Ordered transmission zeros

Zeros-real part:

4.739296228e-002

Zeros-imaginary part:

1.003910923e+000

++

INPUT IMPEDANCE Variant Zin=RS*(1+RO)/(1-RO)

Numerator:

2.119845637e+000 1.094785925e+000 1.000000000e+000

Denominator:

-9.967916851e-002 -9.052140754e-001 1.000000000e+000

++

EXTRACTION OF THE CELLS

++

k=1 (th) A PAIR OF COMPLEX ZEROS (all_pass) s=4.739296228e-002 w=1.003910923e+000

Variant with one transformer , implemented in the SPICE description

L1=2.987010957e-004 C4=8.395494585e-009 C3=7.467527394e-011
L2=3.358197834e-002 N= -1.000000000e+000 C1=0.000000000e+000
C2=0.000000000e+000

Node_i=1000 Node_o=1004

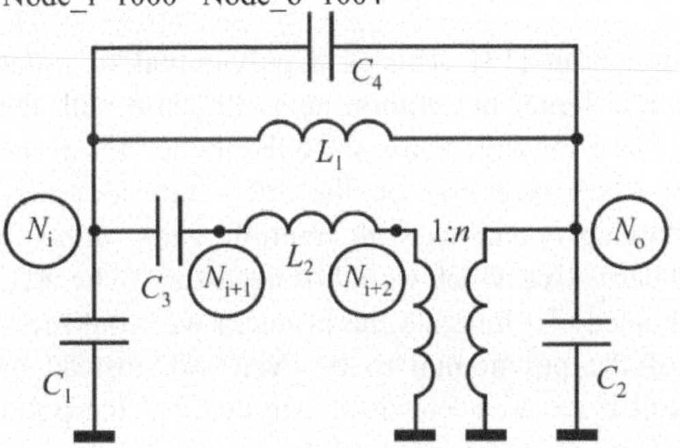

-----------------Residual load impedance--------------------

Residual is resistor (Implemented in the SPICE description) and R=1.000000000e+003

Node_end=1004

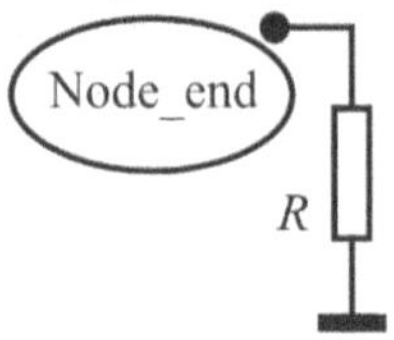

Here ends the synthesis process

End.E

14.7 Developer's Corner

Despite the fact that the subject of passive cascade synthesis is settled for long time now, developing a new program for synthesis imposes several challenges. These are due to the large variety of possible circuit implementations for the same type of transmission zero, to the building of capability of controlling the order of extraction, and, of course, to the numerical accuracy.

It is up to the software developer's talent and skills to cope with combinations and variants either of circuit or of order of extraction nature. That is not the case, however, when the numerical accuracy is considered. There are two braking points where accuracy may be jeopardized.

The first one is related to the extraction of the zeros of the reflection coefficient. The problem is implicitly stated by (14.14) which is rewritten below

$$R(j\omega)\cdot R(-j\omega) = D(\mathrm{j}\omega)\cdot D(-\mathrm{j}\omega) - 4\cdot\left(\frac{2}{G(0)} - 1\right)\cdot N(\mathrm{j}\omega)\cdot N(-\mathrm{j}\omega) = 0.$$

It is known as the Feldkeller Equation [13]. This is a polynomial of order $2n$ which, per se, is a challenge if n is large. In addition, as is the case with the Elliptic or Chebyshev filters, it may have multiple zeros since the numerator of the characteristic function of the filter (the expression on the left hand side) may represent a squared polynomial. Solving equations with multiple zeros imposes problems with a zero valued second derivative which would be necessary to be used if Newton-Raphson algorithm is planned. To mitigate the problem we would recommend exploiting the property of the polynomial to be even. So, instead of solving it for $s = \mathrm{j}\omega$, one could profit if $z = s^2 = -\omega^2$ is substituted and the polynomial solved for z. After solutions found on may go back by $s = \pm\sqrt{z}$ which in the same time allows for choices among the zeros to form the reflection coefficients as discussed in Paragraph 14.2.

The second aspect of numerical nature is related to the realization of (14.22). Here two problems are to be recognized. The first one is related to subtraction of polynomials which takes place in both the numerator and the denominator every now after a cell is extracted. Subtraction of numbers of similar value (which is very probable after so many actions taken) is the most inaccurate numerical operation. In addition, as is expected, cancelation-out of parts of the numerator and denominator should happen (when second order cells are extracted) since the resulting impedance is of lower order. That has to be programmed carefully to avoid numerical inaccuracies.

So, having these in mind, if one has aspiration to manipulate higher order and highly selective filters one would, probably, adopt multiple precision in the calculation such as long double in C programming.

References

1. Darlington S (1939) Synthesis of reactance four-poles which produce prescribed insertion loss characteristics. J Math Phis 18:257–353
2. Ming NT (1968) Simultaneous realization of the transfer and reflection factors of two-ports and n-ports. IEEE Trans Circuit Theory 15(4):354–370
3. Scanlan JO, Rhodes JDI (1970) Unified theory of cascade synthesis. Proc IEE 117(4):665–669
4. Dimopoulos H (2012) Analog electronic filters, Theory design and synthesis. Springer, Dordrecht
5. Koo RL, Sobral M Jr (1975) On the choice of rejection coefficient zeros of coupling networks. J Frankl Inst 300(3):197–202
6. Koo RL, Sobral M Jr (1975) Pole sensitivity in coupling networks. J Frankl Inst 300 (3):203–207
7. Guillemin EA (1957) Synthesis of passive networks. Wiley, New York
8. Youla D (1961) A new theory of cascade synthesis. IRE Trans Circuit Theory 8(3):244–260
9. Neyrinck J, Tran C (1974) On the extraction order of the attenuation poles in LC filters. In: Proceedings of the IEE European Conference on CT and Design, pp 20–23
10. Litovski VB, Raković BD (1974) Attenuation characteristics and element values of least-squares monotonic passband filters with an arbitrary number of transmission zeros. Publications of Electrical Engineering Faculty, University of Belgrade, Series Electronics, Telecommunications and Automation (ETA) 97-101, pp 67–101
11. Neyrinck J, Cuong DM (1975) Realizability of bandpass filters. Electron Lett 11(12):256–358
12. Litovski V (1979) Synthesis of monotonic passband sharp cutoff filters with constant group delay response. IEEE Trans Circuits Systems CAS 26(8):597–602
13. Oldoni M, Macchiarella G, Bellini S (2013) Accurate computation of poles of a lossless multiport network via cepstrum analysis. IEEE Microw Wirel Compon Lett 23(2):57–59

Chapter 15
Active RC Cascade Circuit Synthesis

The term *active* is used to denote the presence of electronic operational amplifiers in the filter's structure. That allows elimination of the need for inductors and transformers. In addition, realization of the filter in a form of cascaded cells (of active electronic circuits) which does not load each other is enabled which eliminates the numerical problems encountered in passive cascade realization. To that purpose a second order function named *biquad* is introduced. To create a cascade of biquads, however, one has to solve the "pole-zero pairing" and the "order of extraction" problem. In addition, most of the cells (physical realizations of the biquads) have variants. In that way the number of cell types and their order in the cascade becomes enormous if high order filters are to be synthesized. In this chapter we will recommend an exhaustive list of types of cells encompassing every type of transmission zero and corresponding pair (or single) of poles. We will also recommend order of extraction leading to reduced noise and nonlinear distortions which are specifics of the active technology. A short study of the influence of the imperfection (limited gain) of the operational amplifier to the frequency response will be given.

15.1 Introduction

Several problems are encountered when using passive filters among which the dominating is the presence of inductors. The inductor has several disadvantages. It needs volume and it needs magnetic material which, in time and under temperature stresses, loses its properties. The former is especially exposed when low frequency filters are to be designed when not only large values of inductances are needed but large values of capacitances, too. Voluminous, heavy and expensive becomes the filter. The solution of first level was enabled by introduction of active components (operational amplifiers -OAs) into the circuit which allowed for circuits without inductors. That was a small revolution in the low-pass filtering especially in the

V. Litovski, *Electronic Filters*, Lecture Notes in Electrical Engineering 596,
https://doi.org/10.1007/978-981-32-9852-1_15

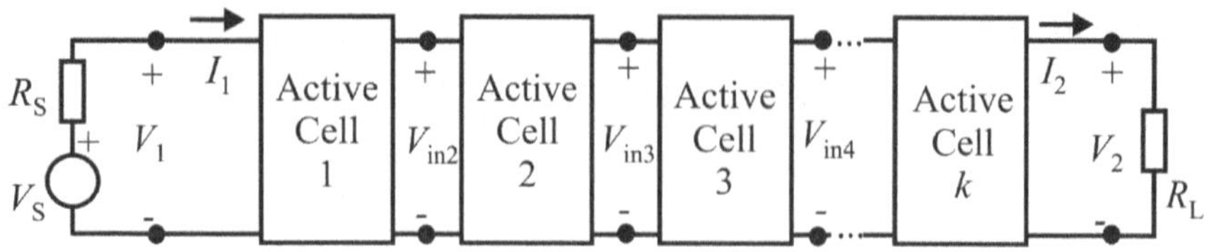

Fig. 15.1 Structure of the filter and enumeration of the port nodes

telephone baseband transmission. The resulting circuit structure contains operational amplifiers, resistors and capacitors hence the term active RC filters.

Operational amplifier [1–3] is not a perfect component. Apart of the need for a supply voltage (sometimes more than one) it performs not perfectly. Among others one may expect frequency dependent gain, and drift of the DC voltage values. In addition, having resistors and transistors incorporated they produce noise which is one of the most unwanted phenomenon in telecommunication. Nevertheless, in this chapter we will consider them perfect in the sense that they will be modeled by the simplest model of operational amplifiers of all: the voltage controlled voltage source with extremely large (infinite) gain. The influence of the imperfections of OAs will be partly considered below.

In this chapter the synthesis of active RC filters in a form of a cascade of cells will be described. The structure of the resulting schematic is shown in Fig. 15.1. It is supposed that the output impedances of the cells (being equal to the output impedances of the operational amplifier which must be located at the output of the cell) are negligible which means that the cell is behaving as an ideal voltage source and no mutual loading of the cells happens. From circuit-theoretical point of view this is the most important difference between passive and active circuits. It allows creating a system whose gain is calculated in a simple manner as

$$A = \frac{V_2}{V_1} = \frac{V_2}{V_{\text{in\_}k}} \cdot \ldots \cdot \frac{V_{\text{in\_}4}}{V_{\text{in\_}3}} \cdot \frac{V_{\text{in\_}3}}{V_{\text{in\_}2}} \cdot \frac{V_{\text{in\_}2}}{V_1} = A_k \cdot \ldots A_3 \cdot A_2 \cdot A_1. \tag{15.1}$$

During creation of the physical structure, however, there are several choices to be resolved.

First, decomposition of the transfer function must be performed. This means the gains of the cells, A_l, in the following

$$H(s) = A_0 \cdot \frac{\prod_{i=1}^{m}(s - z_i)}{\prod_{i=1}^{n}(s - p_i)} = A = \frac{V_2}{V_1} = \prod_{l=1}^{k} A_l \tag{15.2}$$

are to be found. Here the transfer functions A_l are usually restricted to be of first and second order. The latter are usually referred to as biquads.

To complete the information we will introduce the concept of the Q of a pole. It will allow for choosing among alternatives of biquads with the same type of transfer function.

Let the transfer function of a single biquad be rewritten in the form (the index l is omitted for convenience)

$$A = \frac{N}{D} = \frac{(s-z)(s-z^*)}{(s-p)(s-p^*)} = \frac{s^2 + s \cdot g + q}{s^2 + s \cdot a + b} = \frac{s^2 + s \cdot g + q}{s^2 + s \cdot \frac{\sqrt{b}}{Q} + b} \tag{15.3}$$

where $p = \alpha + \mathrm{j} \cdot \beta$, $z = \gamma + \mathrm{j} \cdot \delta$, $a = -2 \cdot \alpha$, $b = \alpha^2 + \beta^2$, $g = -2 \cdot \gamma$, and $q = \gamma^2 + \delta^2$. Then, the quantity

$$Q = \sqrt{b}/a \tag{15.4}$$

is named the pole's or the cell's Q-factor. It is related to the peak value of the gain of the cell which occurs approximately at the frequency $\omega^2 = b$. It is obvious that for a fixed modulus of the pole, the peak value of the gain of the cell is directly dependent on the real part of the pole. The nearer the pole to the ω-axis the higher the peak of the amplitude characteristic of the biquad is. By pushing the pole towards the higher frequencies, keeping a fixed distance from the ω-axis, the peak is rising, too. So, selective low-pass filters having poles near the ω-axis and, in the same time, near to the cut-off frequency will need biquads with very high Q-factors to be synthesized. The reader is advised to go back to the examples of synthesized functions in the previous chapters to see how large the difference may be between Q-factors within a single filter and Q-factors of filters produced under different approximation criteria.

To create (15.2) decisions are to be made as to which pair of poles is to be paired with which zero (or pair of zeros), and which will be the order of the cell in the cascade (order of extraction).

As a result of that activity which will be described below in more details, a sequence of second order transfer functions is obtained with possibly one first order cell needed in case of odd order filter.

For realization of every single second (and first) order transfer function, choices are to be made based on the type of the function (low-pass, high-pass, band-pass, band-stop and all-pass), on the type of the complex zero if there is any, and on the value of the pole's Q factor (small and large). These choices are obvious and proper circuit will be chosen based on the literature e.g. [4].

Every cell structure described below and the element values calculation is based on the literature e.g. [5–9]. Alike other texts, in the sequel we are not suggesting alternatives despite the fact that there are many. If any alternative is suggested it will be justified by proper reasons.

15.2 Order of Extraction and Gain Assignment

Given at transfer function intended to be realized as a cascade of second order cells (plus one of first order when odd order transfer functions are dealt with) beside the choice of the cell's structure two additional problem are to be solved. Namely, it is known that, by proper

- Pairing poles and zeros, and
- Ordering the extraction of the cell.

One may improve several important properties of the final solution such as noise, linearity, and range of element values i.e. total silicon area [10–12].

A transfer function of the form

$$H(s) = A_0 \cdot \frac{\prod_{i=1}^{m}(s - z_i)}{\prod_{i=1}^{n}(s - p_i)} \tag{15.5}$$

which, for the sake of simplicity, in the case of n- and m-even, may be rewritten as

$$H(s) = \begin{cases} A_0 \cdot \dfrac{1}{\prod_{i=m/2+1}^{n/2} D_i} \cdot \prod\limits_{i=1}^{m/2} \frac{N_i}{D_i} & \text{for} \quad n > m \\ A_0 \cdot \prod\limits_{i=1}^{m/2} \frac{N_i}{D_i} & \text{for} \quad n = m \end{cases} \tag{15.6}$$

where $D_i = (s - p_i)(s - p_i*)$ represents a factor of the denominator related to a pair of conjugate poles while $N_i = (s - z_i)(s - z_i*)$ is the same for the numerator, may be realized in a large number of variants depending on the pairing of Ds and Ns into biquads and depending on the ordering of biquads so obtained.

According to [3], for example, in the case of $n = m = 8$, there are 18 possible combination to create biquads while for $n = m = 12$, one may create 1350 combinations. Of course, in the similar way rises the number of filter structures due to the ordering of the biquads in the cascade.

A procedure is implemented within the *RC_cascade* program of the *RM* software enabling pairing in order to get the optimal biquads and ordering of the biquads in order to get optimum from linearity and noise point of view which is based on the literature. Choices are made as to which final circuit should be the most appropriate for realization according to the following:

- Pairing the transfer function poles having highest imaginary part with the attenuation poles having minimal frequency;
- High-Q sections should be in the middle;
- First sections should be low-pass or band-pass, to suppress incoming high-frequency noise;
- All-pass sections should be near to the input;
- Last stages should be high-pass or band-pass to avoid output DC offset.

Let now address the issue of gain assignment. The idea is to distribute the gain of the cells so that the amplitudes of the signals do not force the operational amplifiers into saturation i.e. do not produce distortions for the maximum prescribed input signal.

To appreciate the problem Fig. 15.2 depicts the properties of the cells of a 16th order band-pass filter. Its zeros and poles are given in Table 15.1. As can be seen, there are two bandpass cells which realize the zeros at infinity and at the origin; two cells realizing the pairs of complex right half plane zeros; two high-pass notch cells realizing the ω-axis transmission zeros of the lower stopband; and two low- pass notch cells realizing the ω-axis transmission zeros of the upper stopband.

The gain being given in dB, all these add to produce the overall gain as depicted in Fig. 15.2e. All diagrams are obtained by SPICE simulation and the results partially post-processed for convenience. Since all the cells were chosen to have transfer functions of the form of (15.3) no cell is representing A_0 of (15.5).

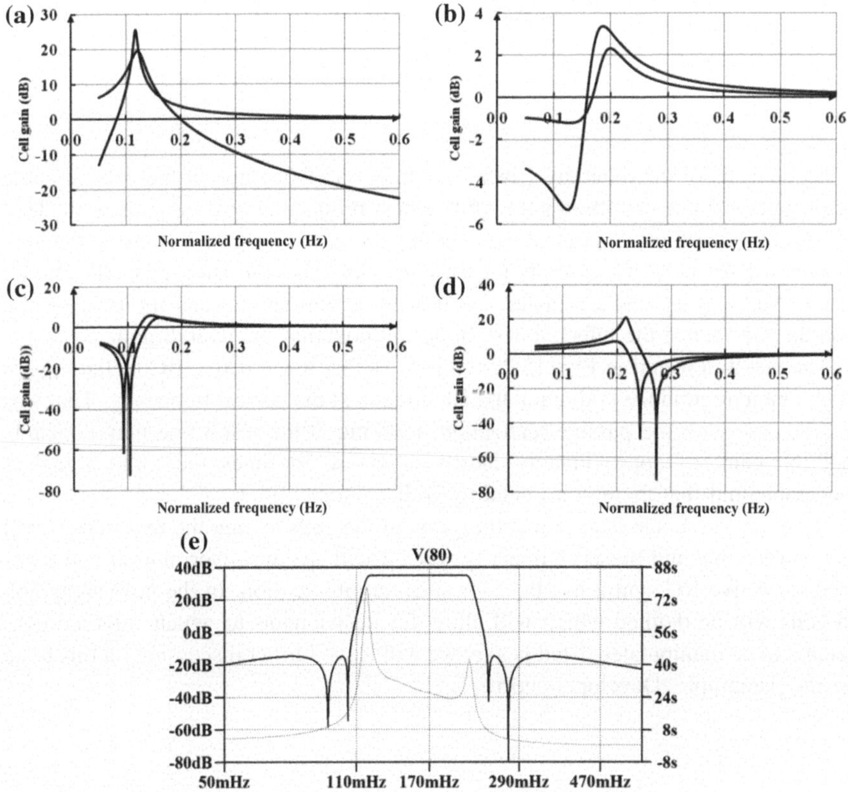

Fig. 15.2 The attenuation characteristic of a 16th order bandpass filter decomposed. **a** Two band-pass cells (zeros at the origin and at infinity), **b** Two cells realizing complex transmission zeros, **c** Two high-pass notch cells, **d** Two low-pass notch cells, and **e** The overall gain obtained by SPICE simulation

Table 15.1 Zeros and poles of the example filter

Zeros		Poles	
Real part	Imag. part	Real part	Imag. part
0.000000	1.521987	−0.036475	1.360313
0.000000	−0.657036	−0.019697	−0.734597
0.000000	−1.521987	−0.019697	0.734597
0.000000	0.657036	−0.036475	−1.360313
0.000000	1.712502	−0.119414	1.302120
−0.000000	−0.583941	−0.069842	−0.761574
0.000000	−1.712502	−0.069842	0.761574
0.000000	0.583941	−0.119414	−1.302120
0.165947	−0.880263	−0.171450	1.164513
0.206813	1.097036	−0.123747	−0.840509
0.206813	−1.097036	−0.123747	0.840509
0.165947	0.880263	−0.171450	−1.164513
0.000000	0.000000	−0.173698	1.053741
0.000000	0.000000	−0.152294	−0.923895
		−0.152294	0.923895
		−0.173698	−1.053741

By analysis of the diagrams given, one may find that some of the cells at some frequencies exhibit very high gain while others produce almost no gain at all. That leaves the opportunity to control the overall gain by reducing or increasing the gain of some of the cells by multiplying their transfer function using properly chosen factor. Multiplying cell's transfer function by a constant means multiplying the overall gain, hence the effectiveness in accommodating the overall gain.

The result depicted in Fig. 15.2e reminds us that when active RC technology is used, one may produce both gain and attenuation at the central frequency. That was not the case with the passive realization. Looking at the result one may conclude that the gain, of approximately 34 dB or about 50 times the input signal is acceptable and that there is no need to further interventions.

If not satisfied, one is to adjust the gains of the cells so that the required overall gain is obtained and preserve linearity. As we will see, however, not all cell types used for active RC synthesis allow for such an intervention. In the next paragraph all cells will be defined which will allow for decisions as to which cells are convenient to be manipulated. That is why we will leave further discussion on this issue for the paragraph "Developers corner".

15.3 Cells Used for Cascade Synthesis of Active RC Filters

In this paragraph complete set of cell will be listed and procedures for element value calculation will be given. The cells are chosen among alternatives from literature which, by the author's opinion, are the most appropriate ones. The reader is free to use alternatives described in the references already stated above.

In creating this paragraph we made rank for every cell based on

- Type of the function (low-, band-, high-, all-pass or notch),
- Order of the function (first or second),
- Sign of the gain (inverting or non-inverting).

The maximum number of transfer function parameters, as can be seen from (15.3), is 4. Most of the circuit discussed below will have larger number of circuit elements, however. Because of that, very frequently, the values of some (chosen by the designer) elements will be fixed in advance. So, solutions realizing the same transfer function and using the same cell may lead to different set of element values.

The frequency and element values discussed below will be normalized.

15.3.1 Low-Pass Cells

15.3.1.1 First Order Low-Pass Inverting Cell

The following transfer function is to be realized

$$H(s) = \frac{g}{s+a}. \tag{15.7}$$

For this purpose we use the circuit of Fig. 15.3a. After circuit analysis one gets:

$$H(s) = A(s) = -\frac{1}{C_1R_1} \cdot \frac{1}{s+1/(R_2C_1)} \tag{15.8}$$

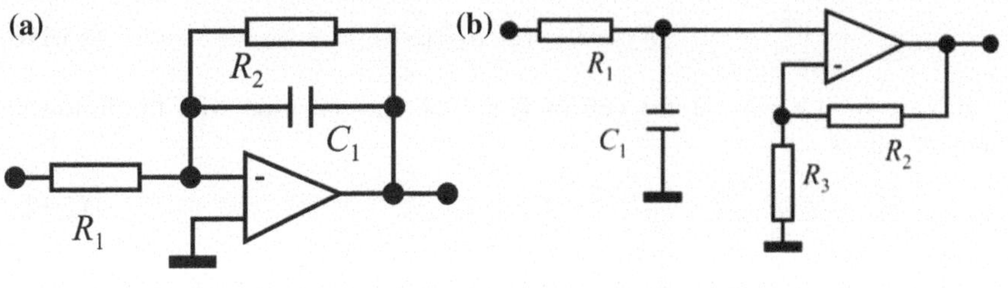

Fig. 15.3 The circuit realizing first order low-pass cell. **a** Inverting and **b** non-inverting

To design this cell one has to input: g and f_c and to adopt the value of C_1. Design equations are

$$g = -1/(C_1 R_1)$$
$$a = 2 \cdot \pi \cdot f_c = 1/(R_2 C_1) \tag{15.9a}$$

By choosing the capacitance to be fixed to unity one has the following element values:

$$C_1 = 1; \tag{15.9b}$$

$$R_1 = 1/(C_1 \cdot |g|); \tag{15.9c}$$

$$R_2 = 1/(C_1 \cdot a); \tag{15.9d}$$

15.3.1.2 First Order Low-Pass Non-inverting Cell

Here we use again (15.6) being aware that $g > 0$.

When $g/a > 1$ for this purpose we use the circuit of Fig. 15.3b. After circuit analysis one gets

$$H(s) = \left(1 + \frac{R_2}{R_3}\right) \cdot \frac{1}{R_1 C_1} \cdot \frac{1}{s + \frac{1}{R_1 C_1}}. \tag{15.10}$$

To design this cell one has to input: g and f_c and to adopt the values of C_1 and R_3. The design equations are

$$g = \left(1 + \frac{R_2}{R_3}\right) \cdot \frac{1}{R_1 C_1}$$
$$a = 2 \cdot \pi \cdot f_c = 1/(R_1 C_1) \tag{15.11a}$$

Accordingly we first choose $C_1 = 1$ and $R_3 = 1$ which allows for

$$R_1 = \frac{1}{a \cdot C_1} \tag{15.11b}$$

$$R_2 = R_3 \cdot (g \cdot R_1 C_1 - 1). \tag{15.11c}$$

In case of $0 < g/a < 1$ the cell of Fig. 15.4 may be used with the following element values

$$C_1 = 1; \tag{15.12a}$$

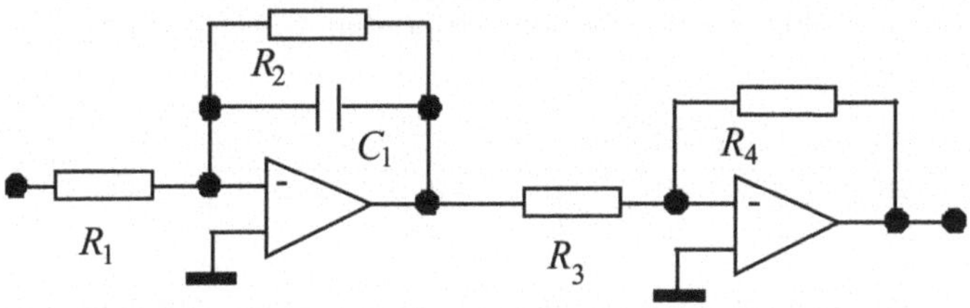

Fig. 15.4 Low-pass non-inverting cell with $0 < g < 1$

$$R_1 = 1/(C_1 \cdot |g|); \tag{15.12b}$$

$$R_2 = 1/(C_1 \cdot a); \tag{15.12c}$$

$$R_3 = 1.0; \tag{15.12d}$$

$$R_4 = 1.0; \tag{15.12e}$$

15.3.1.3 Second Order Low-Pass, Low-Q Cell

The function is

$$H(s) = \frac{g}{b + a \cdot s + s^2}. \tag{15.13}$$

The circuit realizing this function with negative gain ($g < 0$) is depicted in Fig. 15.5. Since it uses single operational amplifier it is referred to as single-amplifier biquad (SAB).

After circuit analysis one gets

$$H(s) = \frac{-\frac{1}{C_1C_2R_1R_2}}{s^2 + s\frac{1}{C_1}\left(\frac{1}{R_1} + \frac{1}{R_2} + \frac{1}{R_3}\right) + \frac{1}{C_1C_2R_3R_2}}. \tag{15.14}$$

Fig. 15.5 Multiple feedback second order low-pass low-Q cell

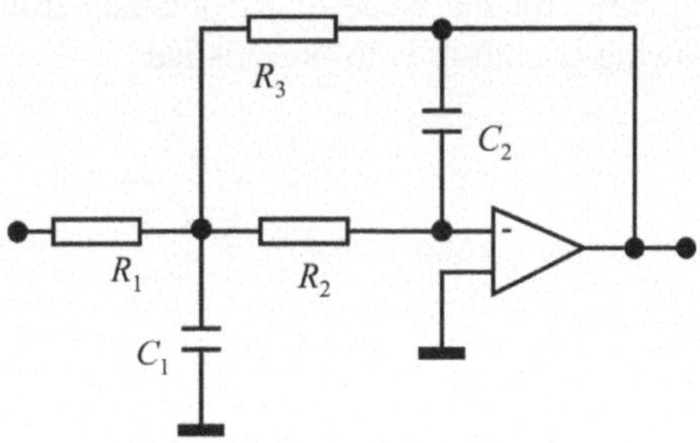

The Q-factor of this cell may be found according to (15.4) to be

$$Q = \frac{\sqrt{b}}{a} = \frac{\sqrt{\frac{1}{C_1 C_2 R_3 R_2}}}{\frac{1}{C_1}\left(\frac{1}{R_1} + \frac{1}{R_2} + \frac{1}{R_3}\right)} = \sqrt{\frac{R_1^2 R_2 R_3 C_1/C2}{(R_1 R_2 + R_1 R_3 + R_2 R_3)^2}}. \tag{15.15}$$

If we suppose that all resistors are equal and the same for the capacitors, the Q-factor will be $Q = 1/3$, which places the circuit in the category of low Q.

Design equations are

$$b = 1/(C_1 \cdot C_2 \cdot R_2 \cdot R_3) \tag{15.16a}$$

$$a = (1/R_1 + 1/R_2 + 1/R_3)/C_1 \tag{15.16b}$$

$$g = -1/(C_1 \cdot C_2 \cdot R_1 \cdot R_2). \tag{15.16c}$$

There are three degrees of freedom and five circuit elements which means that two must be adopted.

First the value of R_1 will be extracted to be

$$2 \cdot R_1 = -\frac{a}{g \cdot C_2} \pm \sqrt{\frac{a^2}{(g \cdot C_2)^2} - 4 \cdot \frac{b - g}{g^2 C_1 C_2}}. \tag{15.17a}$$

Then, combining (15.16a) and (15.16c) one gets

$$C_1 C_2 R_3 \cdot \left(-\frac{1}{C_1 C_2 R_1 g}\right) = \frac{1}{b} \tag{15.17b}$$

which results into

$$R_3 = -R_1 \cdot g/b. \tag{15.17c}$$

Further, from (15.16c) one gets:

$$R_2 = -1/(C_1 C_2 R_1 g). \tag{15.17d}$$

Now, for the value of R_1 obtained from (15.17a) to be a real number the following condition is to be satisfied

$$\frac{a^2}{(g \cdot C_2)^2} - 4 \cdot \frac{b - g}{g^2 C_1 C_2} > 0. \tag{15.17e}$$

For that reason one must choose

$$C_1 > 4 \cdot C_2 \cdot (b - g)/a^2. \tag{15.17f}$$

If twice as big a value of C_1 is chosen than the right hand side of this expression is, after adopting the value of C_2, one will have

$$C_1 = 8 \cdot C_2 \cdot (b - g)/a^2. \tag{15.17g}$$

If, however, C_1 was adopted first, following the same consideration, C_2 is to be calculated from

$$C_2 = \frac{C_1}{8 \cdot (b - g)/a^2}. \tag{15.17h}$$

The program *RC_cascade* of the *RM* software adopts $C_1 = 1$.

We will proceed now with another cell which may be successfully used for the same purpose as the one depicted in Fig. 15.5. It is the Tow-Thomas cell which uses three operational amplifiers and is popular due its small sensitivity to parasitics which is very important when integrated CMOS active RC filters are produced. The schematic of the cell is depicted in Fig. 15.6 and its transfer function obtained after circuit analysis is

$$H(s) = -\frac{\frac{1}{C_1C_2R_1R_2}}{s^2 + s\frac{1}{C_1R_3} + \frac{1}{C_1C_2R_4R_2}} \tag{15.18}$$

If adopted: $C_1 = C_2 = C$ and $R_1 = R_2 = R$, the design equations will be

$$\begin{aligned} g &= 1/\left(R^2 \cdot C^2\right) \\ a &= 1/(R_3 \cdot C) \\ b &= 1/\left(R \cdot R_4 \cdot C^2\right). \end{aligned} \tag{15.19}$$

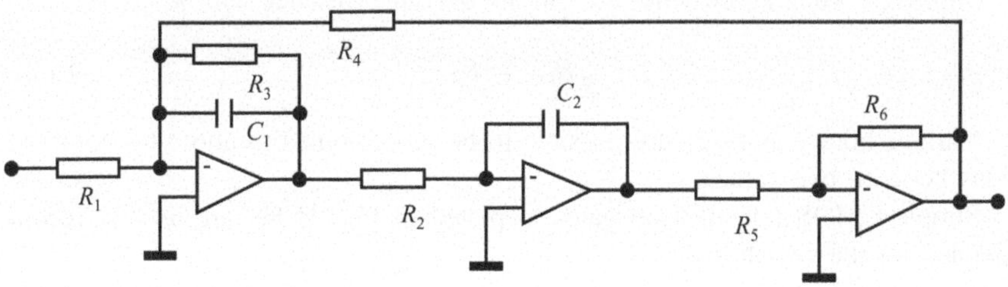

Fig. 15.6 A version of Tow-Thomas biquad implementing low-Q low-pass filter

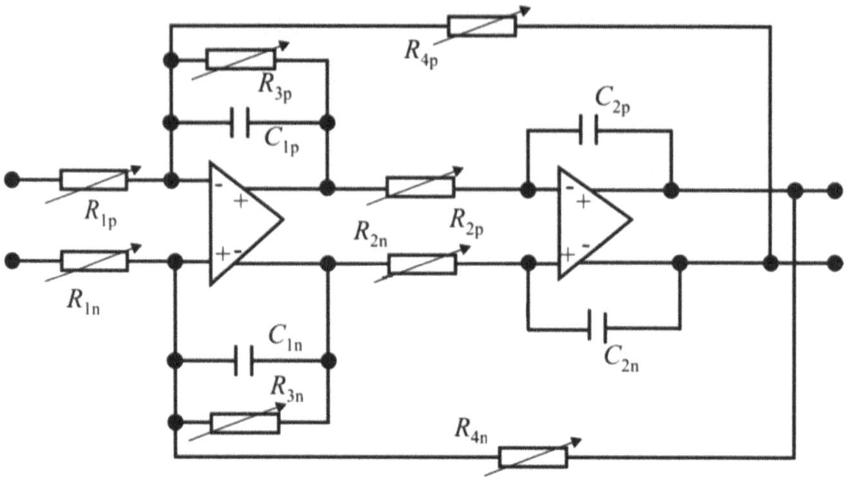

Fig. 15.7 A symmetric variant of the Tow-Thomas low-pass cell

Accordingly for the element values one has: $C = 1$; $R = 1/(C \cdot \sqrt{|g|}$; $R_1 = R$; $R_2 = R$; $R_3 = 1/(a \cdot C)$; $R_4 = 1/(R \cdot C^2 \cdot b)$; $C_1 = C$; $C_2 = C$; $R_5 = 1$; $R_6 = 1$.

A symmetric variant for the Tow-Thomas low-pass cell, convenient for integration, is depicted in Fig. 15.7. Since it manipulates the signal in a differential mode one may expect full elimination of the effects of the temperature drift of the DC regime within the operational amplifiers. Note, the last inverter of Fig. 15.6 is omitted since the inversion is obtained by interchanging the output signals of the second operational amplifier.

Arrows are drawn over the resistors to mark that these are adjustable. The way how adjustments work is depicted in Fig. 15.8a. Here 5 resistors are connected in series. Their values decrease by a factor of two. Four of the resistors have switches in parallel and if all switches are open the overall resistance would be

$$R_{\max} = R_0 \cdot \sum_{i=1}^{5} 2^{-(i-1)}. \tag{15.20}$$

Similarly, when all switches are on, the overall resistance will be

$$R_{\min} = R_0. \tag{15.21}$$

Various binary numbers are possible to be programmed to allow for combinations between two values.

Simple CMOS transmission gates, depicted in Fig. 15.8b, are usually recommended for the switches.

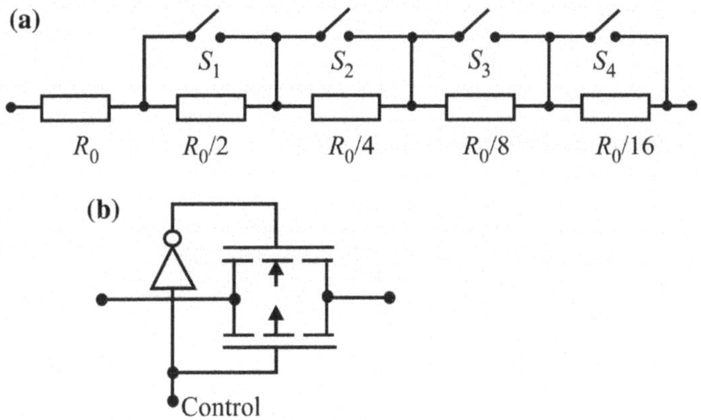

Fig. 15.8 The adjustable integrated CMOS resistor. **a** the resistor and **b** The switch

15.3.1.4 Second Order Low-Pass, High-Q Positive Gain Cell

The schematic of the SAB is depicted in Fig. 15.9. As can be seen the operational amplifier has its own negative feedback which stretches its frequency response and makes the cell applicable for higher frequencies than usually.

The function is again

$$H(s) = \frac{g}{b + a \cdot s + s^2} \tag{15.22}$$

with $g > 0$.

After circuit analysis one gets

$$H(s) = \frac{H \cdot \frac{1}{C_1 C_2 R_1 R_2}}{s^2 + s\left[\frac{1}{C_1}\left(\frac{1}{R_1} + \frac{1}{R_2}\right) + \frac{(1-H)}{R_2 C_2}\right] + \frac{1}{C_1 C_2 R_1 R_2}} \tag{15.23a}$$

Fig. 15.9 Sallen and Key low-pass high-Q cell

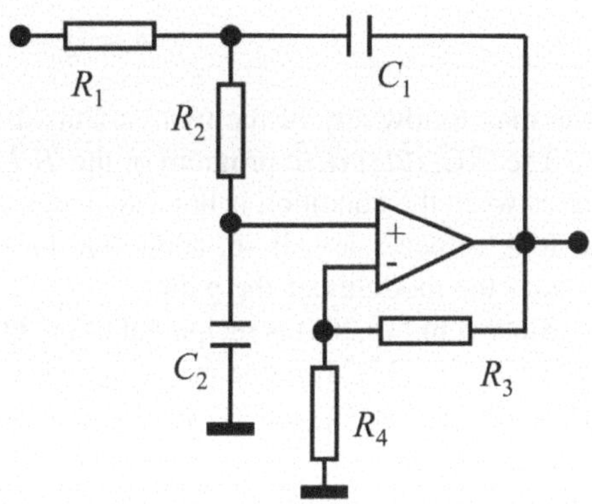

where

$$H = 1 + R_3/R_4. \tag{15.23b}$$

The Q-factor of this cell may be obtained as

$$Q = \frac{\sqrt{b}}{a} = \frac{\sqrt{\frac{1}{C_1 C_2 R_1 R_2}}}{\left[\frac{1}{C_1}\left(\frac{1}{R_1} + \frac{1}{R_2}\right) + \frac{(1-H)}{R_2 C_2}\right]}. \tag{15.24a}$$

Putting $R_1 = R_2$ and $C_1 = C_2$ it simplifies into

$$Q = \frac{1}{(3-H)}. \tag{15.24b}$$

Now, since $H > 1$, by proper choice of H one may produce very high Q. The design equations are

$$H \cdot \frac{1}{C_1 C_2 R_1 R_2} = g \tag{15.25a}$$

$$\frac{1}{C_1 C_2 R_1 R_2} = b \tag{15.25a}$$

and

$$\frac{1}{C_1}\left(\frac{1}{R_1} + \frac{1}{R_2}\right) + \frac{(1-H)}{R_2 C_2} = a \tag{15.25a}$$

To design, after use of

$$\alpha = a/\sqrt{b} \tag{15.26a}$$

and

$$m = H - 1 + \alpha^2/4 \tag{15.26b}$$

one may follow any of the two variants depicted in Table 15.2.

The *RC_cascade* program of the *RM* software implements Variant A and in cases when the condition is not satisfied, it creates a loop in which g is divided by 2 as long as necessary. If so, additional inverter is added in cascade to the SAB to restore the modulus of the gain.

Since α in (15.26a) is reciprocal to the Q-factor and H has limited value which is suggested by (17.24b), one may conclude that the value of m is not large a number. This means that C_1 and C_2 in the Variant A do not differ significantly. Similar

Table 15.2 Design equation for the high-Q low-pass SAB

Variant A	Variant B
If values for C_1 and R_3 are adopted the design equations become $k = 2 \cdot \pi \cdot f_c C_1$ (15.27a) $R_4 = R_3/(H - 1)$ (15.27b) $C_2 = m \cdot C_1$ (15.27c) $R_1 = 2/(k \cdot \alpha)$ (15.27d) $R_2 = \alpha/(2 \cdot m \cdot k)$ (15.27e) The condition is $H > 1\text{-}\alpha^2/4$	If one adopts $C_1 = C_2$ (15.28a) one gets $R_1 = 1/(b \cdot C_1 C_2 R_2)$ (15.28b) and $R_2 = \sqrt{\frac{H \cdot C_1 - C_1 - C_2}{b \cdot C_1 C_2^2}}$ (15.28b) The condition is $H > 2$.

conclusion may be developed for the resistors R_1 and R_2 meaning that, in this SAB, the spread of the element values is small which is an advantage from the physical realization point of view.

On the other side the value of H being part of the definition of the real part of the pole makes this cell sensitive to great extent to its value.

15.3.2 *High-Pass Cells*

15.3.2.1 First Order High-Pass Inverting Cell

The transfer function to be realized is

$$H(s) = \frac{g \cdot s}{a + s} \tag{15.29}$$

For this purpose we use the circuit of Fig. 15.10.

After circuit analysis one gets

$$H(s) = -\frac{\frac{R_2}{R_1} \cdot s}{s + \frac{1}{R_1 \cdot C_1}} \tag{15.30}$$

The design equations are

$$g = -R_2/R_1 \tag{15.31a}$$

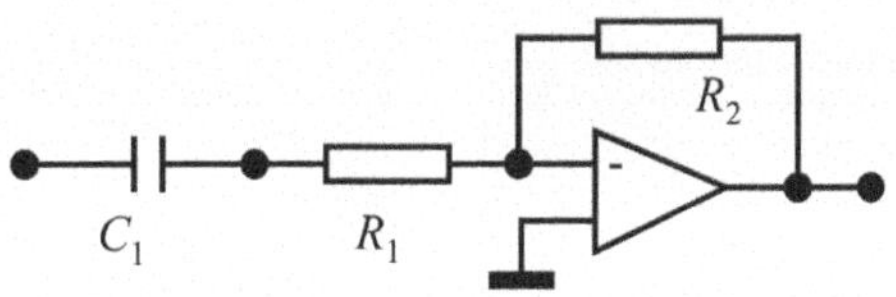

Fig. 15.10 First order inverting high-pass cell

and

$$a = 1/(R_1 C_1) \tag{15.31b}$$

Given a and g, one may adopt $C_1 = 1$ and find:

$$R_1 = 1/(a \cdot C_1) \tag{15.32a}$$

$$R_2 = -g \cdot R_1. \tag{15.32b}$$

15.3.2.2 First Order High-Pass Non-inverting Cell

For this purpose we use the circuit of Fig. 15.11.

After circuit analysis one gets:

$$H(s) = \frac{\left(1 + \frac{R_2}{R_3}\right) \cdot s}{s + \frac{1}{R_1 \cdot C_1}} \tag{15.33}$$

The design equations are

$$g = 1 + \frac{R_2}{R_3} \tag{15.34a}$$

and

$$a = \frac{1}{R_1 \cdot C_1}. \tag{15.34b}$$

Given a and g, choose $C_1 = 1$ and $R_3 = 1$ to obtain:

$$R_1 = 1/(a \cdot C_1); \tag{15.35a}$$

$$R_2 = (g - 1) \cdot R_3; \tag{15.35b}$$

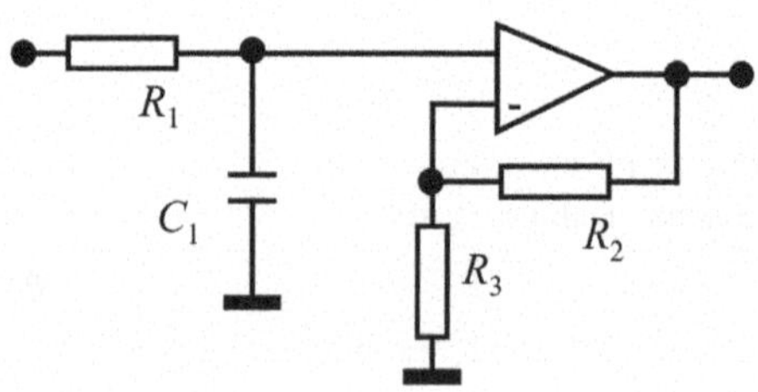

Fig. 15.11 High-pass, First order non-inverting cell

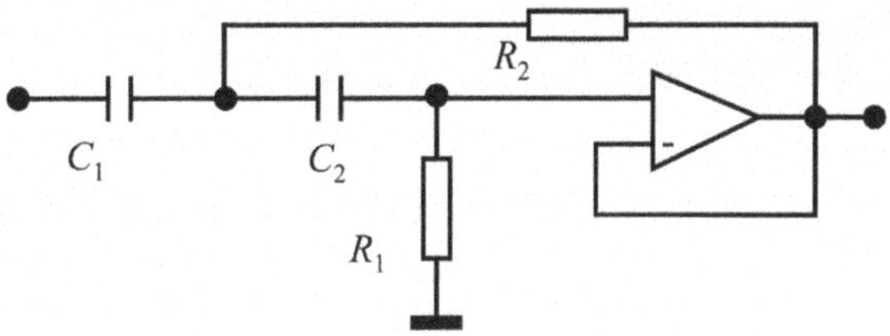

Fig. 15.12 Unity gain Sallen-Key high-pass low-Q filter cell

15.3.2.3 Second Order High-Pass Non-inverting Low-Q Cell

The transfer function to be realized is

$$H(s) = \frac{g \cdot s^2}{s^2 + a \cdot s + b}. \tag{15.36}$$

For this purpose we use the circuit of Fig. 15.12.
After circuit analysis one gets

$$H(s) = \frac{s^2}{s^2 + 2/(R_1 \cdot C) \cdot s + 1/(R_1 \cdot R_2 \cdot C^2)} \tag{15.37}$$

Design equations are:

$$a = 2/(R_1 C) \tag{15.38a}$$

$$b = 1/(R_1 R_2 C^2). \tag{15.38b}$$

Note this cell insists that $g - 1$. For a different value it will be needed an additional amplifier to be added to the cascade whose gain and topology will depend on value and the sign of the original g.

The element values are

$$C = 1.0; \tag{15.39a}$$

$$R_1 = 2/(a \cdot C); \tag{15.39b}$$

$$R_2 = 1/(b \cdot R_1 \cdot C^2); \tag{15.39c}$$

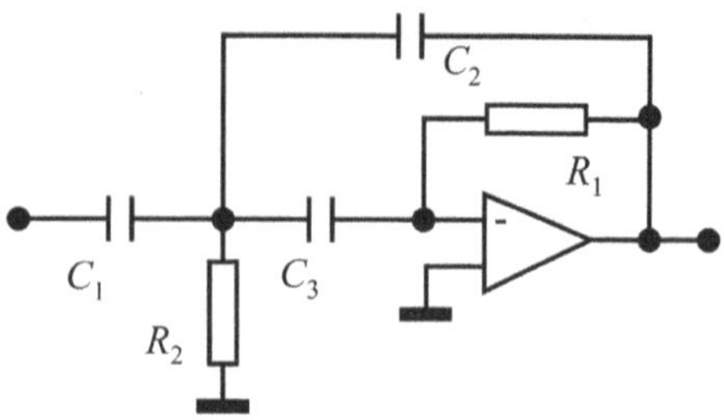

Fig. 15.13 Second order multiple feedback high-Q high-pass cell

15.3.2.4 Second Order High-Pass Inverting High-Q Cell

For this purpose we use the circuit of Fig. 15.13.

After circuit analysis one gets

$$H(s) = \frac{-s^2 \cdot (C/C_2)}{s^2 + s\frac{2 \cdot C + C_2}{C \cdot C_2 R_1} + \frac{1}{C \cdot C_2 R_1 R_2}} \quad (15.40)$$

The design equations are

$$a = \frac{2 \cdot C + C_2}{C \cdot C_2 R_1} \quad (15.41a)$$

$$b = \frac{1}{C \cdot C_2 R_1 R_2} \quad (15.41b)$$

$$g = -C/C_2 \quad (15.41c)$$

After adopting $C = 1$ for the element values one has

$$C_2 = -C/g; \quad (15.42a)$$

$$R_1 = (2/C_2 + 1/C)/a; \quad (15.42b)$$

$$R_2 = 1/(b \cdot C \cdot C_2.R_1); \quad (15.42c)$$

15.3.3 *Band-Pass Cells*

The transfer function is

$$H(s) = \frac{s \cdot g}{s^2 + as + b}. \quad (15.43)$$

15.3.3.1 Band-Pass Low-Q Non-inverting Cell

For this purpose we use the circuit of Fig. 15.14.

After circuit analysis one may obtain

$$H(s)\frac{\frac{1+K}{R_1 \cdot C_2}\cdot s}{s^2+s\left[\frac{1}{R_2}\left(\frac{1}{C_1}+\frac{1}{C_2}\right)-\frac{K}{R_1 C_2}\right]+\frac{1}{C_1 C_2 R_1 R_2}} \tag{15.44}$$

The design equations are

$$a=(C_1+C_2)/(R_2 \cdot C_1 \cdot C_2)-K/(R_1 \cdot C_2) \tag{15.45a}$$

$$b=1/(R_1 \cdot R_2 \cdot C_1 \cdot C_2) \tag{15.45b}$$

$$g=(1+K)/(R_1 \cdot C_2) \tag{15.45c}$$

$$K=R_{\mathrm{a}}/R_{\mathrm{b}}. \tag{15.45d}$$

We first compute $Q=\frac{\sqrt{b}}{a}$; $t=1.5/(4 \cdot Q^2)$; $n=Q \cdot (q+1)\sqrt{t/q}-1$ with q = 1; and $K=(n/q)\cdot\sqrt{t/q}$.

Then we adopt the value of C = 1 and R_{a} = 1.

The element values are calculated from

$$C_1=C_2=C, \tag{15.46a}$$

$$R_1=(1+K)/(g \cdot C_2), \tag{15.46b}$$

$$R_2=\frac{1}{C_1}\cdot\sqrt{\frac{q}{t \cdot b}}, \tag{15.46c}$$

and

$$R_{\mathrm{b}}=R_{\mathrm{a}}/K. \tag{15.46d}$$

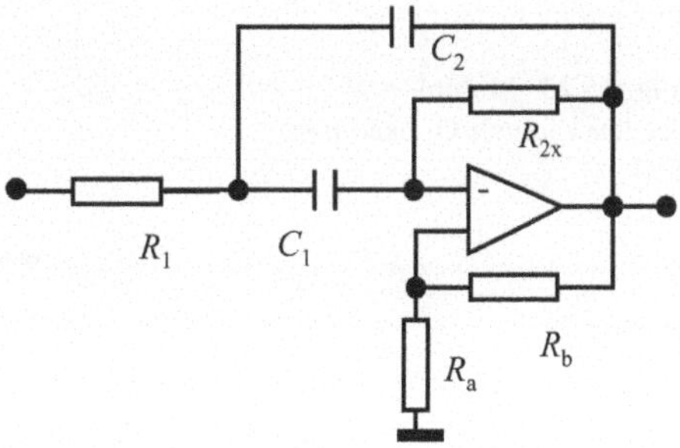

Fig. 15.14 The band-pass low-Q non-inverting cell

15.3.3.2 Band-Pass High-Q Inverting Cell

For this purpose we use the circuit of Fig. 15.15.

After circuit analysis one gets

$$H(s) = \frac{-\frac{1}{R_1 \cdot C} \cdot s}{s^2 + \frac{2}{R_2 C} s + \frac{R_1 + R_3}{R_1 R_2 R_3 C^2}} \tag{15.47}$$

which leads to the following design equations:

$$a = 2/(R_2 \cdot C); \tag{15.48a}$$

$$b = (R_1 + R_3)/(R_1 \cdot R_2 \cdot R_3 \cdot C^2); \tag{15.48b}$$

$$g = 1/(R_1 \cdot C); \tag{15.48c}$$

After adopting $C = 1$ and compute:

$$R_2 = 2/(C \cdot a). \tag{15.49}$$

If

$$(b/(a \cdot |g|) - 1 < 0 \tag{15.50}$$

a loop is formed in which | g | is halved until this condition is not valid. A warning is issued by the *RC_cascade* program informing how many times the operation took place (how many times g was divided by 2).

The rest of the elements (with corrected value of g, if necessary) are obtained from

$$R_1 = 1/(C \cdot |g|); \tag{15.51}$$

$$R_3 = R_1/((\mathrm{b}/(\mathrm{a} \cdot |g|) - 1). \qquad (15.51)$$

If g was changed it will be needed to add to the cascade an additional amplifier whose gain and topology will depend on value and sign of the original g.

Fig. 15.15 Multiple feedbacks, high Q, band-pass cell

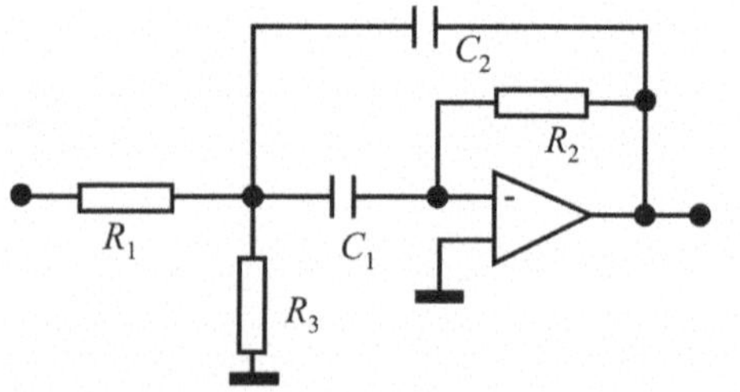

15.3.4 *All-Pass Cells*

15.3.4.1 First Order All-Pass Cell

Variant A: Zero at the Right Half-Plane

The transfer function to be synthesized is

$$H(s) = \frac{s-b}{s+a}. \tag{15.52}$$

The case $b > 0$ will be considered only while discussion on the case of $b < 0$ will be postponed. This means a right-half plane zero will be realized. For all-pass solution one needs $b = a$.

For this purpose we use the circuit of Fig. 15.16.

After circuit analysis one gets

$$T(s) = \frac{s - R_2/(R_1 R_3 \cdot C_1)}{s + 1/(R_3 \cdot C_1)} \tag{15.53a}$$

To create a first order all-pass cell we choose $R_1 = R_2 = R_3 = R$ and $C_1 = C$. The resulting transfer function is.

$$T(s) = \frac{s - 1/(R \cdot C)}{s + 1/(R \cdot C)}. \tag{15.53b}$$

Design equation is

$$a = 1/(R \cdot C) \tag{15.54a}$$

After $C = 1.0$ is adopted one gets

$$R = 1/(a \cdot C). \tag{15.54b}$$

We will use the opportunity of the existing context to consider the case when right-half plane zero at the real axis ($b \neq a$) is created. This happens in odd order

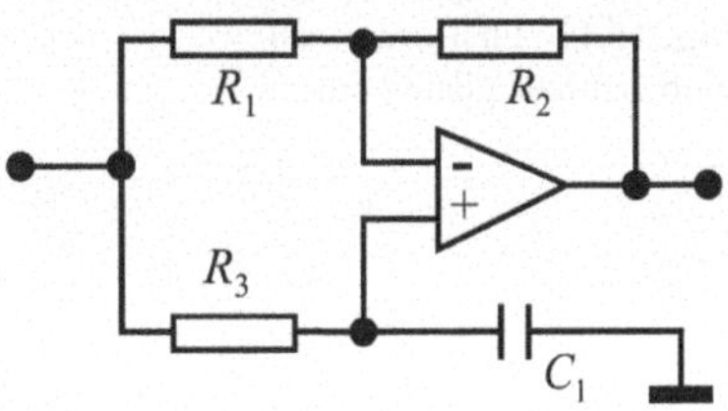

Fig. 15.16 First order all-pass cell with right-half plane zero

filters. In such a case, if we choose $R_1 = R_3 = R$ and $C_1 = C$, for the element values will have

$$R_3 = 1/(a \cdot C) \tag{15.54c}$$

and

$$R_2 = b/\left(Ca^2\right). \tag{15.54d}$$

Variant B: Zero at the Left Half-Plane

The transfer function to be synthesized is

$$H(s) = \frac{s-b}{s+a}. \tag{15.56}$$

Here again $a > 0$ but in order the zero to be transferred to the left half plane $b < 0$ is needed. Now, if

$$b = -a \tag{15.57a}$$

cancellation occurs so that it makes no sense to have an all-pas with pole and zero overlapping.

In the next we will, again, use the opportunity of the existing context to consider the case when

$$b \neq -a \tag{15.57b}$$

which allows for a zero in the left half plane (being different from any real pole) to be realized. These will not produce an all-pass, of course. Odd order filters are considered only.

For this purpose we use the circuit of Fig. 15.17.

After circuit analysis one gets

$$H(s) = \frac{s + R_2/(R_1R_3 \cdot C)}{s + 1/(R_3 \cdot C)}. \tag{15.58}$$

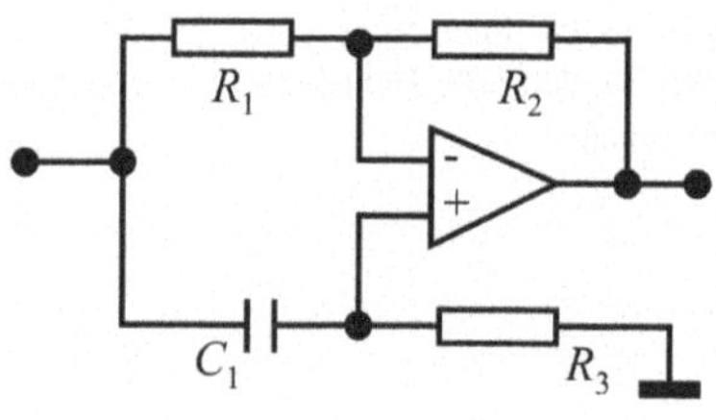

Fig. 15.17 First order cell with left-half plane zero

The condition (15.57b) now becomes

$$R_1 \neq R_2. \tag{15.59}$$

A wide range of positions of the left-half plane zeros is possible to be produced by changing the ratio R_1/R_2.

The design equations are

$$a = 1/(R_3 C_1) \tag{15.60a}$$

and

$$-b = R_2/(R_1 R_3 C_1). \tag{15.60b}$$

If $R_3 = 1$ is adopted we get

$$C_1 = 1/(a \cdot R_3) = 1/a \tag{15.61a}$$

and

$$\frac{R_2}{R_1} = -b \cdot R_3 \cdot C_1 = -b/a. \tag{15.61b}$$

15.3.4.2 Second Order All-Pass Cell

The transfer function is

$$H(s) = \frac{s^2 + s \cdot g + q}{s^2 + s \cdot a + b}, \tag{15.62}$$

with $q = b$ and $g = -a$.

For this purpose we use the circuit of Fig. 15.18.

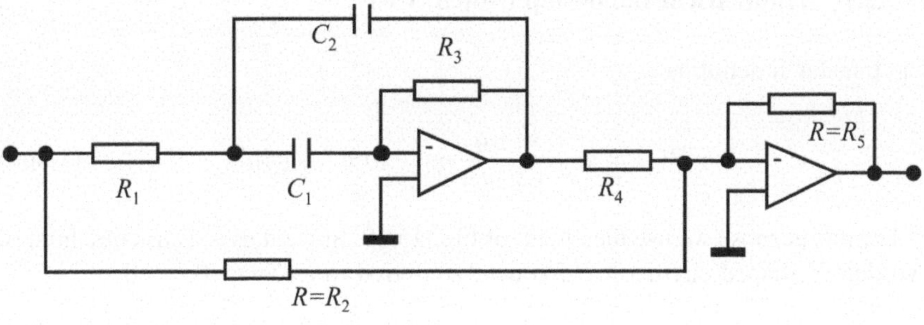

Fig. 15.18 Second order all-pass filter cell

After circuit analysis one gets

$$H(s) = \frac{s^2 + s \cdot \frac{(2 \cdot R_1 - R \cdot R_3/R_4)}{R_1 R_3 C} + \frac{1}{R_1 R_3 C^2}}{s^2 + s \cdot \frac{2}{R_3 C} + \frac{1}{R_1 R_3 C^2}} \tag{15.63}$$

Design equations are:

$$a = 2/(R_3 \cdot C), \tag{15.64a}$$

$$b = 1/(R_1 \cdot R_3 \cdot C^2), \tag{15.64b}$$

$$g = (2 \cdot R_1 - R \cdot R_3/R_4)/(R_1 \cdot R_3 \cdot C), \tag{15.64c}$$

and

$$q = b. \tag{15.64d}$$

If $C_1 = C_2 = C = 1$ and $R_2 = R_5 = R = 1$ is adopted, the following element values will be obtained

$$R_3 = 2/(a \cdot C); \tag{15.65a}$$

$$R_1 = 1/(b \cdot R_3 \cdot C^2) = a/(2 \cdot b \cdot C); \tag{15.65b}$$

$$R_4 = R \cdot R_3/(2 \cdot R_1 + a \cdot R_1 \cdot R_3 \cdot C). \tag{15.65c}$$

15.3.5 Band-Stop Cells

Only second order cells may be created by the nature of the problem. Two variants will be considered the symmetrical and asymmetrical one.

15.3.5.1 Symmetrical Band-Stop (Notch) Cell

The transfer function is

$$H(s) = g\frac{s^2 + q}{s^2 + s \cdot a + b}, \quad \text{with} \quad b = q! \tag{15.66}$$

For this purpose we use the circuit of Fig. 15.19. Since it uses a structure built of two dual T-shaped circuits it is frequently referred to as Twin-Tee cell.

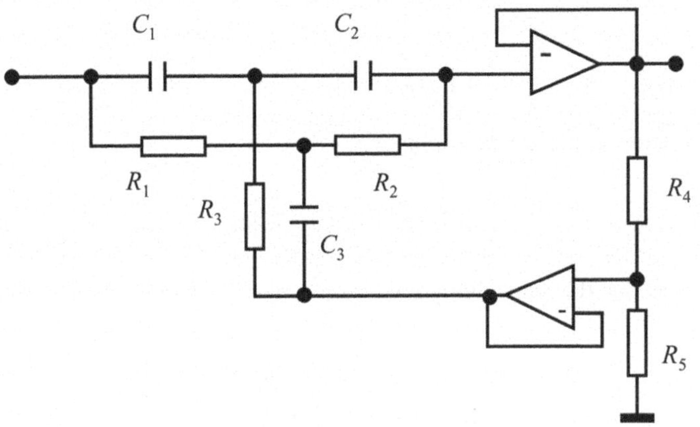

Fig. 15.19 A band-stop (notch) Twin-Tee cell

After circuit analysis with $R = R_1 = R_2 = 2 \cdot R_3$ and $C = C_1 = C_2 = C_3/2$ [13], one gets:

$$H(s) = \frac{s^2 + \frac{1}{R^2C^2}}{s^2 + 4 \cdot s \cdot \frac{1-K}{R \cdot C} + \frac{1}{R^2C^2}} \tag{15.67}$$

where $K = R_5/(R_4 + R_5)$.

Design equations are:

$$b = 1/(R^2 \cdot C^2); \tag{15.68a}$$

$$q = b; \tag{15.68b}$$

$$a = 4 \cdot (1 - K)/(R \cdot C); \tag{15.68c}$$

$$g = 1. \tag{15.68d}$$

The incomplete list of element values is

$$C = 1 \tag{15.69a}$$

$$R = \frac{1}{C \cdot \sqrt{b}} \tag{15.69b}$$

$$R_1 = R_2 = 2 \cdot R_3 = R \tag{15.69c}$$

and

$$C_1 = C_2 = C_3/2 = C. \tag{15.69d}$$

To complete the list we first find

$$K = 1 - a/\left(4 \cdot \sqrt{b}\right) \tag{15.70}$$

This means that the realizability condition is $K < 1$ which is always satisfied. We choose

$$R_4 = (1 - K) \cdot R_1 \tag{15.71a}$$

and

$$R_5 = K \cdot R_1 \tag{15.71b}$$

In case of violation of the realizability condition the program *RC_cascade* skips the cell and issues a warning.

15.3.5.2 Asymmetrical Band-Stop (Notch) Cell

The transfer function is

$$H(s) = g\frac{s^2 + q}{s^2 + s \cdot a + b}. \tag{15.72}$$

For this purpose we use the circuit of Fig. 15.20. After circuit analysis one has

$$H(s) = -\frac{C_1}{C} \cdot \frac{s^2 + \frac{1}{CC_1RR_2}}{s^2 + s \cdot \frac{1}{CR_4} + \frac{1}{C^2R^2}} \tag{15.73}$$

Design equations are

$$g = -C_1/C \tag{15.74a}$$

$$q = \frac{1}{CC_1RR2} \tag{15.74b}$$

$$a = \frac{1}{CR_4} \tag{15.74c}$$

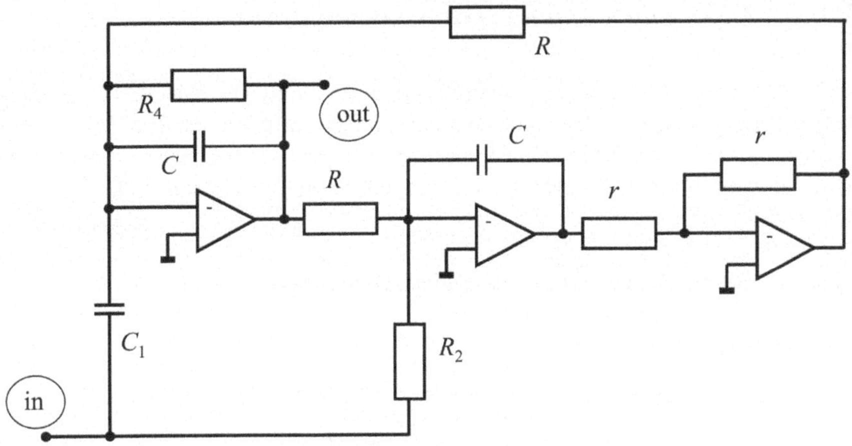

Fig. 15.20 The version of Tow-Thomas cell used for realization of a notch function

and

$$b = \frac{1}{C^2 R^2}. \tag{15.74d}$$

Having in mind that

$$Q_p = \frac{\sqrt{\sigma_0^2 + \omega_0^2}}{-2\sigma_0} = \frac{\sqrt{b}}{a} \tag{15.74d}$$

where σ_0 and ω_0 are the coordinates of the pole, the following parameter values may be obtained

$$C = \text{arbitrary e.g. } C = 1, \tag{15.75a}$$

$$C_1 = |g| \cdot C \tag{15.75b}$$

$$R = r = \frac{1}{C \cdot \sqrt{b}} \tag{15.75c}$$

$$R_4 = R \cdot Q_p \tag{15.75d}$$

$$R_2 = \frac{R}{-g} \cdot \frac{b}{q} \tag{15.75e}$$

15.3.6 Cells with Complex Transmission Zeros

In this paragraph second order cell will be considered only since the first order ones were elaborated earlier in sections "Variant A: Zero at the Right Half-Plane" and "Variant B: Zero at the Left Half-Plane". In addition, second order all-pass cells (since already introduced in Sect. 15.3.3.2) will be not considered, too.

15.3.6.1 Second Order Cell with a First Order Zero

The transfer function is

$$H(s) = g\frac{s+q}{s^2+as+b}. \tag{15.76}$$

For this purpose we use the circuit of Fig. 15.21.
After circuit analysis one gets

$$H(s) = -\frac{1}{R_4C_1}\cdot\frac{s+\frac{R_8R_5-R_6R_2}{R_2R_3R_5C_2}R_4}{s^2+\frac{1}{R_1C_1}s+\frac{R8}{R_2R_3R_7C_1C_2}} \tag{15.77}$$

Design equations are

$$a = 1/(R_1C_1) \tag{15.78a}$$

$$b = R_8/(R_2R_3R_7C_1C_2) \tag{15.78b}$$

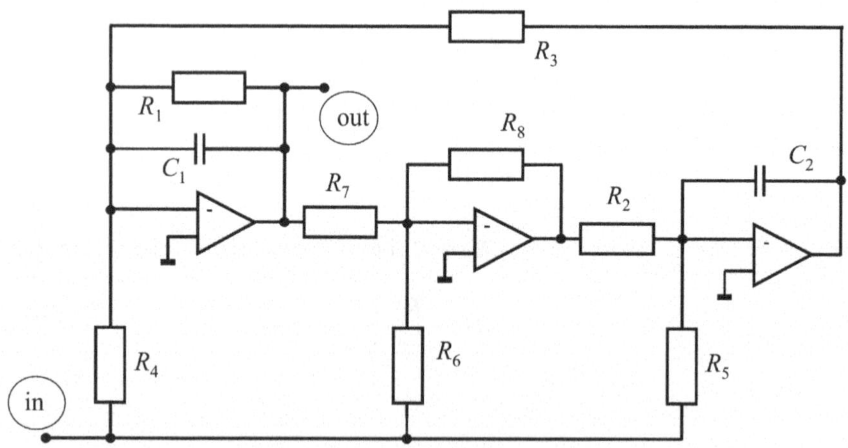

Fig. 15.21 The Tow-Thomas biquad cell realizing zero at the real axis

$$q = \frac{R_4(R_5R_8 - R_2R_6)}{R_2R_3R_5R_6C_2} \tag{15.78c}$$

and

$$g = -1/(C_1R_4). \tag{15.78d}$$

First we adopt the values of $C_1 = C_2 = C = 1$. Then, we adopt the values of $R_2 = R_3 = R_7 = R = 1$. With that set, one may calculate

$$R_1 = 1/(a \cdot C) \tag{15.79a}$$

$$R_4 = 1/(|g| \cdot C) \tag{15.79b}$$

$$R_8 = b \cdot C^2R^3 \tag{15.79c}$$

and

$$\begin{array}{ll} R_5 = R \text{ and } R_6 = R_8/[R \cdot (q \cdot R \cdot C/R_4 + 1/R_5)] & \text{for} \quad q > 0 \\ R_6 = R \text{ and } R_5 = 1/[R_8/(R \cdot R_6) - q \cdot R \cdot C/R_4] & \text{for} \quad q < 0. \\ R_5 = R \text{ and } R_6 = R_5 \cdot R_8/R_2 \quad \text{for} \quad q = 0 & \end{array} \tag{15.79d}$$

Note, in the above case ($g < 0$) both left and right half-plane zeros at the real axis are possible i.e. $q > 0$ and $q < 0$ is allowed. To keep this property, when positive gain i.e. when $g > 0$ is required, we introduce inverters for every such cell.

15.3.6.2 Second Order Cell with a Second Order Complex Zero

The transfer function is now

$$H(s) = \frac{s^2 + g \cdot s + q}{s^2 + as + b}. \tag{15.80}$$

For this purpose we use the circuit of Fig. 15.22.

After circuit analysis one gets

$$H(s) = \frac{R_8}{R_6} \frac{s^2 + \frac{1}{R_1C_1} \cdot \left(1 - \frac{R_1R_6}{R_4R_7}\right)s + \frac{R_6}{R_3R_5R_7C_1C_2}}{s^2 + \frac{1}{R_1C_1}s + \frac{R_8}{R_2R_3R_7C_1C_2}}. \tag{15.81}$$

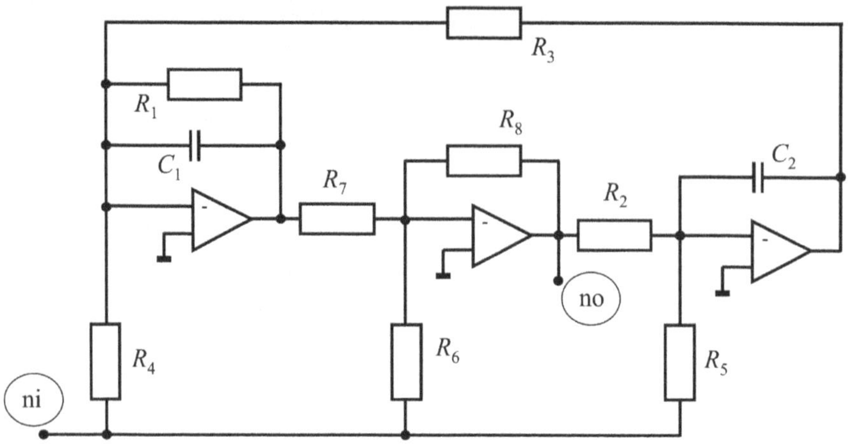

Fig. 15.22 The Tow-Thomas biquad cell realizing complex pair of zeros

Design equations are:

$$a = \frac{1}{R_1 C_1} \tag{15.82a}$$

$$b = \frac{R_8}{R_2 R_3 R_7 C_1 C_2} \tag{15.82b}$$

$$q = \frac{R_6}{R_3 R_5 R_7 C_1 C_2} \tag{15.82c}$$

and

$$g = \frac{1}{R_1 C_1} \cdot \left(1 - \frac{R_1 R_6}{R_4 R_7}\right). \tag{15.82d}$$

We adopt $C_1 = C_2 = 1.0$ and $R_8 = 1.0$ so that the rest of the elements are

$$R_1 = \frac{1}{C_1 a} \tag{15.83a}$$

$$R_2 = \frac{1}{C_2 \cdot \sqrt{b}} \tag{15.83b}$$

$$R_3 = \frac{1}{C_1 \sqrt{b}} \tag{15.83c}$$

$$R_4 = \frac{1}{C_1 \cdot (a - g)} \tag{15.83d}$$

$$R_5 = \sqrt{\frac{b}{q \cdot C_2}} \tag{15.83e}$$

$$R_6 = R_8 \tag{15.83f}$$

and

$$R_7 = R_8. \tag{15.83g}$$

15.4 Design Example

This example is representing a 10th order band-stop filter obtained by transformation of a modified elliptic prototype. The passband attenuation was a_{max} = 0.1 dB while the stopband attenuation was a_{min} = 50 dB. The band-stop configuration was chosen to demonstrate the need for both high-Q asymmetrical notch cells, and a symmetrical notch cell which is characteristic for the band-stop filter having transmission zero at the central frequency of the stopband.

In the next a copy of the full .html synthesis report is given. Figure 15.23 represents simulation results obtained by running the SPICE file created atomically by the *RC_cascade* program.

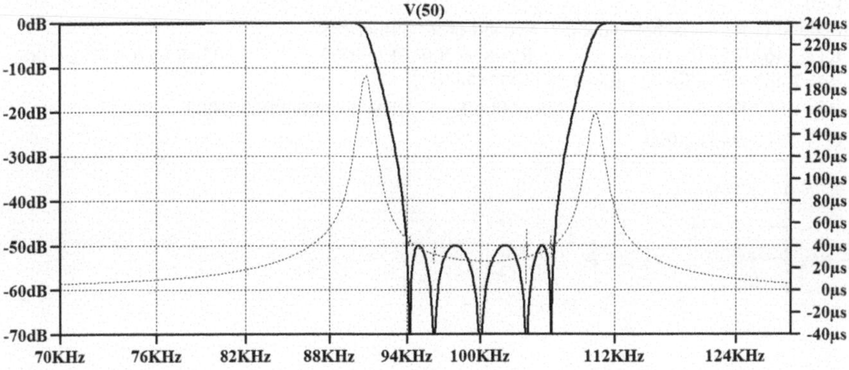

Fig. 15.23 Frequency response of the example filter

Start.E

Program RC-CASCADE, SYNTHESIS OF THE CASCADE ACTIVE RC FILTER

Project name: BS_RC_5_4_01db_50dB_10PC

Input data on the transfer function

Order of the numerator, n=10 Order of the denominator, m=10

Before normalization zeroes:
sigma=[1]=0.000000000e+000 omega=[1]=1.061929327e+000
sigma=[2]=-0.000000000e+000 omega=[2]=-9.416822522e-001
sigma=[3]=0.000000000e+000 omega=[3]=-1.061929327e+000
sigma=[4]=0.000000000e+000 omega=[4]=9.416822522e-001
sigma=[5]=0.000000000e+000 omega=[5]=1.040157159e+000
sigma=[6]=-0.000000000e+000 omega=[6]=-9.613931816e-001
sigma=[7]=0.000000000e+000 omega=[7]=-1.040157159e+000
sigma=[8]=0.000000000e+000 omega=[8]=9.613931816e-001
sigma=[9]=0.000000000e+000 omega=[9]=1.000000000e+000
sigma=[10]=0.000000000e+000 omega=[10]=-1.000000000e+000
Before normalization poles:
sigma=[1]=-1.037790338e-002 omega=[1]=9.073618725e-001
sigma=[2]=-1.260351731e-002 omega=[2]=-1.101951969e+000
sigma=[3]=-1.260351731e-002 omega=[3]=1.101951969e+000
sigma=[4]=-1.037790338e-002 omega=[4]=-9.073618725e-001
sigma=[5]=-1.705427371e-001 omega=[5]=9.853502803e-001
sigma=[6]=-1.705427371e-001 omega=[6]=-9.853502803e-001
sigma=[7]=-6.695855835e-002 omega=[7]=1.110701359e+000
sigma=[8]=-5.407991550e-002 omega=[8]=-8.970718175e-001
sigma=[9]=-5.407991550e-002 omega=[9]=8.970718175e-001
sigma=[10]=-6.695855835e-002 omega=[10]=-1.110701359e+000

Gain at nominal frequency (0, 1, or infinity)=1.000000000e+000

Cell No. 1
Second order Fleischer–Tow cell Second order numerator
R1=7.467305335e+003 R2=8.987004412e+002 R3=8.987004412e+002
R4=7.467305335e+003 R5=9.867197382e+002
R6=1.000000000e+003 R7=1.000000000e+003 R8=1.000000000e+003
C=1.591549431e-009

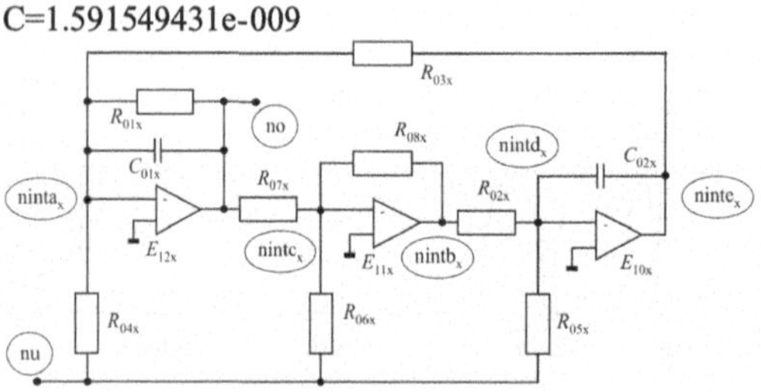

Cell No. 2
Second order Fleischer–Tow cell Second order numerator
R1=3.967146533e+004 R2=9.074212190e+002 R3=9.074212190e+002
R4=3.967146533e+004 R5=1.018575310e+003
R6=1.000000000e+003 R7=1.000000000e+003 R8=1.000000000e+003
C=1.591549431e-009

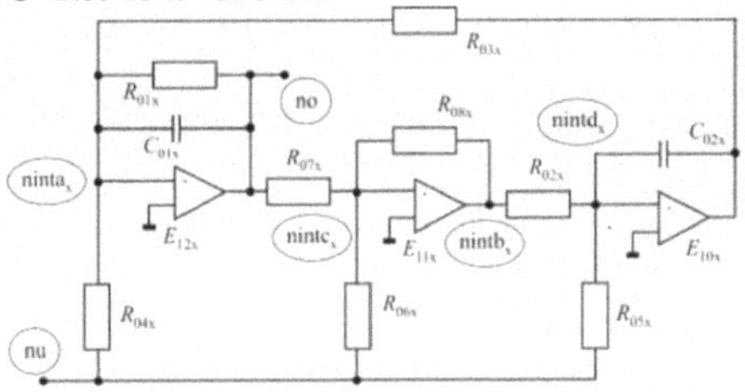

Cell No.=3 Second order Notch symmetrical

c=1.591549431e-009 r=1.000000000e+003 r4=8.527136855e+001 r5=9.147286314e+002

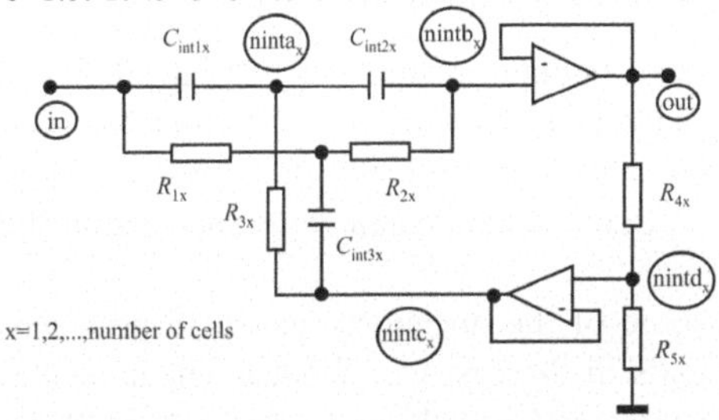

Cell No. 4
Second order Fleischer–Tow cell Second order numerator
R1=4.817928839e+004 R2=1.102024043e+003 R3=1.102024043e+003
R4=4.817928839e+004 R5=9.817634396e+002
R6=1.000000000e+003 R7=1.000000000e+003 R8=1.000000000e+003
C=1.591549431e-009

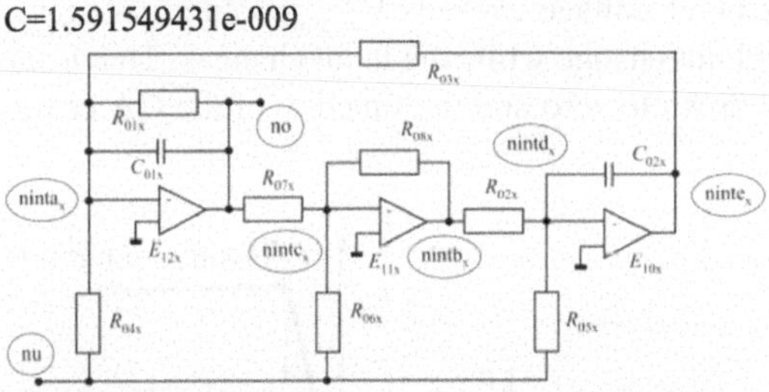

Cell No. 5
Second order Fleischer–Tow cell Second order numerator
R1=9.245576577e+003 R2=1.112717825e+003 R3=1.112717825e+003
R4=9.245576577e+003 R5=1.013459001e+003
R6=1.000000000e+003 R7=1.000000000e+003 R8=1.000000000e+003
C=1.591549431e-009

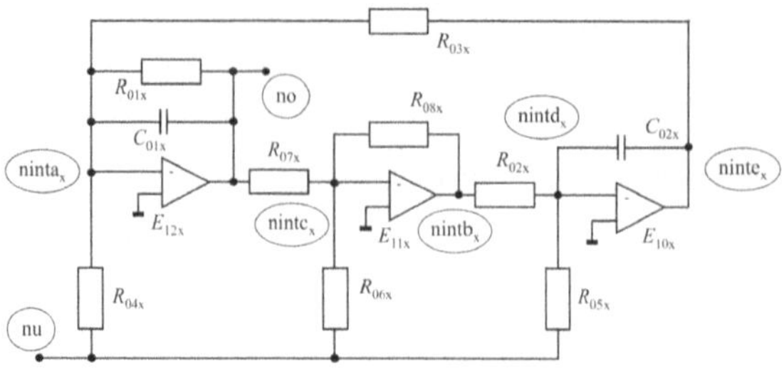

End of the synthesis procedure

End.E

15.5 Developer's Corner

The reader could recognize from the very beginning that the natural similarity between a transfer function represented in a factored form and the cascaded connection of cells which do not load each other, is a lure which imposes a large amount of uncertainty as to what is to be done in the design steps succeeding the factoring.

In Sect. 15.2 we already answered some of the questions related to pole-zero pairing and the mutual location of the cells in the cascade. What is left is to find recommendation for the linearity to be preserved and the noise to be kept at the lowest possible level.

To remind the reader, the operational amplifier (OA) which is always on the output of a cell (and handling the largest signal within it) is a device with limited linearity. Figure 15.24 depicts a direct current (DC) or static transfer characteristic of the OA. As can be seen, if the input voltage exceeds V_{in_max} saturation occurs which means the output voltage will not change while the input changes. That is the (dynamic) gain of the OA will fall down to zero and the function of the OA ceases

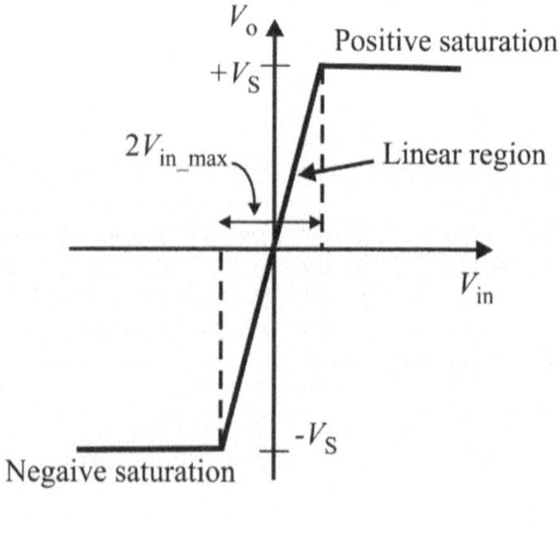

Fig. 15.24 The DC transfer characteristic of an operational amplifier and the definition of the maximum input voltage amplitude

to exist. The output waveform, during a short time when saturation occurs, is constant as if a negative signal was added to the output which makes the response to a sinusoidal input to be clamped. The presence of such an additional and unwanted signal is, of course, distortion.

It is obvious that the gain of the cell should be limited so as, for the prescribed input signal, the OA never enters saturation. In the same time, to keep the signal-to-noise ratio low one needs as large a gain as possible. So, scaling the dynamic range is necessary to get an optimum from both points of view.

What is usually recommended in the literature listed below is to find the maximum swing in the frequency domain by sweeping the frequency of an input sine-wave to the filter, and compare V_o with the maximum swing of the output OA.

Information needed for such analysis for the bandpass filter of Sect. 15.2 is depicted in Fig. 15.2. There, for convenience, logarithmic scale was used. For given amplitude of the input signal, these diagrams allow for calculation of the maximum amplitude of the input and the output voltages. There from one may conclude whether saturation occurs (For a given voltage supply V_s and given OA characteristic). If unsatisfactory, one may try to reduce the input signal by reducing the gain of the previous cell or to change the order of the cells in the cascade. In the former case, before deciding to reduce the gain, one is to study the properties of the cell. Namely, as we could see from Sect. 15.3, not all cells allow for change of the gain. If the preceding cell belongs to that category one is to search backwards until proper cell is found. If there is no such cell in the complete search back to the input, rearrangement of the cells is the only solution.

Figure 15.2e suggests that the overall passband gain may be a specific design parameter which may or may not be strictly connected to the transfer function's poles and zeros. Namely, if the passband gain is a design parameter, and a freedom exists for the gain of some cells to be accommodated, one may use the opportunity to adjust the overall gain without adding additional cells. Of course, the adjustment should be distributed evenly to as many cells as possible.

It is especially important, in cases when reduction of the overall gain is necessary, not to use resistive voltage dividers at the input of the cascade. That will drastically increase the noise.

It is well known that reducing the resistance values leads to reduction of the thermal noise in the system. That is especially important for the input circuitry of the cascade being connected in series with the internal resistance of the source. Reducing the resistances will certainly reduce the noise and the overall area of the chip (if integrated filter is planned) but it will, in the same time, give rise to the power consumption. Optimum should be sought from this point of view, too. In any case, when designing the first cell, the low noise criterion must prevail.

All together, defining the optimum order of the cells in a large cascade of active RC cells is a tedious, time consuming, and uncertain task.

Now we will briefly visit another aspect of the design, the imperfection of the OA. It is known that the open loop gain (OLG) of the OA is not infinite (as used in the previous analyses) and it is frequency dependent (decreasing with the rise of the frequency). To make things short we will here consider only the nominal value of

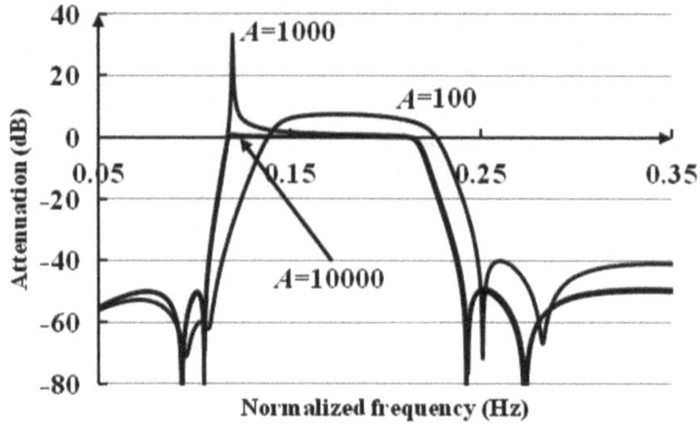

Fig. 15.25 Normalized gain of the filter as a function of frequency for various values of the OA's nominal OLG denoted by *A*

the OLG or what is known as its value for zeroth frequency. For simplicity, the OLG will be considered frequency independent.

Figure 15.25 depicts the (normalized) gain of the filter discussed in Paragraph 15.2 as a function of frequency for three values of the nominal OLG of all OAs used for realization. The largest value of OLG depicted in Fig. 15.25 is $A = 10^4$. Further increasing the OLG gives the same results as in the case of infinite amplification. This characteristic is almost perfect with miniscule distortion at the beginning of the passband. Reduction of the OLG by a factor of ten produces an unexpected spike in the gain characteristic at the lower edge of the passband. Its value is significant since it reaches almost 40 dB (100 times). Further reduction of the OLG by additional 10 times produces a fully distorted amplitude characteristic which may be stated as unacceptable.

We may conclude that, for this filter, the amplitude characteristic will be preserved as long as the OLG of the OA, in the frequency range of interest, remains above 10^4.

Since the example considered is complex enough and since it is dealing with an amplitude characteristic of relatively high selectivity, we may conclude that the cascade physical realization leads to a solution being not much sensitive to the value of the OA's OLG. Namely, $A = 10^4$ is much lower of the OLGs of standard OAs available as discrete components. That is not to be taken as granted when integrated CMOS operational amplifiers are to be used since that value of OLG is not easily obtainable. One is not to forget however, the frequency dependence of OLG no matter the technology.

From the software development point of view one is to note that there are cells declared by some set of attributes which may be substituted by others described by different attributes. For example, the all-pass cell given in Sect. 15.3.3.2 may be substituted by the Tow-Thomas cell described in Sect. 15.3.5.2. It is up to de software developer to decide which one to implement.

Finally, since the factorization of the transfer function is essential for the implementation of the method, here we may see ones again that representation of the transfer function in a factored form is crucial not only from transfer function analysis but for system synthesis point of view, too.

References

1. Laker KR, Sansen WMC (1994) Design of analog integrated circuits and systems. McGraw-Hill Education (ISE Editions), New York
2. Gray PR, Hurst PJ, Lewis SH, Meyer RG (2009) Analysis and design of analog integrated circuits. Wiley, New York
3. Fonderie MJ, Huijsing J (2013) Design of low-voltage bipolar operational. Springer, Heidelberg
4. Mohan PVA (2013) VLSI analog filters: active RC, OTA-C, and SC, modeling and simulation in science, engineering and technology. Springer Science + Business Media New York
5. Huelsman LP (1968) Theory and design of active RC circuits. McGraw-Hill, New York
6. Sallen RP, Key EL (1955) A practical method of designing RC active filters. IRE Trans Circuit Theory CAS 2(1):74–85
7. Fleischer PE, Tow J (1973) Design formulas for biquad active filters using three operational amplifiers. Proc IEEE, Proc Lett 61(5):662–663
8. Thomas LC (1971) The biquad: Part I-Some practical design considerations. IEEE Trans Circuits Syst CAS 18(3):350–357, and The Biquad: Part II-A multipurpose active filtering system. IEEE Trans Circuits Syst CAS 18(3):358–361
9. Tow J (1968) Active RC filters-a state-space realization. Proc IEEE 56(6):1137–1139
10. Hospodka J (2006) Optimization of dynamic range of cascade filter realization. Radioengineering 15(3):31–34
11. Xuexiang C, Sánchez-Sinencio CE, Geiger RL (1987) Pole-zero pairing strategies for cascaded switched-capacitor filters. IEE Proc G-Electron Circuits Syst 134(4):199–204
12. Chiou C-F, Schaumann R (1981) Refined procedure for optimizing signal-to-noise ratio in cascade active-RC filters. IEE Proc 128, Pt. G(4):181–191
13. Winder S (1997) Analog and digital filter design. Newness (Elsevier), Oxford

Chapter 16
Parallel Active-RC Circuit Synthesis

Parallel realization of active RC filters is dual to the cascade. It is an alternative which is based on partial fraction expansion of the transfer function and asks for residue calculation. The rest of the activities in the physical synthesis are straightforward since the cells are connected in parallel and their outputs are simply summed. Only one biquad type and one first order cells type are in use so dramatically simplifying the synthesis. No order of extraction and pole-zero pairing is necessary. As compared to the cascade this method needs additional circuitry for summation while offering extreme synthesis simplicity. To our knowledge no parallel synthesis as a general method was elaborated in the literature. After describing the procedure and giving proper example of a circuit of very high complexity, a short tolerance analysis is performed on a properly selected example, based on the Monte Carlo method, to demonstrate the how-to-do and the general necessity for this kind of verification of the physical synthesis of any filter.

16.1 Introduction

While parallel implementation is frequent in design of digital IIR and switched capacitor filters, to our knowledge, there was a single attempt to implement parallel realization of active RC circuits [1]. There, an effort was made to produce a realization implemented in hybrid technology (with limitations on the capacitance values) more economical than the cascade one on the expense of reduced sensitivity. To achieve that the author created a specific (multiple input and multiple output) cells with reduced number of capacitors which we find the main reason for this method not to be further investigated.

In these proceedings we will develop a straightforward procedure for parallel synthesis based on repetitive use of one type of widely known and frequently used second order cell [2]. To our knowledge this is the first complete description of the

V. Litovski, *Electronic Filters*, Lecture Notes in Electrical Engineering 596,
https://doi.org/10.1007/978-981-32-9852-1_16

procedure. To come to a base for comparison with a cascade solution, high order band-pass filters will be synthesized.

There are some apparent advantages of the parallel synthesis of active RC circuits as compared with the solution taking the form of cascade connection of cells.

First, when using parallel synthesis the types of transfer functions of the cells (circuits) is reduced to two: one for the first order cell and another for the second order cell. When the realization of the cell is sought, the later one may take two forms (low- and high-Q) which altogether is incomparably smaller in number of possible variants than in the case of cascade synthesis.

Then, there is no amplification of the noise generated by the cells since there is no cascade which (in case of a higher order filter) may become prohibitive. This issue is further elevated if the shape of the amplitude characteristics of the cells used is considered. When parallel synthesis is taken into account, both the second order and the first order cell (as will be demonstrated) are never high-pass which means the noise bandwidth is limited even when high-pass filters as a whole are to be synthesized.

Then, there is no need for pole-zero pairing in order to reduce nonlinear distortions and noise since, as will be seen below, only poles and residues are used. In addition, there is no need for assigning specific gain values to the cells in order to keep the signals within prescribed limits. Furthermore, when parallel (in case of low-pass, band-stop and all-pass filters) there is no amplification of the input DC offset.

The price paid for all these advantages is the cost of the additional circuitry needed to perform the summing at the output of the filter.

In the sequel we will first describe the synthesis technique and subsequently we will use an example to demonstrate the design sequence implemented within the $\mathcal{RM}$ software for filter design.

16.2 Decomposition of the Transfer Function

The continuous time transfer function may be represented in a form of sum of partial fractions as follows [3],

$$\begin{aligned} H_a(s) &= A_0 \cdot H(s) = A_0 \cdot \frac{\prod_{i=1}^{m}(s - z_i)}{\prod_{i=1}^{n}(s - p_i)} \\ &= A_0 \cdot \sum_{i=1}^{n} \frac{r_i}{(s - p_i)} = \begin{cases} \sum\limits_{i=0}^{n/2} H_e(s), n - \text{even} \\ H_o(s) + \sum\limits_{i=1}^{n/2} H_e(s), n - \text{odd} \end{cases}. \end{aligned} \tag{16.1}$$

where, to get unity nominal gain:

$$A_0 = \frac{\prod_{i=1}^{n}(0 - p_i)}{\prod_{i=1}^{m}(0 - z_i)} \tag{16.2a}$$

for low-pass filters and band-stop filters having the lower part of the pass-band as a normalization region;

$$A_0 = \frac{\prod_{i=1}^{n}(1 - p_i)}{\prod_{i=1}^{m}(1 - z_i)} \tag{16.2b}$$

for band-pass filters;

and

$$A_0 = 1 \tag{16.2c}$$

for high-pass filters and band-stop filters having the upper part of the pass-band as a normalization region.

In the above s stands for the complex frequency as independent variable, p_i stands for the ith pole, z_i stands for the ith zero, n is the order of the filter, and m is the order of the numerator of the transfer function. Index e is used to denote a second order fraction constructed by a complex pair of poles while o denotes a first order fraction constructed by a simple real pole. **r** are the complex residues in the poles. Note, some programs within the $\mathcal{RM}$ software are restricted to simple poles.

Note (16.1) is valid for $n > m$ only. In the case $n = m$ polynomial long division must be carried out first (as explained in [1]) which leads to

$$H_a(s) = A_0 \cdot [1 + H(s)]. \tag{16.3}$$

The structure of the resulting filter is depicted in Fig. 16.1. As can be seen an auxiliary unity gain path is allowed for filters having $n = m$.

The residues needed for the realization of the above computations are obtained (for the case of simple poles) as follows,

$$r_i = \lim_{s \to p_i} \{[(s - p_i) \cdot H(s)]\} = [(s - p_i) \cdot H(s)]_{|s=p_i} \tag{16.4a}$$

or

$$r_i = A_0 \cdot \frac{\prod_{j=1}^{m}(p_i - z_j)}{\prod_{j=1,\, j \neq i}^{n}(p_i - p_j)} \tag{16.4b}$$

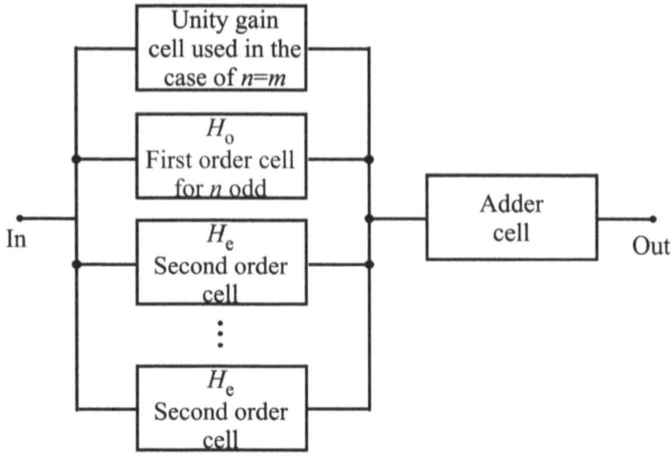

Fig. 16.1 Structure of the filter

The summands in (16.1) in the case of a pair of conjugate poles may be collected into partial fractions as follows

$$H_e(s) = G_i \frac{s + b_{0,i}}{s^2 + a_{1,i}s + a_{0,i}} \tag{16.5}$$

with

$$G_i = 2 \cdot A_0 \cdot \mathrm{re}\{r_i\}, \tag{16.6a}$$

$$b_{0,i} = \begin{cases} -\left(\mathrm{re}\{p_i\} + \frac{\mathrm{im}\{r_i\} \cdot \mathrm{im}\{p_i\}}{\mathrm{re}\{r_i\}}\right) & \text{if} \quad \mathrm{im}\{r_i\} \cdot \mathrm{im}\{p_i\} > 0 \\ \left(\frac{\mathrm{im}\{r_i\} \cdot \mathrm{im}\{p_i\}}{\mathrm{re}\{r_i\}} - \mathrm{re}\{p_i\}\right) & \text{if} \quad \mathrm{im}\{r_i\} \cdot \mathrm{im}\{p_i\} < 0 \end{cases}, \tag{16.6b}$$

$$a_{1,i} = -2\mathrm{re}\{p_i\}, \tag{16.6c}$$

$$a_{i0} = |p_i|^2, \tag{16.6d}$$

while in the case of a simple pole on the real axis one has

$$H_o(s) = G_o \frac{1}{s + a_o}, \tag{16.7}$$

with $G_o = r_o$, and $a_o = -p_o$. In the above “re” stands for “real part” and “im” for “imaginary part”.

We will denote $p_i = \sigma_i + j\omega_i$, and $z_i = \alpha_i + j\beta_i$ and $r_i = \mu_i + j\xi_i$. Accordingly (16.5), (16.6a) and (16.7) may be rewritten as:

$$H_e(s) = \frac{G_i \cdot s + G_i \cdot b_{0,i}}{s^2 + a_{1,i}s + a_{0,i}} \tag{16.8}$$

with

$$G_i = 2 \cdot A_0 \cdot \mu_i, \tag{16.9a}$$

$$G_i \cdot b_{0,i} = \begin{cases} -2 \cdot A_0 \cdot (\mu_i \sigma_i + \xi_i \cdot \omega_i) & if \quad \xi_i \cdot \omega_i > 0 \\ 2 \cdot A_0 \cdot (\xi_i \cdot \omega_i - \sigma_i \cdot \mu_i) & if \quad \xi_i \cdot \omega_i < 0 \end{cases}, \tag{16.9b}$$

$$a_{1,i} = -2 \cdot \sigma_i, \tag{16.9c}$$

$$a_{i0} = |p_i|^2 = \sigma_i^2 + \omega_i^2, \tag{16.9d}$$

while

$$H_o(s) = G_o \frac{1}{s + a_o}, \tag{16.10}$$

with $G_o = A_0 \cdot r_o = A_0 \cdot \mu_0$, and $a_o = -p_o = -\sigma_0$.

The developments expressed so far are (apart of the notation) equal to the ones used in [1]. The difference and, accordingly, the novelty we are introducing, is in the use of standard and universally accepted circuits (cells) which are realizing (16.8) and (16.10) in the place of "multiple entry" cells used in [1]. Since two types of cell transfer functions are in view, only two types of circuit cells will be involved. Note that the second order cell has one zero at infinity and another on the real axis of the frequency plane being not restricted to any part of the real axis.

16.3 Choice of the Cells and Physical Realization of the Filter

In this section the two filter cells and the summing circuit will be discussed. The reader may recognize the fundamental difference as compared with cascade RC realization where the number of different cells was almost two dozen.

16.3.1 Second Order Cell

According to the literature, e.g. [4], there are several concepts for creation of second order cells mainly intended to be used in cascade synthesis of active RC filters. The specifics of parallel synthesis may be seen from (16.5) which represent a second order cell with a zero at the real axis of the complex frequency plane. Note, again, the zero is not restricted to any part of the real axis.

To avoid the choice of a specific cell and to go for well established methods, in [1] further decomposition of the function was performed by parting the second order cell into two: a band-pass and a low-pass cell as follows:

$$H_e(s) = G_i \frac{s}{s^2 + a_{1,i}s + a_{0,i}} + G_i \frac{b_{0,i}}{s^2 + a_{1,i}s + a_{0,i}}. \tag{16.11}$$

This solution is feasible with the main drawback that it in fact needs two second order cells and, usually, needs at least four different capacitors per cell which in cases when programmable transfer function is sought, may be difficult to control.

There are only several cells that are qualified for implementation in a case of transfer function containing a zero at the real axis. In [5] a single amplifier biquad (SAB) is proposed. It is known in the literature as the Friend's circuit. A two operational amplifier biquad was proposed in [6] by Soliman, while in [2] a three amplifier biquad is described. It is referred here to as Tow-Thomas (TT) biquad and due to its low sensitivities to parasitics is very frequently in use. As will be shown later on in this text, when using the TT cell, all capacitors in the filter (not only in a cell) may have the same value. The Tow-Thomas cell was adopted for synthesis here.

The schematic of the version of the TT biquad cell used in *RC-parallel* is depicted in Fig. 16.2.

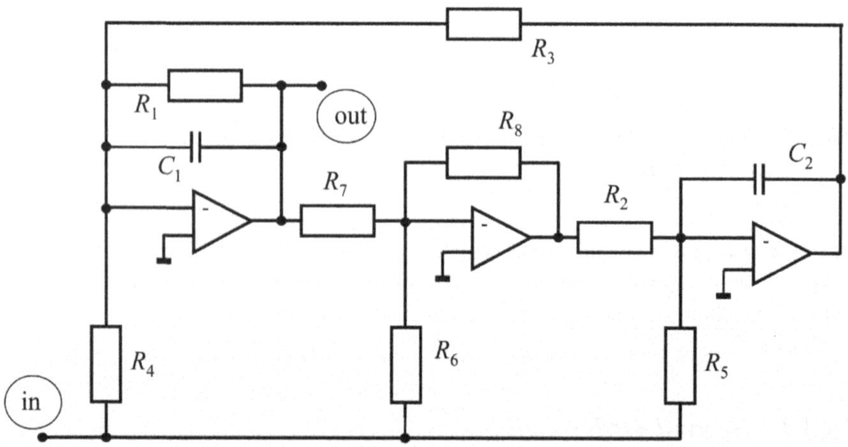

Fig. 16.2 The Tow-Thomas biquad cell and node notation. "ni" stands for "input node" while "no" for the "output node"

To get the design equations we first simplify the notation. One may find easily by analogy that (16.8), for a given cell, may be rewritten as

$$H_e(s) = g\frac{s+q}{s^2+as+b}. \tag{16.12a}$$

Now, after circuit analysis one gets

$$a = 1/(R_1C_1) \tag{16.12b}$$

$$b = R_8/(R_2R_3R_7C_1C_2) \tag{16.12c}$$

$$q = \frac{R_4(R_5R_8 - R_2R_6)}{R_2R_3R_5R_6C_2} \tag{16.12d}$$

and

$$g = -1/(C_1R_4). \tag{16.12e}$$

Since there are more circuit elements than degrees of design freedom (expressed by the number of coefficients in 16.12a) some of the element values are to be adopted as design constants. Speaking in normalized element values we first adopt $C_1 = C_2 = C = 1$. Then, we adopt the $R_2 = R_3 = R_7 = R = 1$. With that set, one may calculate

$$R_1 = 1/(a \cdot C) \tag{16.13a}$$

$$R_4 = 1/(|g| \cdot C) \tag{16.13b}$$

$$R_8 = b \cdot C^2R^3 \tag{16.13c}$$

and

$$\begin{array}{lll} R_5 = R \text{ and } R_6 = R_8/[R \cdot (q \cdot R \cdot C/R_4 + 1/R_5)] & \text{for} & q > 0 \\ R_6 = R \text{ and } R_5 = 1/[R_8/(R \cdot R_6) - q \cdot R \cdot C/R_4] & \text{for} & q < 0\ . \\ R_5 = R \text{ and } R_6 = R_5 \cdot R_8/R_2 & \text{for} & q = 0 \end{array} \tag{16.13d}$$

Note, in the above case ($g < 0$) both left and right half-plane zeros at the real axis are possible i.e. $q >$ and $q < 0$ is allowed. To allow for this property to be tractable, when positive gain i.e. when g > 0 is required, we introduce inverters for every such cell.

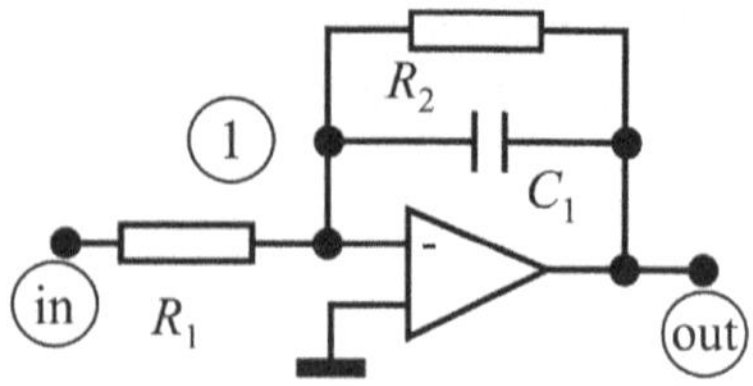

Fig. 16.3 Inverting first order filter and node notation

16.3.2 First Order Cell

The first order cell is a simple inverting circuit as depicted in Fig. 16.4. Note we suggest using an additional inverter if necessary to accommodate to the sign of the first order fraction of the filter function. Namely, the positive-gain first order cell exhibits a gain that is larger than unity so that additional circuitry will be necessary in cases when $G_0 < 1$. Having that in mind we restricted our procedure to either one (for negative gain) or two in cascade (for positive gain) inverting amplifiers.

The transfer function of the circuit depicted in Fig. 16.3 may be expressed in the form

$$H_o = g/(s + a), \tag{16.14}$$

where, by analogy, for the normalized element values we have

$$C_1 = 1, \tag{16.15a}$$

$$R_1 = 1/(C_1 \cdot \mathrm{g}), \tag{16.15b}$$

$$R_2 = 1/(C_1 \cdot \mathrm{a}). \tag{16.15c}$$

Note, since there are no restrictions to the value of C_1, it may be equal to the capacitances used within the TT cells.

16.3.3 Summing

Figure 16.4 represents how the summing is performed within *RC-parallel*.

Here, apart from the auxiliary path which will be used in the case of high-pass, all-pass, and band-stop filters, this schematic has as many paths as is the number of filtering cells connected in parallel denoted by k_{norm} in the figure. Part of these paths (k_{inv} of them) have to use inverters which are collected together and realized as a separate unity gain inverting summing amplifier.

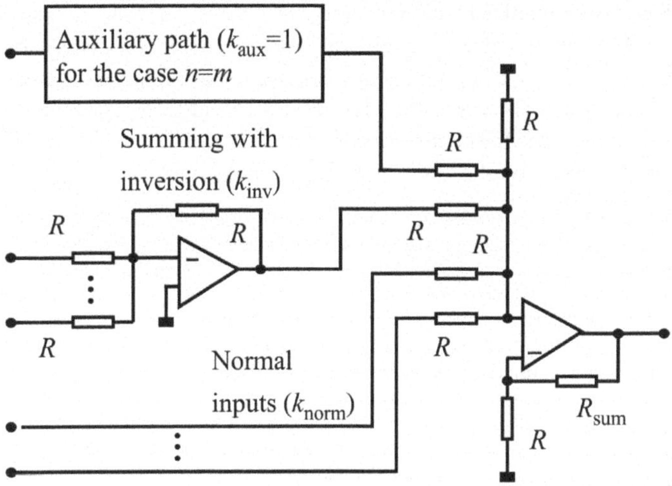

Fig. 16.4 Realization of the summing (adder cell)

To avoid the summing amplifier to change the value of the overall gain, the value of R_{sum} is calculated as follows

$$R_{sum}/R = \begin{cases} k_{aux} + k_{norm} - k_{inv} + 1 & \text{for} \quad k_{inv} > 0 \\ k_{aux} + k_{norm} & \text{for} \quad k_{inv} = 0 \end{cases}. \tag{16.16}$$

R is the normalization resistance supplied by the system designer.

16.4 Synthesis Example

In this example a band-pass filter will be synthesized with the following prerequisites: Central frequency f_0 = 100 kHz, Upper frequency of the pass-band f_h = 120 kHz, Lower frequency of the pass-band f_l = 80 kHz, Maximum pass-band attenuation a_{max} = 3 dB and Minimum stopband attenuation a_{min} = 45 dB. The normalization resistance was supplied to be R_0 = 1 kΩ while the capacitances are denormalized by the use of the formula

$$C = C_{norm}/(2 \cdot \pi \cdot R_0 \cdot f_0) \tag{16.17}$$

where C_{norm} is adopted to be equal to unity as stated above.

The LSM polynomial low-pass prototype of order 8 was first selected and synthesized. Then its transfer function was extended with 6 transmission zeros to increase its selectivity. The so obtained rational low-pass function was transformed into a band-pass one. These results were used as input to the physical realization program. Following are (In order to save space the tables are partly edited) excerpts from the report created by the *RC_parallel* system synthesis program being part of the *RM* software.

Start.E

RC-Parallel, Program for synthesis of active RC-circuits ,
in the form of parallel connection of Tow-Thomas biquads

Project name: BP_SIM_14_8_2_20p

The parameters for the analysis read from the input file

Order of the numerator, n=10 Order of the denominator, m=14

Before normalization
zeros:
sigma[1]=0.000000e+000 omega[1]=0.000000e+000
sigma[2]=0.000000e+000 omega[2]=0.000000e+000
sigma[3]=0.000000e+000 omega[3]=1.284856e+000
sigma[4]=0.000000e+000 omega[4]=-1.284856e+000
sigma[5]=0.000000e+000 omega[5]=1.406737e+000
sigma[6]=0.000000e+000 omega[6]=-1.406737e+000
sigma[7]=0.000000e+000 omega[7]=7.137798e-001
sigma[8]=0.000000e+000 omega[8]=-7.137798e-001
sigma[9]=0.000000e+000 omega[9]=6.164873e-001
sigma[10]=0.000000e+000 omega[10]=-6.164873e-001
poles:
sigma[1]=-1.300627e-001 omega[1]=7.141267e-001
sigma[2]=-1.966151e-001 omega[2]=8.064309e-001
sigma[3]=-2.294117e-001 omega[3]=8.923336e-001
sigma[4]=-2.393232e-001 omega[4]=9.761194e-001
sigma[5]=-2.287358e-001 omega[5]=1.059705e+000
sigma[6]=-1.954219e-001 omega[6]=1.144987e+000
sigma[7]=-1.288047e-001 omega[7]=1.236174e+000
sigma[8]=-1.300627e-001 omega[8]=-7.141267e-001
sigma[9]=-1.966151e-001 omega[9]=-8.064309e-001
sigma[10]=-2.294117e-001 omega[10]=-8.923336e-001
sigma[11]=-2.393232e-001 omega[11]=-9.761194e-001
sigma[12]=-2.287358e-001 omega[12]=-1.059705e+000
sigma[13]=-1.954219e-001 omega[13]=-1.144987e+000
sigma[14]=-1.288047e-001 omega[14]=-1.236174e+000

Partial fraction expansion

Residues in the poles
pl[1]= -1.594843e+000 q1[1]= 8.821194e-001
pl[2]= 2.899842e+001 q1[2]= -2.144391e+001
pl[3]= -1.232130e+002 q1[3]= 8.874356e+001
pl[4]= 2.032878e+002 q1[4]= -1.335745e+002
pl[5]= -1.461327e+002 q1[5]= 8.363654e+001
pl[6]= 4.128998e+001 q1[6]= -1.931854e+001
pl[7]= -2.635629e+000 q1[7]= 9.428247e-001
pl[8]= -1.594843e+000 q1[8]= -8.821194e-001
pl[9]= 2.899842e+001 q1[9]= 2.144391e+001
pl[10]= -1.232130e+002 q1[10]= -8.874356e+001
pl[11]= 2.032878e+002 q1[11]= 1.335745e+002
pl[12]= -1.461327e+002 q1[12]= -8.363654e+001
pl[13]= 4.128998e+001 q1[13]= 1.931854e+001
pl[14]= -2.635629e+000 q1[14]= -9.428247e-001

Gain at nominal frequency (0, 1, or infinity)=20.004103 ipb=1 (m-n)is equal:0

Residues after division with the gain
pl[1]= -7.972580e-002 q1[1]= 4.409692e-002
pl[2]= 1.449623e+000 q1[2]= -1.071976e+000
pl[3]= -6.159386e+000 q1[3]= 4.436268e+000
pl[4]= 1.016230e+001 q1[4]= -6.677354e+000
pl[5]= -7.305138e+000 q1[5]= 4.180969e+000
pl[6]= 2.064075e+000 q1[6]= -9.657288e-001
pl[7]= -1.317544e-001 q1[7]= 4.713157e-002
pl[8]= -7.972580e-002 q1[8]= -4.409692e-002
pl[9]= 1.449623e+000 q1[9]= 1.071976e+000
pl[10]= -6.159386e+000 q1[10]= -4.436268e+000
pl[11]= 1.016230e+001 q1[11]= 6.677354e+000
pl[12]= -7.305138e+000 q1[12]= -4.180969e+000
pl[13]= 2.064075e+000 q1[13]= 9.657288e-001
pl[14]= -1.317544e-001 q1[14]= -4.713157e-002

Ordered vector of transfer functions to be parallelized T(s)=(g*s+q)/(s**2+a*s+b)

g	q	a	b
-1.5945159e-001	5.2505141e-001	2.6012541e-001	5.2689321e-001
2.8992468e+000	7.9295909e-001	3.9323020e-001	6.8898827e-001
-1.2318772e+001	8.7211069e-001	4.5882334e-001	8.4888898e-001
2.0324610e+001	8.8070287e-001	4.7864648e-001	1.0100848e+000
-1.4610275e+001	8.3523971e-001	4.5747165e-001	1.1752950e+000
4.1281507e+000	7.3113251e-001	3.9084378e-001	1.3491853e+000
-2.6350884e-001	5.7101250e-001	2.5760943e-001	1.5447176e+000

SYNTHESIS OF THE PARALLEL ACTIVE RC FILTER

Cell No. 1
Second order Fleisher-Tow cell
Elements of the RC circuit: i-inver=0
Additional inverter required? Look obove: i-invr=0, no; i-invr=1, yes;
Input node is ni=1, output node is no=7
R1=3.844299557e+003 R2=1.000000000e+003 R3=1.000000000e+003
R4=6.271495914e+003 R5=1.000000000e+003
R6=4.861893037e+002 R7=1.000000000e+003 R8=5.268932092e+002
C1=1.591549433e-009 C2=1.591549433e-009

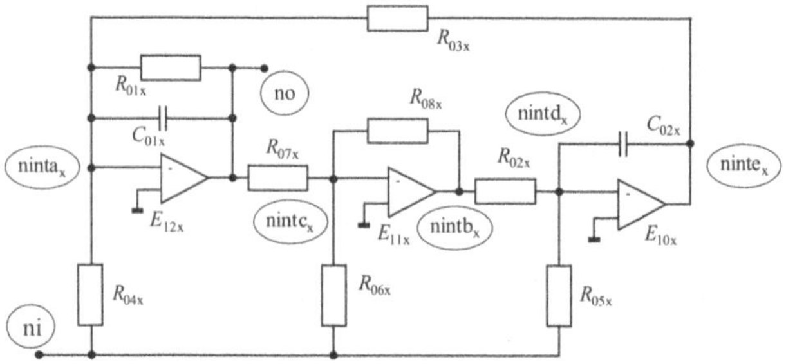

-------------------End of the 1-th cell description ---------------------------

.
.
.

-------------------End of the 6-th cell description ---------------------------

Cell No. 7
Second order Fleisher-Tow cell
Elements of the RC circuit: i-inver=0
Additional inverter required? Look obove: i-invr=0, no; i-invr=1, yes;
Input node is ni=1, output node is no=127
R1=3.881845550e+003 R2=1.000000000e+003 R3=1.000000000e+003
R4=3.794939063e+003 R5=1.000000000e+003
R6=1.342687613e+003 R7=1.000000000e+003 R8=1.544717579e+003

C1=1.591549433e-009 C2=1.591549433e-009

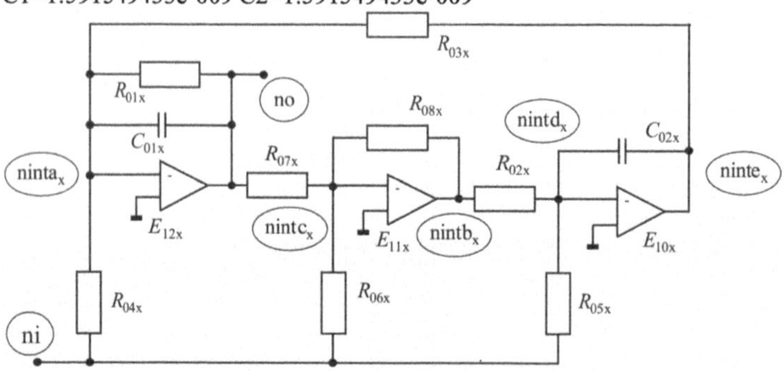

------------------End of the 7-th cell description --------------------------

•
•
•

HERE ENDS THE SYNTHESIS PROCESS

The resistance of the summing resistor was calculated to be R_{sum}=0+7-4+1=4 kΩ since 4 out of 7 cells need additional inversion of their output signals and no auxiliary path was necessary.

End.E

The SPICE net-list created by *RC_parallel* was run by the LTspice program. The simulation results are depicted in Fig. 16.5. Note the linear scale for the *f*-axis.

16.5 Developer's Corner

The procedure of parallel synthesis of active RC circuits seems to be the simplest of all.

Having the poles and zeros of the transfer function one has to do the following

1. Find the residues of the poles. This activity is relatively simple and it is supposed to be already done for the calculation of the time domain responses of the low-pass filters.
2. Normalize the residues so that they are divided by the gain at the central frequency. The trick here is that one needs to use the temporary partial fraction expansion for calculation of the gain at the central frequency of the passband.
3. Create the partial fractions. Here one simply finds pairs of conjugate poles and their residues and manipulate them to get the biquads. It is important to extract the sign of the cell in order to allow for proper summation.

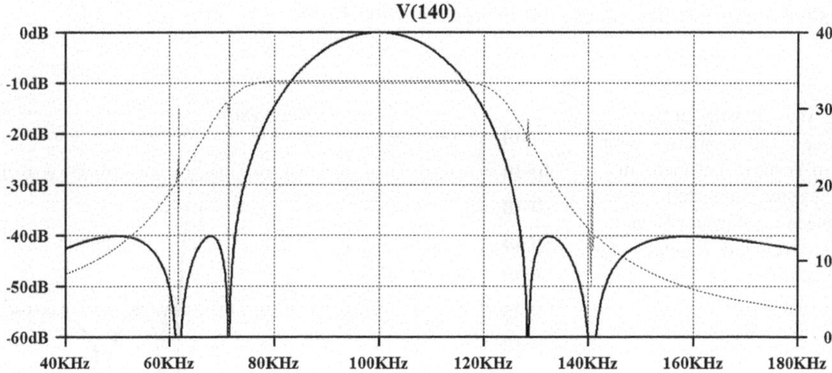

Fig. 16.5 Frequency response of the "Band_pass_CMAC_LSM" filter

4. Calculate the element values of the cells.
5. Calculate the resistance in the summing amplifier.

In the previous chapter we suggested an action within the verification of the feasibility of the circuit obtained: search for the minimum acceptable open loop gain. That, of course is not the only action which is necessary to be taken before proceeding towards designing the printed circuit board or the integrated circuit. Here we will study the influence of the tolerances of the elements to the amplitude characteristic of the filter. As a carrier we will use a 7th order Modified Elliptic filter exhibiting 0.1 dB of passband attenuation and 50 dB of stopband attenuation. The reason for this is the fact that the sensitivity of the attenuation characteristic is proportional to its derivative as expressed with (7.20) which means the example depicted in Fig. 16.5 is not a representative one since the derivatives of the attenuation are not large enough to expose the influence of the circuit parameter's tolerances.

Monte Carlo analysis available in the LTSpice software (used throughout this book) was used to activate the parameter variations. To avoid over-crowded and unpainted images only five samples of the variable parameter were used to create a family of amplitude characteristics.

Three experiments were undertaken.

In the first one we imposed maximum variation of 2% to the resistor connecting the output of the second cell and the input of the summing OA. To make it easier to the reader that is one of the bottom horizontal resistors in Fig. 16.4. The Overall gain characteristic of this filter obtained by Monte Carlo analysis is depicted in Fig. 16.6. As can be seen the passband response is not noticeably affected while the transition region and the stopband are. The main reason for that is the fact that the second cell which was purposely chosen is responsible for the realization of the transmission zero nearest to the passband. In any case, one may conclude that with 2% tolerance of this resistor (if it is the only source of variations) the design is not seriously jeopardised.

The second experiment is related to the resistor R_8 (Fig. 16.2) in the second cell of the parallel connection. Obviously, as seen from Fig. 16.7, even with ten times smaller maximum tolerance, the design is at the brink to be non-acceptable.

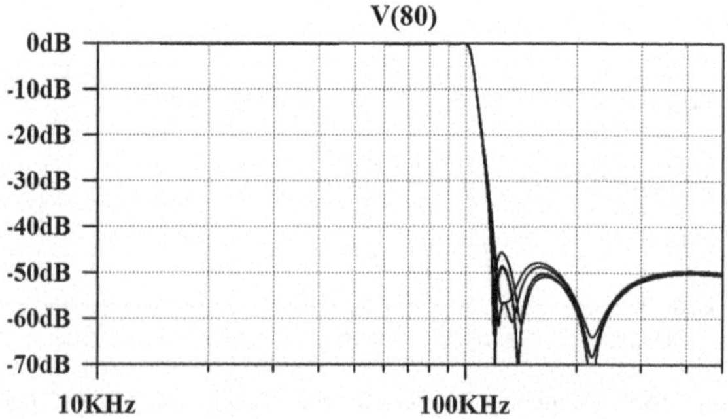

Fig. 16.6 Results of the Monte Carlo simulation with 2% maximum variations of the resistor connected between the output of the second cell and the input of the summing OA

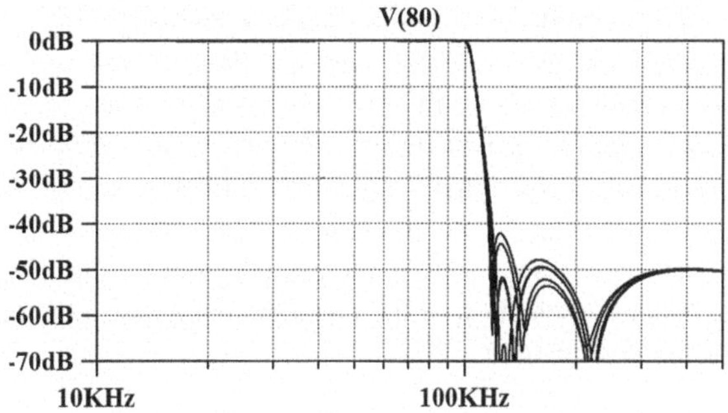

Fig. 16.7 Results of the Monte Carlo simulation with 0.2% maximum variations of the resistor R_8 of the second cell

Finally, the capacitance C_1 (Fig. 16.2) of the second cell connected in parallel was taken under scrutiny. Here again 2% tolerance was imposed. The resulting amplitude characteristic obtained after Monte Carlo analysis is depicted in Fig. 16.8. Figure 16.8a represents the overall gain as a function of frequency. It reveals that 2% is a large value for the tolerance of this element if one intends to preserve the width of the transition region. Figure 16.8b depicts the distortions of the gain in the passband. As already said, this cell is not affecting much the passband. This conclusion is not to be generalized, however. Namely, there is another cell which is realising a pole falling into the passband. That one is to be considered for the passband distortions.

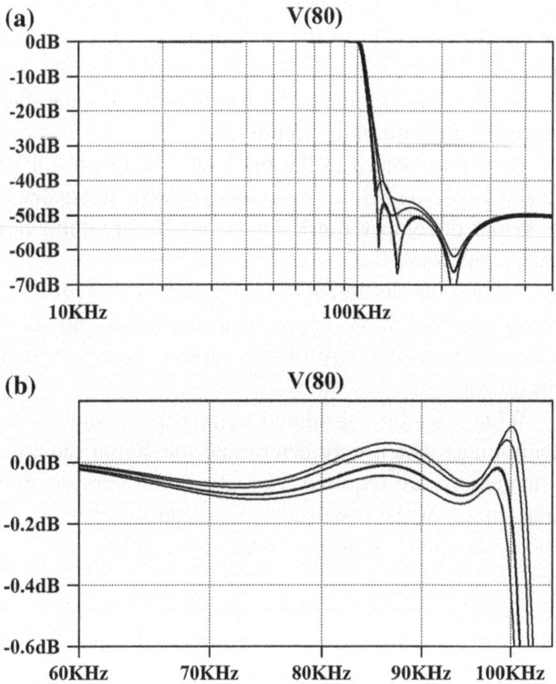

Fig. 16.8 Results of the Monte Carlo simulation with 2% maximum variations of the resistor connected between the output of the second cell and the input of the summing OA. **a** Overall gain and **b** passband gain

To conclude, a thorough tolerance analysis must be performed before a design is proclaimed acceptable. It was not done above but simultaneous variation of the most influential circuit elements must be introduced and statistically simulated. Note, there is a possibility offered by the LTSpice program for worst-case (or corner) analysis. That would be useful but it will take enormous time if several parameters are to be toleranced. Our advice is to use natural intelligence and experience to save time in the design process. The choices done above may be used as hints for selection of the elements being the most influential and so for reduction of the design time.

16.6 Conclusion

Parallel active RC filtering circuits are rarely found in practice. We suppose that is by inertia since cascaded synthesis looks natural when looking to the structure of the transfer function. Namely, as shown above, the transfer function is defined as ratio of *products* while cascade connection of cells is doing *product* of signals. In addition, the need for a cell having transmission zero on the real axis (necessary for parallel realization) seems to be a repelling factor for many designers. It is our experience that even zeros at the imaginary axis are not welcomed by many.

Nevertheless, parallel realization has important advantages. To appreciate the extreme simplicity of the synthesis algorithm we will mention the following. First, there is no need for pole-zero pairing algorithms. Second, there is no need for order of extraction (order of the cells in the cascade) algorithm. Finally, there is no need for gain distribution (to avoid nonlinear distortions) concerns. A special advantage is that the noise produced by a cell within the parallel connection is amplified only once (by the summing amplifier).

As it was shown in the previous chapter, in the case of cascade synthesis, one needs a specific cell for every type of zero and every type of biquad which becomes a serious developer's task. One is not to forget the variations existing for practically every type of cell.

Among the disadvantages of parallel connection is the need for summing which introduces two more operational amplifiers and several resistors. Having in mind the overall complexity of the system, however, that seems not to be a decisive argument.

What may be considered as possible problem is the accuracy of the addition (summing) which may deteriorate the signal outside of the passband if very high attenuations are required. Namely, when very small signals are to be obtained the inaccuracy of the resistances may lead to *relatively* large subtraction (which is the most inaccurate arithmetic operation) error.

References

1. Moran PL (1978) A low-cost parallel implementation for active filters. Electron Circuits Syst 2 (1):21–25
2. Fleisher PE, Tow J (1973) Design formulas for biquad active filters using three operational amplifiers. Proc IEEE 61(5):662–663
3. Williams AB, Taylor FJ (2006) Electronic filter design handbook. McGraw-Hill, New York
4. Deliyannis TL, Yichuang S, Kel Filder J (1999) Continuous-time active filter design. CRC Press LCC, Boca Raton, Florida
5. Friend JJ, Harris CA, Hilberman D (1975) STAR: an active biquadratic filter section. IEEE Trans Circuits Syst 22(2):115–121
6. Soliman AM (1974) A new active rc configuration for realizing nonminimum phase transfer functions. Circuit Theory Appl 2:307–315

Chapter 17
Gm-C Filter Synthesis Based on LC Prototypes

Two main problems are encountered when high frequency integrated analog filters are to be synthesized: the limitations of the operational amplifiers and the properties of the integrated resistor. The gain of the operational amplifier strongly depends on the frequency so deteriorating the properties of the overall system. As for the resistor it occupies large area on the silicon chip, its value is subject to large tolerances and it is accompanied by a large parasitic capacitance. To avoid that, operational transconductance amplifier (OTA) was introduced which in conjunction with capacitors may perform well enough at high frequencies. Hence the name Gm-C. To avoid all burdens introduced with cascaded synthesis one of the variant of this type of filters is to develop a cell which is simulating the inductance and use this equivalent circuit to substitute the inductors in an already synthesized LC filter. The simulated inductor is based on a specific negative impedance converter (NIC) named gyrator. In this chapter we are developing the analogy and demonstrate how an LC filter cell is transformed into a Gm-C one. The main imperfection of the OTA is its finite output resistance. Its influence to the overall amplitude characteristic is studied.

17.1 Introduction

One of the problems encountered in high frequency analog integrated filter synthesis is the area needed to produce an inductor. It is realized in a form of a flat spiral line the inductance of which is limited not only by the area but by its huge parasitic capacitance. One such inductor realized in CMOS was reported in [1].

The layout of one inductor of this kind is depicted in Fig. 17.1a. One may see that the wires are twisted to reduce the parasitic capacitance. This inductor is specific in the sense that it has a tap terminal allowing specific uses. The most important electrical parameters of an integrated inductor are the inductance (L), its resistance (R_s), its parasitic capacitance and its Q-factor (Q) as a secondary

V. Litovski, *Electronic Filters*, Lecture Notes in Electrical Engineering 596,
https://doi.org/10.1007/978-981-32-9852-1_17

parameter. The reactance of this kind of components is linearly rising at low frequencies (several GHz) to exhibit inductive character. The values of the inductances obtained in this way are of order of nH. The inductance is normally accompanied by a series resistance as depicted in Fig. 17.1b where one half of the whole component is modeled. The value of the resistance depends on technologies and it is seen to be of order of several Ω. In that way the Q-factor of the inductor becomes relatively large being of order of 10.

One must have in mind that this component occupies large area on the silicon chip (of the order of 5,000 μm^2) which has as a consequence presence of influential parasitics. The simplified (lumped) version of the parasitic circuit is depicted in Fig. 17.1b and consists of parallel connection of a conductance (G_s) and capacitance (C_s). Their values are of tenth of mS and hundreds of fF, respectively. At higher frequencies the influence of the parasitics comes in fore. Namely, the reactance starts to rise due to resonance effects with the parasitic capacitance. The reactance reaches its maximum at the internal parallel resonance frequency of the tank-circuit created by L and C_S. Above that frequency the overall reactance starts to decrease sharply giving the controlling role to the parasitics and making the component of no use.

The main problem of designing such components are the confronting requirements for large inductance and small parasitics. To avoid that the shape of the wires is investigated (circular, hexagonal and similar) and stacked inductors are used. In the latter case, due to the mutual coupling, difficulties in the design are encountered. Finally, the influence of the magnetic field on the transistors being (possibly) under the inductor is a special issue to investigate.

To avoid such a component, attempts were made to find an active circuit that simulates the inductance good enough to be implemented in high frequency CMOS integrated circuits.

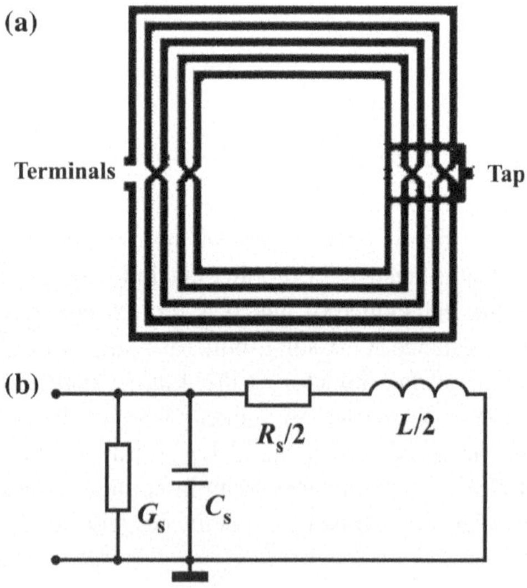

Fig. 17.1 Planar integrated inductor. **a** layout, **b** equivalent circuit of the half of the inductor

There are many solutions offering a simulated inductor using active components and capacitors. These are based on the concept of NIC (negative impedance converter) which will be not elaborated further here as a general circuit theoretical issue. The main difference in all such circuits if high frequencies are planned to be used is related to the active component. Namely, the CMOS operational amplifier (OA) performance is not good enough for these purposes so that the OTA

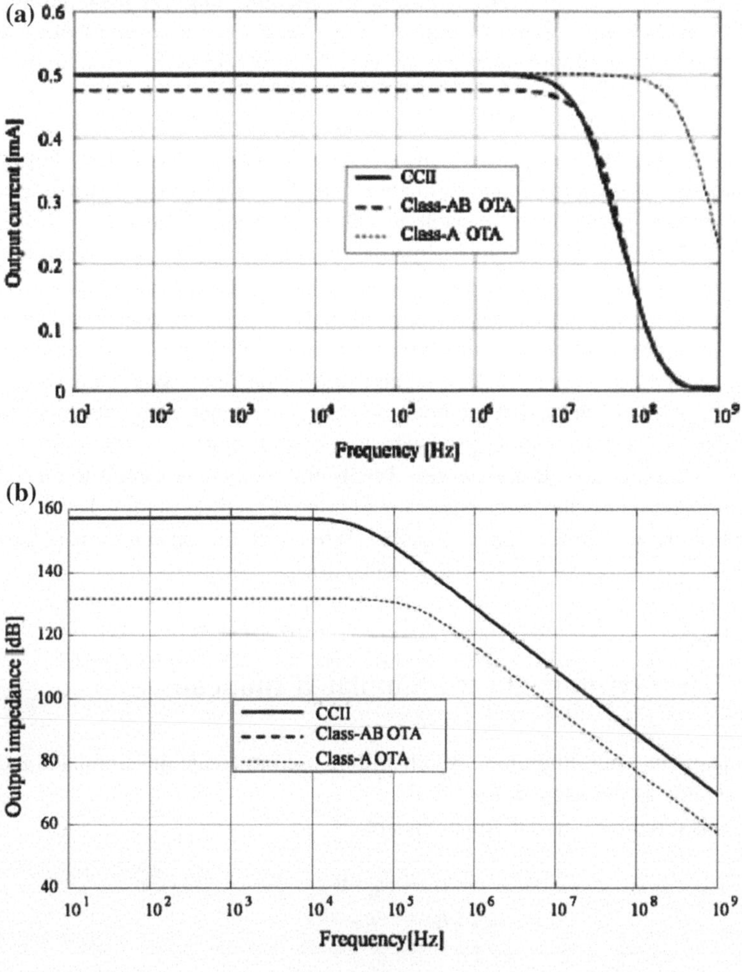

Fig. 17.2 Properties of OTA. **a** Transconductance as a function of frequency and **b** output impedance (modulus) as a function of frequency [7]. (Courtesy of the authors)

(operational transconductance amplifier) took over. Based on that solutions were found for the simulated inductances [2, 3] of which we will later on elaborate the one based on gyrators [4] and described in [5].

The OTA itself is not perfect, too. Its main characteristics are the frequency dependence of its transconductance and the output capacitance. Ideally, one would like to have a perfect OTA which means a component with zero valued output capacitance and frequency independent controllable transconductance. There is no such perfect component, however, despite the fact that improvements are reported almost on daily basis [6]. As an example Fig. 17.2a depicts the frequency dependence of the transconductance of an OTA [7] obtained by simulation while Fig. 17.2b depicts the corresponding output impedance. As can be seen high frequencies are reached (cut-off frequency is claimed to be 567 MHz) the transconductance being much more frequency independent than the output impedance. Unfortunately neither the transconductance nor the output impedance is given in absolute values so one is not capable to extract further conclusions.

The fundamental idea of implementation of simulated inductance is based on the availability of LC cascaded circuits which are synthesized by some other filter synthesis software system or even extracted from an existing catalogue such as [8]. In that way the inductors are substituted by an equivalent circuit containing OTAs and a capacitor and the rest of the filter elements (capacitors) remain the same. That, of course, is a very attractive method and even a designer with extremely limited knowledge of filter design can produce successful designs.

In this chapter we will review very briefly the equivalent circuit to an inductor and demonstrate how the synthesis is functioning. The filter obtained will be tested using ideal OTAs (infinite output impedance) to check for the synthesis process and not for the properties of the practical realization.

17.2 The Gyrator and the Simulated Inductor

The fundamental building block which will be used to create the simulated inductor is the gyrator as depicted in Fig. 17.3.

For this circuit the nodal equations are

$$\begin{aligned} I_1 &= g_m \cdot V_2 \\ I_2 &= -g_m \cdot V_1 \end{aligned} . \tag{17.1}$$

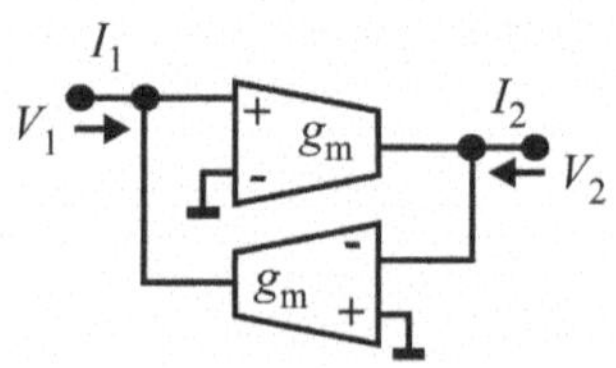

Fig. 17.3 Gyrator realized by a pair of transconductors

When loaded by an impedance Z_L, the output voltage will be

$$V_2 = -Z_L \cdot I_2. \tag{17.2}$$

After substitution in (17.1) one gets

$$I_1 = g_m^2 \cdot Z_L \cdot V_1, \tag{17.3}$$

For $Z_L = 1/(j\omega C)$, one has

$$I_1/V_1 = \frac{g_m^2}{j\omega C} = \frac{1}{j\omega L_e} \tag{17.4}$$

where

$$L_e = C/g_m^2. \tag{17.5}$$

A gyrator loaded by capacitor will behave as an inductor. Since the capacitor has one terminal grounded the resulting simulated inductor will be grounded, too.

It is up to the designer to decide whether to use a fixed value for the transconductance or for the capacitor in order to create the desired value of the inductance. Namely, the transconductance may be voltage controlled while the voltage digitally created which means that, in general, finer granulation may be achieved. The capacitance is programmed by connecting and disconnecting incremental capacitances already available on the chip. From the functionality point of view, however, there may be an additional consideration. Namely, one would prefer smaller transconductances since in that case the output impedance is expected to be larger which makes the component nearer to the perfect OTA. If the transconductance is chosen to be variable this opportunity (probably) will be not available.

17.2.1 Floating Simulated Inductor

The schematic depicted in Fig. 17.4 represents a connection of two gyrators and a capacitor to produce a floating inductor. To show that we will write the nodal equations for the circuit as

$$\begin{aligned} I_1 &= g_m \cdot V_C \\ I_2 &= -g_m \cdot V_C \\ j\omega C \cdot V_C - g_m V_1 + g_m V_2 &= 0 \end{aligned} \tag{17.6}$$

After eliminating V_C from the third equation and having in mind $I_1 = -I_2$, one gets

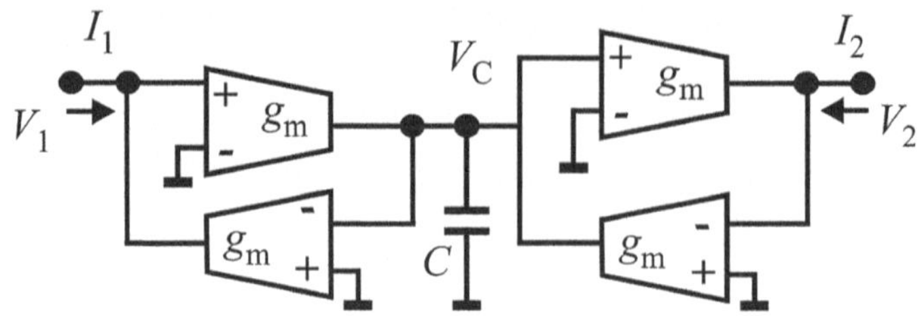

Fig. 17.4 Simulated inductor

$$Z_L = \frac{V_1 - V_2}{I_1} = \frac{j\omega C}{g_m^2}. \tag{17.7}$$

This means that the circuit of Fig. 17.4 behaves as a floating inductor of inductance

$$L_e = C/g_m^2. \tag{17.8}$$

Note, at $\omega = 0$, from (17.6) one gets $V_1 = V_2$ which corresponds to real inductor. The voltage V_C is undefined and so are the currents I_1 and I_2. The last two will be defined by the outer circuit as is the case with the original inductor.

17.2.2 *Simulated Ideal Transformer*

To produce a simulated transformer based on gyrators one may use the circuit of Fig. 17.5.

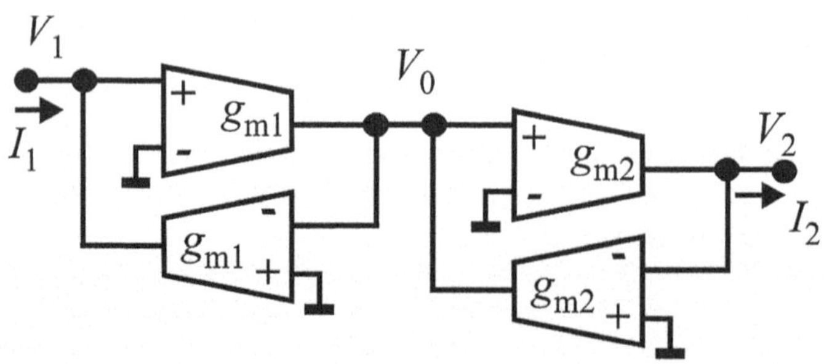

Fig. 17.5 Simulated transformer

Following are the nodal equations of this circuit

$$\begin{aligned} I_1 &= g_{m1} \cdot V_0 \\ I_2 &= -g_{m2} \cdot V_0 \\ -g_{m1} V_1 &+ g_{m2} V_2 = 0 \end{aligned} \quad . \tag{17.9}$$

The transformer's equation is now

$$\frac{V_2}{V_1} = \frac{g_{m1}}{g_{m2}}. \tag{17.10}$$

Note the transformer depicted here has one terminal of both input and output, grounded. That is acceptable for the implementation of the cell realizing complex transmission zero (D-section) depicted in Fig. 14.6b.

Negative "turn-ratio" may be obtained by inverting (interchanging the input terminals) of both transconductance amplifiers in one of the gyrators.

17.3 Circuit Synthesis

The circuit synthesis of this kind of filters is straightforward. One is to synthesize first an LC filter using a conventional synthesis procedure using a conventional synthesis program e.g. *cascade_LP* of the *RM* software. The next step is to substitute the inductors and, if necessary, transformers with their models using OTAs. Here we demonstrate the equivalent circuits to the ones described in Chap. 14. A limited set is given to save space. Nevertheless, this set is satisfactory for most physical realizations especially when low-pass circuits are sought.

Figure 17.6 depicts the equivalent circuit to the grounded inductor. As can be seen from now on the transconductance is considered a constant while the value of the capacitance is evaluated from (17.5) to be

$$C = L \cdot g_m^2. \tag{17.11}$$

At $\omega = 0$ the equivalence is failing since no current flow towards the ground is possible. In the case of synthesis of Gm-C filters based on LC prototypes this problem is usually mitigated by the fact that the inductor is either connected in series with a capacitor or there are two capacitors (to the left and to the right) which disconnect the inductor from DC signals. This will be demonstrated later on by the cell realizing a complex transmission zero without a transformer.

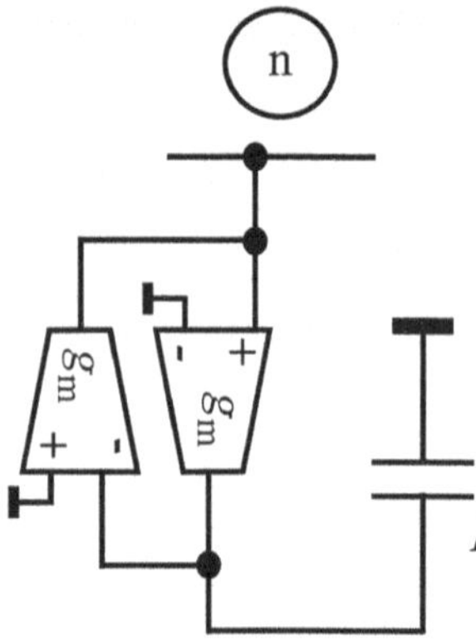

Fig. 17.6 Simulated grounded inductor

Equation (17.11) may be used as a reference when choosing the value of the transconductance. Namely, a small transconductance e.g. $g_m = 10^{-6}$ S, would produce extremely small capacitances. If, for example $L = 100$ μH, one would produce $C = 0.1$ fF which is fairly small value and is in the range of the parasitic capacitances in any CMOS technology. In the opposite case, when large g_m is chosen, the resulting capacitance may become very large. For example, if $g_m = 10^{-1}$ S, and $L = 100$ μH, one gets $C = 10$ nF. It seems that for this inductance a value of $g_m = 10^{-3}$ S, would be preferable. The question is, however, which is the output resistance of such an OTA. If satisfactory, the goal is reached. If not, one must go for a compromise.

Figure 17.7 depicts the equivalent circuit to the one depicted in Fig. 14.4b. It is the equivalent to the Brune's cell. No additional comments are necessary since the capacitance is calculated again from 17.11.

Its alteration realizing transmission zero at the real axis with a cell using a transformer (as depicted in Fig. 14.5b) is depicted in Fig. 17.8. Note, since for an ideal transformer the turn ratio is defined by

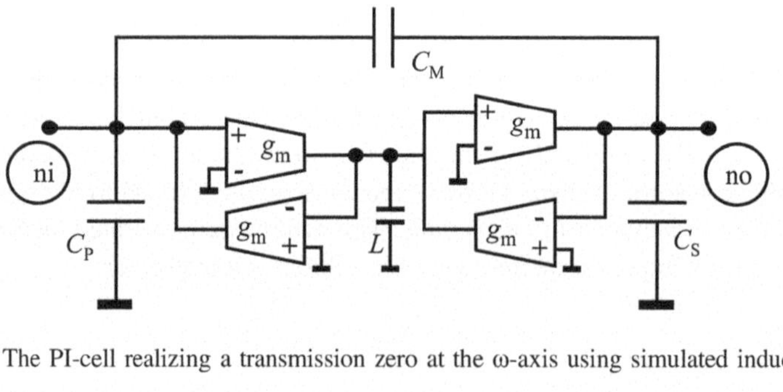

Fig. 17.7 The PI-cell realizing a transmission zero at the ω-axis using simulated inductor

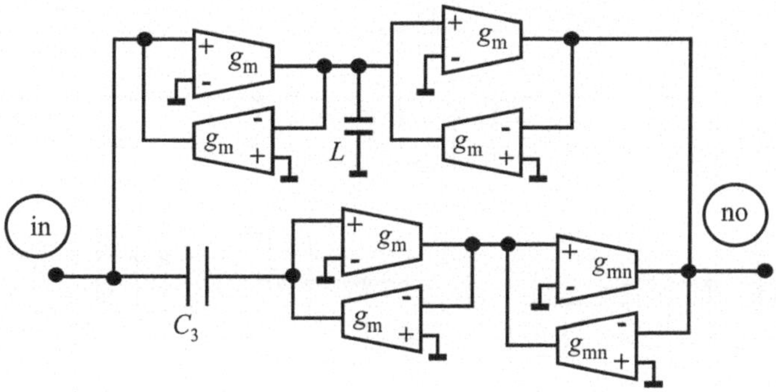

Fig. 17.8 Equivalent circuit to the version of the Brune's cell depicted in Fig. 14.5b

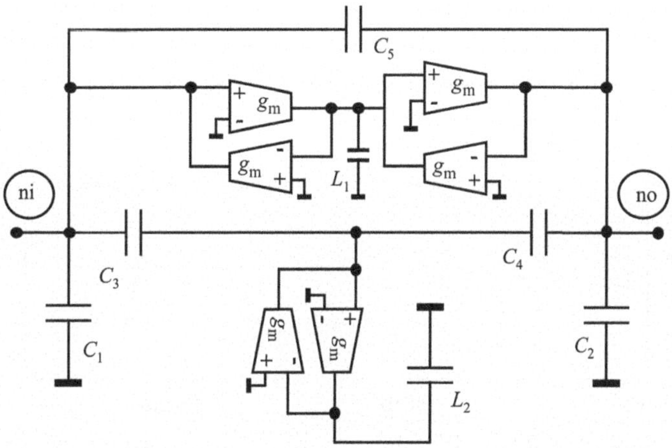

Fig. 17.9 The cell realizing a complex transmission zero without a transformer

$$V_2/V_1 = 1/n, \tag{17.12}$$

having in mind the notation of Fig. 17.8, by comparison with (17.10) we have

$$g_{mn} = n \cdot g_m. \tag{17.13}$$

Figure 17.9 and Fig. 17.10 depict the equivalent circuits to the versions of the D-section depicted in Fig. 14.6b and Fig. 14.6c, respectively. The first one is transformerless while the second includes model of a transformer.

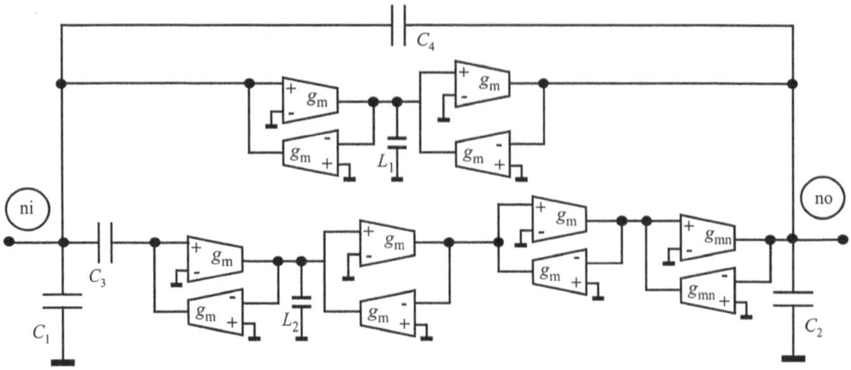

Fig. 17.10 The cell realizing a complex transmission zero using a transformer

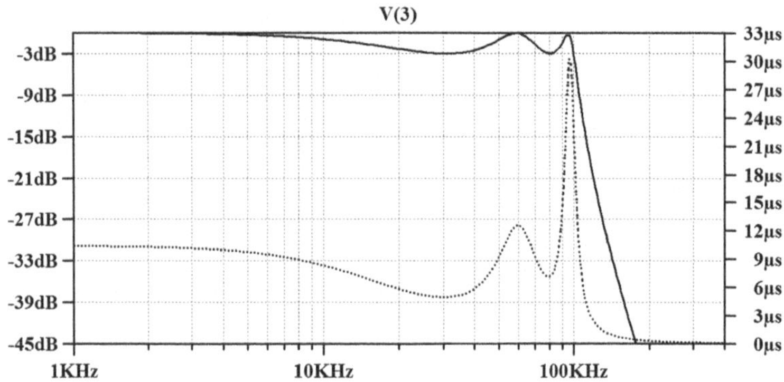

Fig. 17.11 SPICE simulation results for the example

17.4 Design Example

As an example a 5th order Chebyshev filter exhibiting 3 dB attenuation in the passband will be used. In this case, exceptionally, we are delivering the SPICE net-list together with the edited .html report. As can be seen $g_m = 10^{-4}$ S was used. Figure 17.11 depicts the SPICE simulation results.

Start.E

```
Program: Gm_LC
        SYNTHESIS OF LOW-PASS Gm-CFILTERS

        Project name: GM_C_6_CH3
++++++++++++++++++++++++++++++++++++++++++++++++++++++++++++++++++++
Input data on the transfer function
------------------------------------------------------------
Order of the numerator n=0;   Order of the denominator m=5

You entered the poles of the transfer function as follows (m=5):

sigma[1]=-1.775302716e-001    omega[1]=0.000000000e+000
sigma[2]=-1.436250067e-001    omega[2]=-5.969760109e-001
sigma[3]=-1.436250067e-001    omega[3]=5.969760109e-001
sigma[4]=-5.485987094e-002    omega[4]=9.659274761e-001
sigma[5]=-5.485987094e-002    omega[5]=-9.659274761e-001
++++++++++++++++++++++++++++++++++++++++++++++++++++++++++++++++++++
All polynomials in ascending order of s or w**2

The input transfer function
--------------------------------------------------
Poles-real part:
  -1.775302716e-001   -1.436250067e-001   -1.436250067e-001   -5.485987094e-002
  -5.485987094e-002
Poles-imaginary part:
  0.000000000e+000   -5.969760109e-001   5.969760109e-001   9.659274761e-001
  -9.659274761e-001
++++++++++++++++++++++++++++++++++++++++++++++++++++++++++++++++++++
Ordered transmission zeros
Zeros-real part:
  9.999999990e+009  9.999999990e+009  9.999999990e+009  9.999999990e+009
 9.999999990e+009
Zeros-imaginary part:
  9.999999990e+009  9.999999990e+009  9.999999990e+009  9.999999990e+009
 9.999999990e+009
++++++++++++++++++++++++++++++++++++++++++++++++++++++++++++++++++++
This is a low-pass function
------------------------------------------------------------------------
Zeros of the reflection coefficient
Real part
  0.000000000e+000  0.000000000e+000  0.000000000e+000  0.000000000e+000
```

0.000000000e+000
Imaginary part
5.877845956e-001 -5.877859090e-001 9.510559766e-001 -9.510570535e-001
0.000000000e+000
+++
Coefficients of the reflection coefficient (numerator):
0.000000000e+000 3.124999992e-001
0.000000000e+000 1.249999998e+000 0.000000000e+000 1.000000000e+000

Coefficients of the reflection coefficient (denominator):
6.264858082e-002 4.079663127e-001 5.489371140e-001
1.415025140e+000 5.745000270e-001 1.000000000e+000
+++
INPUT IMPEDANCE Variant Zin=RS*(1-RO)/(1+RO)
Numerator:
-6.264858082e-002 -9.546631352e-002 -5.489371140e-001 -1.650251428e-001
-5.745000270e-001
Denominator:
-6.264858082e-002 -7.204663119e-001 -5.489371140e-001
-2.665025138e+000 -5.745000270e-001 -2.000000000e+000
+++
EXTRACTION OF THE CELLS
+++
k=1 (th) ZERO AT INFINITY: parallel capacitance C=5.5406418e-009
Node_UP=1 Node_Down=0

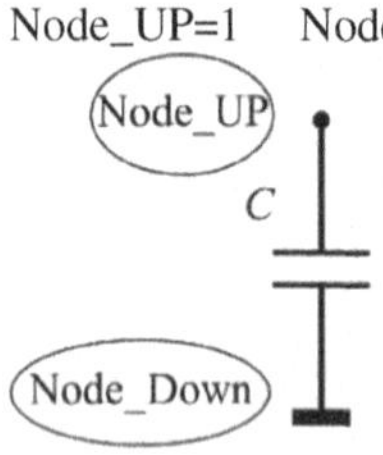

----

k=2 (th) Zero at infinity: SIMULATED series inductance L=1.2126321e-011
ni=1 no=2

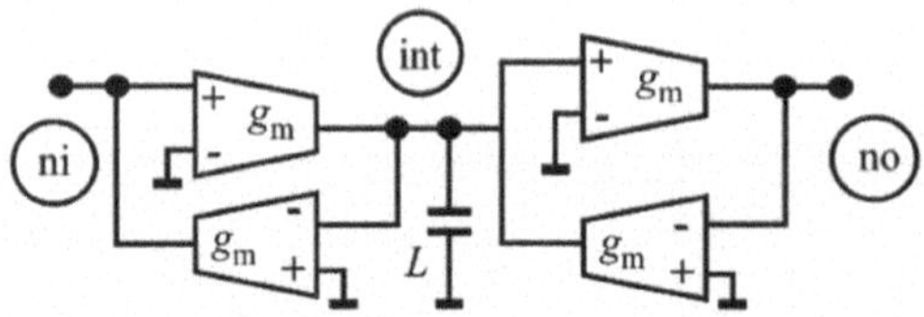

----

k=3 (th) ZERO AT INFINITY: parallel capacitance C=7.2217288e-009
Node_UP=2 Node_Down=0

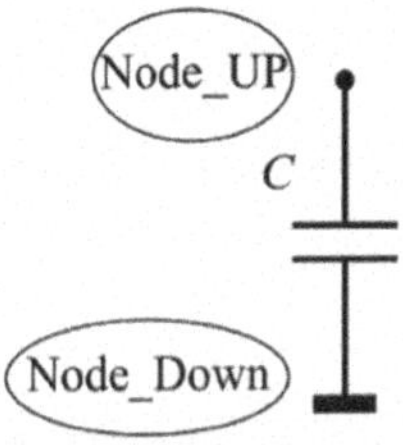

----

k=4 (th) Zero at infinity: SIMULATED series inductance L=1.2126321e-011
ni=2 no=3

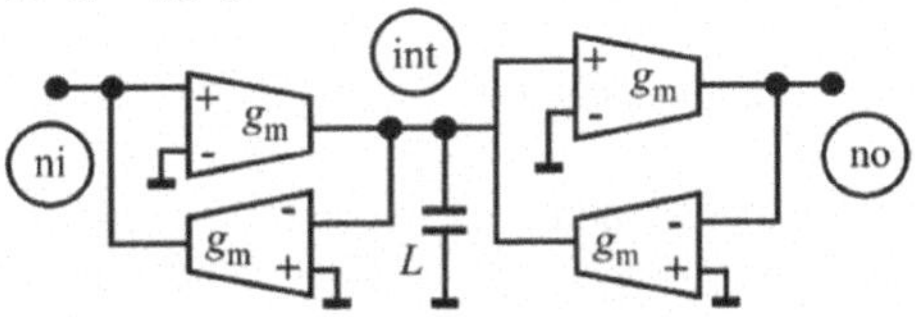

----

k=5 (th) ZERO AT INFINITY: parallel capacitance C=5.5406418e-009
Node_UP=3 Node_Down=0

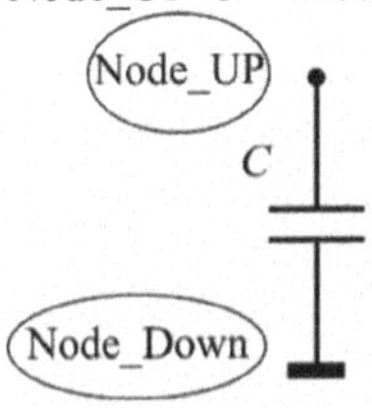

-----------------Residual load impedance--------------------

Residual is resistor (Implemented in the SPICE description) and R=1.0000000e+003
Node_end=3

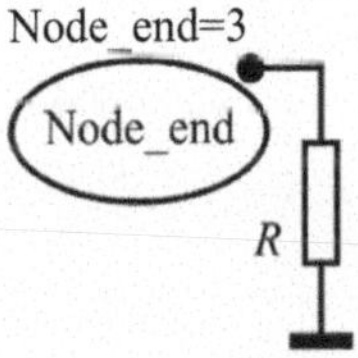

==

Here ends the synthesis process

The following is a copy of the corresponding SPICE net-list.

```
+++++++++++++++++++++++++++++++++
GM_Simulated inductance LC FILTERS.
*PROJECT NAME:          GM_C_6_CH3
*Welcome to the RM software for filter design
v  n0    0       ac      2       sin
Rgen     n0      1       1.00000e+003
*
```

```
* ZERO AT INFINITY
* parallel capacitance
C011     1        0        5.5406418035e-009
*
* ZERO AT INFINITY
* Simulated serial inductance
gleftt2  0        int2     1        0        1.000000000e-004
gleftb2  1        0        int2     0        1.000000000e-004
Rleftb2  1        0        1.000000000e+008
Rint2             int2     0        1.000000000e+008
Csim2             int2     0        1.2126321e-011
grightt2 2        0        0        int2     1.000000000e-004
RrightT2 2        0        1.000000000e+008
grightb2 0        int2     0        2        1.000000000e-004
*
* ZERO AT INFINITY
* parallel capacitance
C013     2        0        7.2217287925e-009
*
* ZERO AT INFINITY
* Simulated serial inductance
gleftt4  0        int4     2        0        1.000000000e-004
gleftb4  2        0        int4     0        1.000000000e-004
Rleftb4  2        0        1.000000000e+008
Rint4             int4     0        1.000000000e+008
Csim4    int4     0        1.2126321e-011
grightt4 3        0        0        int4     1.000000000e-004
RrightT4 3        0        1.000000000e+008
grightb4 0        int4     0        3        1.000000000e-004
*
* ZERO AT INFINITY
* parallel capacitance
C015     3        0        5.5406417802e-009
RP5      3        0        1.0000000000e+003
*AP
* Simulation settings---------------------------
.ac dec  700      1.00000e+003     4.00000e+005
*.print ac v(3)
.end
```

End.E

17.5 Developer's Corner

There is not much to advise related to this subject. The only choice to be done is the value of g_m. As seen from the example $g_m = 10^{-4}$ S leads to capacitances inside the simulated inductor which are by two orders of magnitude lower than the rest of the capacitances. Ten times larger value of g_m would gather the values of all capacitances in a smaller interval but it would ask for higher consumption and smaller output conductance (broader transistors).

To get a rough feeling on the influence of the output resistance of the OTA, the SPICE net-list given above was extended with resistors in parallel to all OTA's outputs. The simulation result is depicted in Fig. 17.12. Not surprisingly, the value of the output resistance of the OTAs has very strong influence on the frequency characteristic of the circuit. Its influence is recognizable even for values as large as 10 MΩ. Lower values are practically destroying the response.

Having in mind Fig. 17.2 where $|Z_{out}| = 10^{(Z_{out\_dB})/20}$, we may conclude that the OTA whose characteristics are depicted in Fig. 17.2 would satisfy the criterion $|Z_{out}| > 10^7$ for frequencies below (approximately) 200 kHz only. Of course, this amplifier was developed for different purposes which is recognizable from its transconductance value.

To produce a large output resistance in [6] the output stage of a transconductance amplifier depicted in Fig. 17.13 was recommended.

Its output resistance is claimed to be given as

$$R_{out} = \frac{A \cdot g_{m2}}{g_{ds1} \cdot g_{ds2}}, \tag{17.14}$$

where A stands for the amplifier's voltage gain, g_{m2} is the transconductance of the amplifying transistor and g_{ds1} and g_{ds2} are the output resistances of the transistors M_1 and M_2, respectively.

The enhancement with respect to $1/g_{ds2}$ would be $A \cdot g_{m2}/g_{ds1}$ times. Since A is supposed to be a large number and since, by no means, $g_{m2} > g_{ds1}$ or even $g_{m2} >> g_{ds1}$, one may expect a really large output resistance.

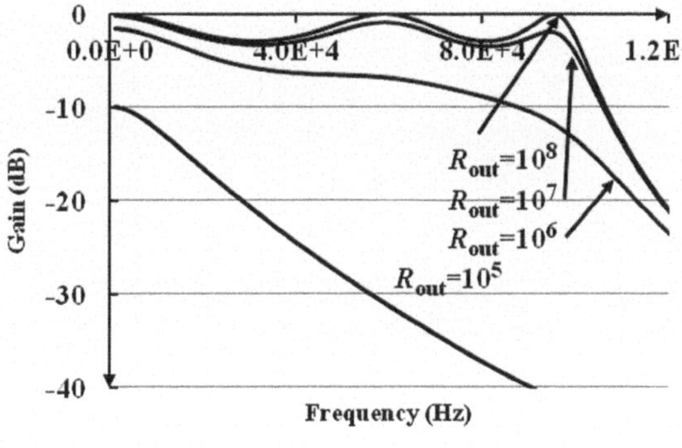

Fig. 17.12 Amplitude characteristic of the example filter for different values of the OTA's output resistance

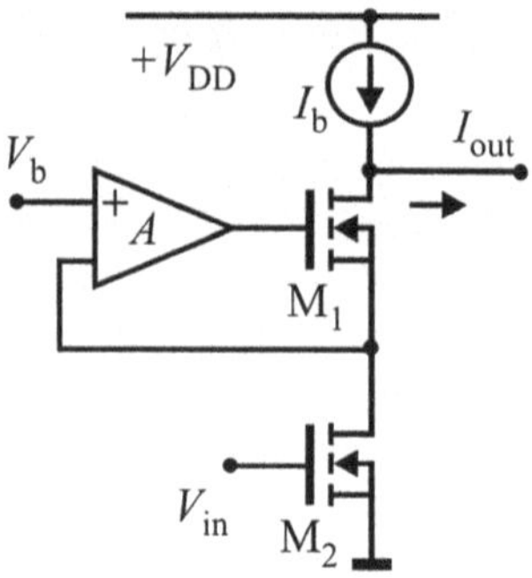

Fig. 17.13 A concept to enhance the output resistance of a transconductance amplifier

References

1. Haobijam G, Palathinkal RP (2014) Design and analysis of spiral inductors. Springer India, New Delhi
2. Tsividis YP, Voorman JO (1993) Integrated continuous-time filters: principles, design, and applications. IEEE, Piscataway, NJ
3. Lawanwisut S, Siripruchyanun M (2012) An electronically controllable active-only current-mode floating inductance simulator. In: Proceedings of the 35th international conference on telecommunications and signal processing (TSP), Prague, Czech Republic, pp 386–389
4. Tellegen BDH (1948) The gyrator, a new electric network element. Philips Res Rep 3:81–101
5. Uzunov IS (2008) Theoretical model of ungrounded inductance realized with two gyrators. IEEE Trans Circuits Syst-II: Express Briefs 55(10):981–985
6. Sanchez-Sinencio E, Silva-Martinez J (2000) CMOS transconductance amplifiers, architectures and active filters: a tutorial. IEE Proc Circuits Devices Syst 147(1):3–12
7. Santos MM, Bertemes-Filho P, Vincence VC (2012) CMOS transconductance amplifier types for low power electrical impedance spectroscopy. XXIII Congresso Brasileiro em Engenharia Biomédica – XXIII CBEB:1382–1386
8. Zverev AI (2005) Handbook of Filter Synthesis. Wiley-Interscience, New York

Chapter 18
Parallel Active SC Circuit Synthesis

Filtering at low frequencies means use of inductors and capacitors of large values. When inductors are excluded, such as in active RC circuits and small acceptable values are assigned to the capacitances, the resistances must have large values. To avoid these, simulated resistors are used in a form of a combination of capacitors and switches. Using high frequency switching along with a tolerable value of a capacitance any practical value of the resistance can be achieved, thus practically eliminating the problem. Since switches and capacitors along with operational amplifiers are used only, the abbreviation SC-filters is in use. The signal here is processed in discretized time instants so that special mathematical apparatus is implemented and new complex variable (z) is introduced to allow for the periodicity of the switching to be represented within the transfer function of the system. The old (s) and new (z) variables are usually related via the *bilinear transform*. Accepting parallel synthesis, a procedure will be described for creation of SC-filters of any complexity using only two (low-Q and High-Q) biquads and one first order cell. Example will be given whose verification will be performed by repetitive SPICE simulations in the time domain.

18.1 Introduction

The active SC (switched capacitor) circuit is built of capacitors, switches, and operational amplifiers. The rationale of introduction of this kind of circuits comes from the fact that by this technology low frequency (audio) analog signal processing is enabled for CMOS integrated circuits. In that way not only inductors are eliminated from the circuit but resistors, too. Resistors are known to occupy large area on the chip, to have large tolerances, and to be very temperature dependent. In addition, large area means also large parasitic capacitances which make audio-frequency integrated active RC circuit practically unfeasible. Now, by the use

V. Litovski, *Electronic Filters*, Lecture Notes in Electrical Engineering 596,
https://doi.org/10.1007/978-981-32-9852-1_18

of SC circuits, not only area is saved but also tolerances and temperature dependences are reduced in a simple and very effective way [1–4].

In addition to the basic well established building elements of this technology (operational amplifiers and capacitors) one needs a source of a pulse train with a stable frequency and quality switches since sharp transients produce unwanted spurious signals which may be considered as a noise to the useful signal [5–7].

The technology available, a new mathematical apparatus allowing for processing discrete signals was introduced [8, 9].

Finally, active SC circuit synthesis (mainly up to second order filter cells) was developed [10] allowing for filter synthesis of any complexity.

In this chapter we will discuss the above aspects of the SC filter synthesis in a form as reduced as possible trying to simplify the developments at most. Nevertheless, at the end, we expect, the reader will gather enough understanding of the phenomenon and knowledge of the subject to be capable do develop his/hers own software for SC filter design or to use existing software if available.

The example given is based on implementation of the routines available within the $\mathcal{RM}$ software for filter design.

18.2 Mathematical Background

A discrete time signal as depicted in Fig. 2.1 may be represented by (2.22) and (2.23) which we repeat here for convenience

$$v_d(t) = v(t) \cdot \sum_{l=-\infty}^{\infty} \delta(t - l \cdot T_s) \tag{18.1}$$

where δ is the delta function

$$\delta(x) = \begin{cases} 1 & \text{for}\, x = 0 \\ 0 & \text{for}\, x \neq 0. \end{cases} \tag{18.2}$$

Implementing the Laplace transform leads to the following s-domain presentation of (18.1):

$$V_d(s) = \sum_{l=-\infty}^{l=\infty} V(l \cdot T_s) \cdot e^{-s \cdot l \cdot T_s} \tag{18.3a}$$

Using the expression for convolution given in Table 3.2, based on [1], one may find the spectrum of $v_d(t)$ to be

$$V_d(s)_{|s=j\omega} = \frac{1}{T_s}\sum_{l=-\infty}^{l=\infty} V\left(j\omega - j\frac{2\pi \cdot l}{T_s}\right). \qquad (18.3b)$$

So, the spectrum of a sampled signal is a sum of shifted copies of the original continuous-time spectrum. If the spectrum of the original signal $v(t)$ is band-limited so that the upper frequency of the spectrum (the cut-off frequency) is lower than $0.5/T_s$ the repeated spectra will not overlap. The term aliasing is used to denote overlapping of spectra. The minimum sampling frequency required to avoid aliasing is called the Nyquist rate, and is equal to two times the signal cut-off frequency. To prevent aliasing, usually, at the input of the sampled data system an anti-aliasing low-pass continuous time filter is positioned. In the next we will consider that the input signal is band limited and no antialiazing occurs due to its bandwidth.

The switching, however, is a real process which takes time and produces unwanted components in the spectrum. To eliminate these effects anti-aliasing filter is unavoidable at the output of the system processing discrete signals.

If in (18.3a, 18.3b) we introduce the so called bilateral z-transform:

$$z = e^{sT_s} \qquad (18.4)$$

it will become

$$V_d(z) = \sum_{l=-\infty}^{l=\infty} V(l \cdot T_s) \cdot z^{-l}. \qquad (18.5)$$

Table 18.1 depicts some characteristic mappings related to the s-to-z transform with normalization implemented so that $T_s = 1$ s.

Given $X(z)$ as a transform of $x(l)$, the $x(l - k)$ maps into $z^{-k}X(z)$. This means that z^{-k} represent a delayed pulse or generally delay. Furthermore, if $y(l) = h(l) \otimes x(l)$ where $\otimes$ denotes convolution, then $Y(z) = H(z) \otimes X(z)$.

Table 18.1 Specific values of the s-to-z transform

s-domain	z-domain
0	1
$\pm j\pi/2$	$\pm j1$
$\pm j\pi$	−1
2π	1
$\pm j\pi/4$	$\sqrt{2}/2 \pm j\sqrt{2}/2$
$\pm j3\pi/4$	$-\sqrt{2}/2 \pm j\sqrt{2}/2$

18.2.1 The Bilinear Transform

Now, given a transfer function in the s-domain a question arise as to how to obtain its counterpart in the z-domain. That is important from the point of view of use of existing continuous time filter function to create discrete time filter functions and consequently discrete time processing systems—filters. To that end we will follow the developments given in [11].

Consider an analog filter whose transfer function has a form:

$$H_A(s) = \frac{\sum_{k=0}^{n} b_k s^k}{\sum_{k=0}^{n} a_k s^k} \tag{18.6}$$

Here s is the complex frequency, while **a** and **b** are vectors of real coefficients of size $n + 1$. We assume $a_n = 1$. It may be represented in the so called canonical controller form as depicted in Fig. 18.1, since its origin is a differential equation of the form

$$\sum_{k=0}^{n} a_k \frac{d^k y(t)}{dt^k} = \sum_{k=0}^{m} b_k \frac{d^k x(t)}{dt^k} \tag{18.7}$$

to which the Laplace transform was implemented.

To transform the above transfer function into a discrete one we need to substitute the integrators i.e. to find the discrete realization of:

$$\dot{g}(t) = f(t) \tag{18.8}$$

where $\dot{g}(t)$ is the input while $f(t)$ is the output of an integrator in Fig. 18.1. Its Laplace transform is

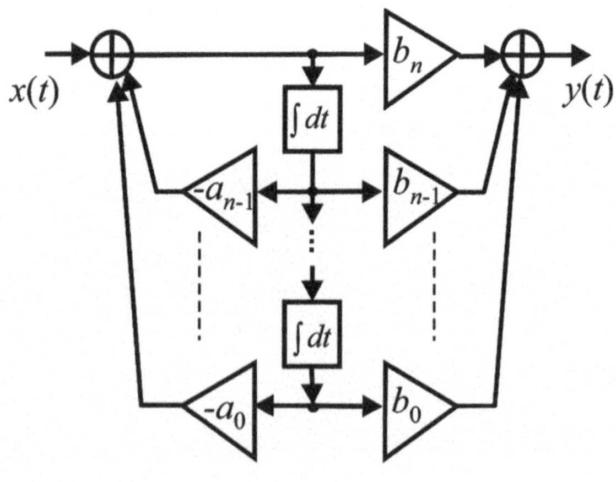

Fig. 18.1 Controller canonical form representation of an analog transfer function

$$s \cdot G(s) = F(s). \tag{18.9}$$

So, to complete the transformation of (18.6) into a discrete transfer function one needs to transform (18.9) into the z-domain.

The following general procedure may be implemented for that purpose. First, do approximate the derivative i.e. do discretize the time and transform the differential equation into a difference one. Second, apply the z-transform to the resulting difference equation, and third using analogy with (18.9), substitute $1/s$ by the approximate expression obtained in the previous step.

The derivative may be expressed using the linear multistep formula given in [12] in the following way:

$$\dot{y}_l = -\frac{\alpha_0}{\beta_0 T_s} y_l - \frac{1}{\beta_0 T_s} \cdot \sum_{j=1}^{r} \left(\alpha_j y_{l-j} + T_s \cdot \beta_j \dot{y}_{l-j} \right) \tag{18.10}$$

where l is the time step counter, meaning the evaluations happens at the moment: $t = l \cdot T_s$, T_s being the discretization (or sampling) period. r is the order of the formula. Two sets of coefficients $\boldsymbol{\alpha}$ and $\boldsymbol{\beta}$ are used to define the formula.

For the applications sought in this book the best already known formulae that may be created from (18.10) are the Euler backward and the trapezoidal as elaborated in [13, 14] and [15, 16]. Here we will briefly describe how these transformations are obtained.

In the case of the Backward Euler (or backward difference) we have: $r = 1$, $\alpha_0 = -1$, $\alpha_1 = 1$, $\beta_0 = 1$, and $\beta_1 = 0$. That leads to

$$\dot{y}_l = \frac{1}{T}(y_l - y_{l-1}) \tag{18.11}$$

or

$$\dot{y}[l \cdot T_s] = \frac{1}{T_s}(y[l \cdot T_s] - y[l \cdot T_s - T_s]). \tag{18.12}$$

After implementation of the z-transform we get

$$\dot{Y} = \frac{1}{T_s}\left(1 - z^{-1}\right)Y \tag{18.13}$$

So, by comparison with (18.9) we get the Backward Euler transform as

$$s = \frac{1}{T_s}\left(1 - z^{-1}\right) = \frac{z - 1}{z \cdot T_s}. \tag{18.14}$$

The trapezoidal rule is obtained if: $r = 1$, $\alpha_0 = -2$, $\alpha_1 = 2$, $\beta_0 = 1$, $\beta_1 = 1$. That leads to

$$\dot{y}_l = -\dot{y}_{l-1} + \frac{2}{T_s}(y_l - y_{l-1}) \tag{18.15}$$

or

$$\dot{y}[l \cdot T_s] = -\dot{y}[l \cdot T_s - T_s] + \frac{2}{T_s}(y[l \cdot T_s] - y[l \cdot T_s - T_s]). \tag{18.16}$$

After implementation of the z-transform we get

$$(1 + z^{-1})\dot{Y} = \frac{2}{T_s}(1 - z^{-1})Y \tag{18.17}$$

or

$$\dot{Y} = \frac{2}{T_s}\frac{(1 - z^{-1})}{(1 + z^{-1})}Y \tag{18.18}$$

So, by comparison with (18.9) we get Trapezoidal transform which is more frequently referred to as the bilinear transform

$$s = \frac{2}{T_s}\frac{(1 - z^{-1})}{(1 + z^{-1})} = \frac{2}{T_s}\frac{(z - 1)}{(z + 1)} \tag{18.19a}$$

or

$$z = \frac{1 + \frac{T_s}{2}s}{1 - \frac{T_s}{2}s}. \tag{18.19b}$$

If we use $s = \sigma + j \cdot \omega$ and $z = \Sigma + j \cdot \Omega$ we may produce

$$\Sigma + j \cdot \Omega = \frac{1 - \frac{T_s^2}{4}(\sigma^2 + \omega^2) + j \cdot T_s\omega}{\left(1 - \frac{T_s}{2}\sigma\right)^2 + \left(\frac{T_s}{2}\omega\right)^2}. \tag{18.20a}$$

Table 18.2 contains mapping of some poles located on important positions while Fig. 18.2 depicts the results in the z-domain.

From Table 18.2 and Fig. 18.2 one may recognize that the imaginary axis of the s-plane is mapped into a unit circle in the z-plane. Points in the left half of the s-plane are mapped onto the inside of the unit circle while the ones from the right side of the s-plane are mapped outside of the unit circle. The origin of the s-plane maps itself into 1 while infinity at both axes in the s-plane is mapped into −1 of the z-plane.

We will use the opportunity here to discuss shortly the issue of stability and minimum-phase. A continuous-time filter is stable if the poles of its transfer

Table 18.2 Specific s-to-z mappings by bilinear transform

Re{s}	Im{s}	Re{z}	Im{z}
0.0e+0	1.0e+0	9.23e−1	3.85e−1
0.0e+0	−1.0e+0	9.23e−1	−3.85e−1
0.0e+0	0.0e+0	1.00e+0	0.00e+0
1.0e+0	0.0e+0	1.50e+0	0.00e+0
−1.0e+0	0.0e+0	6.67e−1	0.00e+0
1.0e+10	0.0e+0	−1.00e+0	0.00e+0
0.0e+0	1.0e+10	−1.00e+0	0.00e+0

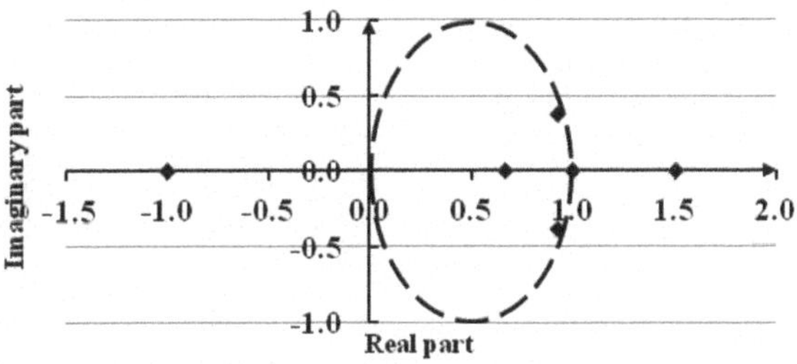

Fig. 18.2 Partial depiction of Table 18.2 (only z is depicted)

function fall in the left half of the complex s-plane. That comes from the need for all exponents in the pulse response to have negative signs which will allow the response to decay to zero at infinite time. A discrete-time filter will be stable if all poles of its transfer function fall inside the unit circle in the z-plane. Having in mind Fig. 18.2, the bilinear transform maps the left half of the complex s-plane to the interior of the unit circle in the z-plane which means that stable filters designed in the s-domain are transformed into filters in z-domain that preserve that stability.

Similarly, a continuous-time filter is minimum-phase if the zeros of its transfer function are in the left half of the s-plane. A discrete-time filter is minimum-phase if the zeros of its transfer function fall inside the unit circle in the complex z-plane. Then, the same mapping property of the bilinear transform provides that continuous-time minimum-phase filters are transformed into discrete-time filters with preserved the property of being minimum-phase.

To examine the properties of the bilinear transform we took two examples a Bessel and Butterworth filter of 6th order. These will be followed throughout the chapter. The first one provides an insight into the distortions of the group delay characteristic, and is based on sixth order Bessel filter as follows (coefficients of both all-pole s-domain transfer functions used in these examples are given in Table 18.3):

$$H(s) = \frac{10395}{s^6 + 21s^5 + 210s^4 + 1260s^3 + 4725s^2 + 10395s + 10395}. \tag{18.20b}$$

Note, in (18.20a, 18.20b) frequency normalization is used. We will consider the coefficients and poles for the beginning.

Table 18.3 Coefficients (a_k) of the transfer functions in *s*-domain ($b_0 = a_0$ and $b_k = 0$, $\forall k > 0$, with reference to (18.6))

k	Bessel, $n = 6$	Butterworth $n = 6$
0	10,395	1
1	10,395	3.864
2	4725	7.464
3	1260	9.142
4	210	7.464
5	21	3.864
6	1	1

Table 18.4 Coefficients of the transfer function in the z-domain obtained by the bilinear transform ($T_s = 0.4$ s)

k	Bessel, $n = 6$ Bilinear		Butterworth, $n = 6$ Bilinear	
	$\hat{a}_k$	$\hat{b}_k$	$\hat{a}_k$	$\hat{b}_k$
0	1.000e+0	1.888e−2	1.000e+0	2.966e−5
1	−2.039e−1	1.133e−1	−4.477e+0	1.780e−4
2	4.658e−1	2.832e−1	8.500e+0	4.449e−4
3	−9.464e−2	3.777e−1	−8.737e+0	5.932e−4
4	4.619e−2	2.832e−1	5.117e+0	4.449e−4
5	−5.504e−3	1.133e−1	−1.616e+0	1.780e−4
6	5.359e−4	1.888e−2	2.148e−1	2.966e−5

After implementation of the bilinear transform of (18.20a, 18.20b) with a normalized value of $T_s = 0.4$ s, new (z-domain) values of the coefficients of transfer function are obtained as depicted in Table 18.4.

To complete the illustration Figs. 18.3 and 18.4 represent the pole locations of both function in the s- and z-domain.

18.2.2 Transfer Function Analysis in the Z-Domain

To allow for a study of the properties of the new transfer function in the z-domain we will first extend the theory of transfer function analysis given in Chap. 3 which was developed for the case of s-domain. To that end we represent the new variable at the imaginary axis by rewriting (18.4) as

$$z = e^{sT_s}_{|s=j\omega} = \cos(x) + j\sin(x). \qquad (18.21)$$

Here, despite the fact that we are in the z-domain, we use the same notation for the angular frequency as we did in the s-domain. No harm will be done as long as we are aware of the domain. In addition, to simplify the development, new variable $x = \omega \cdot T_s$ was introduced.

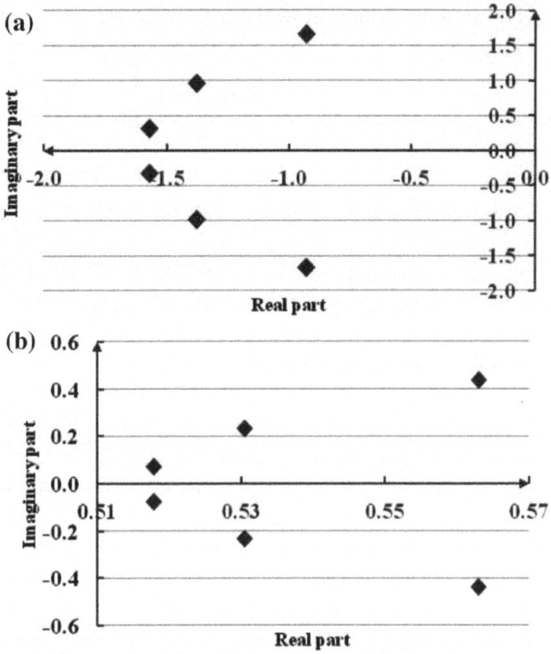

Fig. 18.3 Poles of the 6th order Bessel filter. **a** In the s-plane and **b** in the z-plane mapped with $T_s = 0.4$

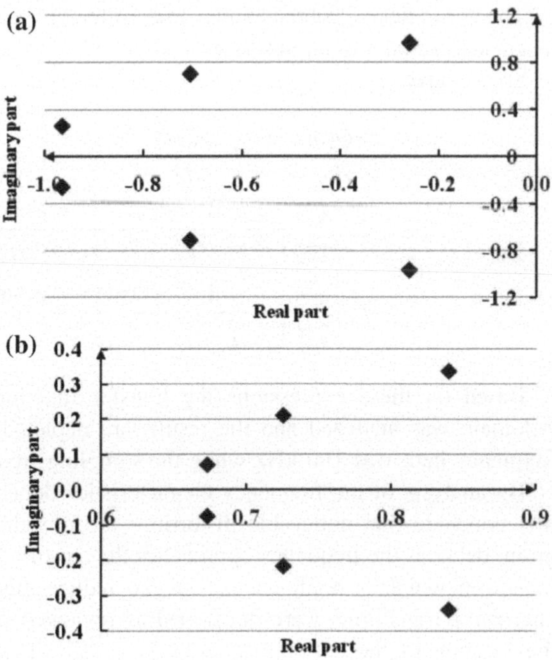

Fig. 18.4 Poles of the 6th order Butterworth filter. **a** In the s-plane and **b** in the z-plane mapped with $T_s = 0.4$

The transfer function in the z-domain may be expressed as

$$H_D(z) = A_0 \frac{\prod_{k=1}^{m} (z - z_k)}{\prod_{k=1}^{n} (z - p_k)}. \tag{18.22}$$

A_0 is here chosen as already explained for the function in the s-domain. After substitution of (18.21) and $z_k = \alpha_k + \mathrm{j} \cdot \beta_k \, p_k = \gamma_k + j \cdot \delta_k$ into (18.22) it becomes

$$H_D(z) = A_0 \frac{\prod_{k=1}^{m} [\cos(x) + j \cdot \sin(x) - (\alpha_k + j \cdot \beta_k)]}{\prod_{k=1}^{n} [\cos(x) + j \cdot \sin(x) - (\gamma_k + j \cdot \delta_k)]} = |H_D(\mathrm{j}\omega)| e^{j\Phi(\omega)}. \tag{18.23}$$

In this expression we use

$$|H_D(\mathrm{j}\omega)|^2 = |A_0|^2 \frac{\prod_{k=1}^{m} \left\{ [\cos(x) - \alpha_k]^2 + [\sin(x) - \beta_k]^2 \right\}}{\prod_{k=1}^{n} \left\{ [\cos(x) - \gamma_k]^2 + [\sin(x) - \delta_k]^2 \right\}} \tag{18.24}$$

and

$$\Phi(\omega) = \arg\{A_0\} + \sum_{k=1}^{m} \mathrm{arctg}\left(\frac{\sin(x) - \beta_k}{\cos(x) - \alpha_k} \right) - \sum_{k=1}^{n} \mathrm{arctg}\left(\frac{\sin(x) - \delta_k}{\cos(x) - \gamma_k} \right). \tag{18.25}$$

The group delay is obtained in a quiet different way since the derivation has to be taken with respect to ω, not to x.

So we have

$$\begin{aligned} \tau_d &= -\frac{d\Phi(\omega)}{d\omega} = -T_s \frac{d\Phi(x)}{dx} \\ &= -T_s \cdot \sum_{k=1}^{m} \frac{1 - \alpha_k \cos(x) - \beta_k \sin(x)}{1 + \alpha_k^2 + \beta_k^2 - 2 \cdot [\alpha_k \cos(x) + \beta_k \sin(x)]} \\ &\quad + T_s \cdot \sum_{k=1}^{n} \frac{1 - \gamma_k \cos(x) - \delta_k \sin(x)}{1 + \gamma_k^2 + \delta_k^2 - 2 \cdot [\gamma_k \cos(x) + \delta_k \sin(x)]}. \end{aligned} \tag{18.26}$$

Based on these expressions the transfer function of the Bessel filter in the z-domain was analyzed and the results are depicted in Fig. 18.5a, b (the cut-off frequency here was 100 kHz while the sampling frequency was 250 kHz).

By analysis of the frequency characteristics depicted in Fig. 18.5a, b one may first conclude that noticeable distortions are introduced in the magnitude of the group delay as the frequency approaches the cut-off. That is inherent to the bilinear transform which is nonlinear per se. To reduce this kind of distortions a set of alternative transforms were developed as reviewed in [11]. In addition, however, the location of the cut-off frequency is shifted. This shift may be found by the following analysis [12].

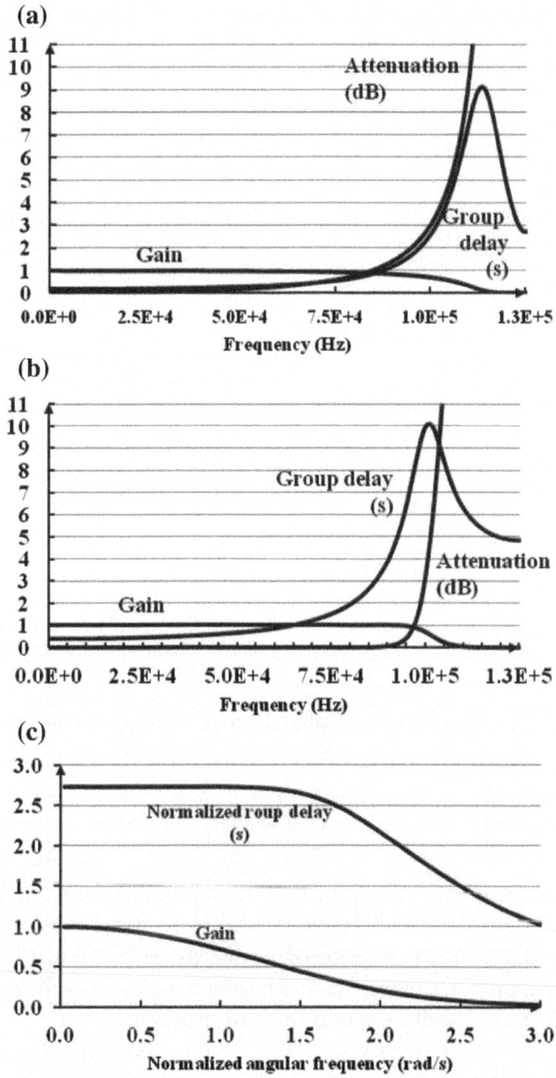

Fig. 18.5 **a** and **b** Gain, attenuation, and group delay of transformed 6th order Bessel and Butterworth filters, respectively, with $T_s = 0.4$. **c** Gain and group delay of transformed 6th order Bessel filter with $T_s = 0.01$

$$H_D(z) = H_A\left(\frac{2}{T_s}\cdot\frac{z-1}{z+1}\right) = H_A\left(\frac{2}{T_s}\cdot\frac{e^{j\omega_D T_s}-1}{e^{j\omega_D T_s}+1}\right) = H_A\left\{j\cdot\frac{2}{T_s}\cdot\tan\left(\omega_D\frac{T_s}{2}\right)\right\} \tag{18.27}$$

From this expression we see that there is nonlinear relation between the angular frequency in the s-domain (ω_A) and the angular frequency in the z-domain (ω_D). This phenomenon is referred to as frequency warping. We should interpret this

Table 18.5 Shift of the cut-off frequency as a consequence of the bilinear transform

Type	Normalized angular frequency (rad/s)	Gain
Bessel	1.0	0.704
Butterworth	1.0	0.691

result as: the gain and phase shift that the discrete-time filter has at frequency ω_D is the same gain and phase shift that the continuous-time filter has at frequency

$$\omega_A = \frac{2}{T_s} \cdot \tan\left(\omega_D \frac{T_s}{2}\right). \tag{18.28}$$

The continuous-time filter should be designed to compensate for this frequency warping by setting in advance ω_A from (18.28) for the cut-off frequency or center frequency. In other words for a given discrete time cut-off frequency the poles and zeros of the continuous time function are supposed to be divided (renormalized) by ω_A from (18.28) before entering the bilinear transformation. This is called pre-warping in the filter design.

The opposite of (18.28) would be

$$\omega_D = \frac{2}{T_s} \cdot \arctan\left(\omega_A \frac{T_s}{2}\right). \tag{18.29}$$

As presented by Table 18.5 at normalized angular frequency $\omega = 1$ rad/s the gain is still smaller of its original value $1/\sqrt{2} = 0.707$ despite the warping being performed. The reason for that is the fact $T_s = 0.4$ is really a large value. The discrepancies seen here, however, may still be considered negligible.

Since $\tan(x)_{|x \ll 1} \approx 1$, for low frequencies, $\omega_A \approx \omega_D$, and the warping is not exposed. In other words if small T_s is chosen no distortions of the amplitude characteristic will be produced. Even the distortions of the group delay will be reduced down to unrecognizable as depicted in Fig. 18.5c. That means that the sampling frequency should be much higher than the cut-off frequency of the system. That is usually referred to as oversampling.

What can we see from Fig. 18.6 is a consequence of the z-transform. Namely the amplitude (and group delay) characteristic of the discrete time filter repeats itself periodically with a period of $2\pi/T_s$. This effect when seen from the continuous time processing point of view is unwanted and before coming back to it one needs to band-limit the response of the discrete-time filter by a low-pass continuous time filter which is again referred to as anti-aliasing. The cut-off frequency of the anti-aliasing filter must be low enough to suppress all "formants" of the discrete time signal and large enough for the filter not to interfere with the properties in the main lobe of the frequency characteristic of the resulting output signal.

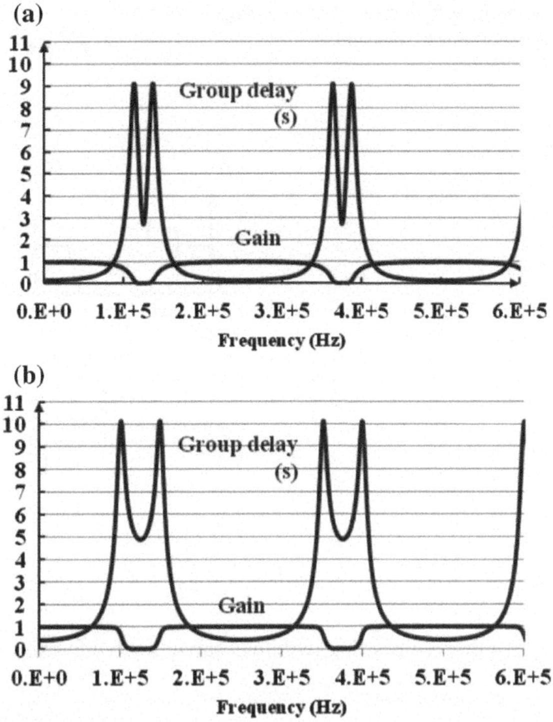

Fig. 18.6 Gain and group delay of transformed 6th order filters. $T_s = 0.4$. **a** Bessel filter and **b** Butterworth filter

18.3 Basics of SC Circuits

Consider the schematic depicted in Fig. 18.7a. It is a capacitor which is connected to two nodes via a switch. It is supposed that the position of the switch, either "1" or "2", is defined by two non-overlapping pulse trains with the same frequency $f_s = 1/T_s$ as depicted in Fig. 18.7b. T_s is, as earlier, the sampling frequency while the pulse trains are denoted ϕ_1 and ϕ_2. We will refer to this circuit element as the switched capacitor.

If connected between two nodes with voltages v_1 and v_2 the following may be observed. When ϕ_1 is high the switch is thrown left and the capacitor is charged to a value of $q_1 = C_R \cdot v_1$. In the next half of the period ϕ_2 goes high and the switch connects the capacitor to node 2. The new value of the voltage is v_2 and the charge is $q_2 = C_R \cdot v_2$. The corresponding change of the charge is $\Delta q = q_1 - q_2 = C_R(v_1 - v_2)$. Now, if the frequency of switching f_s is much larger than the frequency of the signal to be processed f so that the values of v_1 and v_2 remain unchanged during sampling, then the average current transferred by the capacitor will be

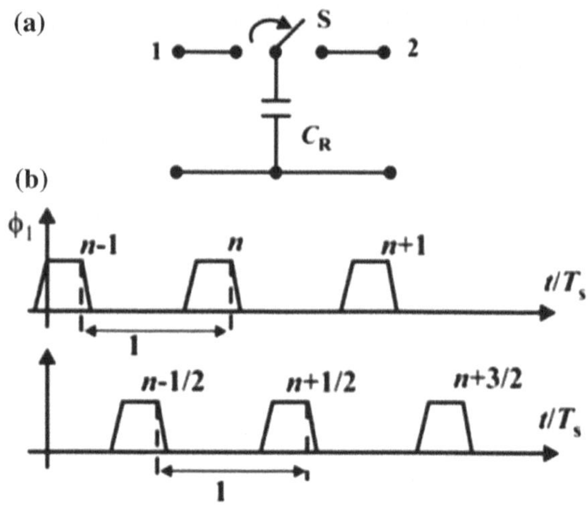

Fig. 18.7 **a** Switched capacitor and **b** controlling pulse trains

$$i_C = \Delta q/T_s = \frac{C_R}{T_s}(v_1 - v_2) = \frac{1}{R}(v_1 - v_2). \tag{18.30a}$$

In other words the switched capacitor behaves as a resistor. Its resistance is

$$R = \frac{T_s}{C_R}. \tag{18.30b}$$

Since the values of C_{R} and T_{s} are not mutually coupled a large span of resistances may be obtained. For example if $C_{\text{R}} = 1$ pF and $f_{\text{s}} = 1$ MHz, the resistance will be $R = 1$ MΩ which is a really large value for an integrated polysilicon resistor. To get the picture, since a polysilicon resistors exhibits a sheet resistance of about 25 Ω/□ (This is read as: 25 ohms per square.), one will need 40,000 □ to produce 1 MΩ. That would be a very long resistor. In the case of n-well resistors the situation would be better from the dimensions point of view (it will need 100 □ only) but it will be a very temperature sensitive one. There is another advantage of this resistor which is related to the fact that in SC circuits the filter's coefficients are expressed as quotients of capacitances which not only exploits the better temperature and tolerance properties of the integrated capacitor but also enables mutual compensation of the simultaneous variations of the capacitances. This issue will be not elaborated any further in this book.

Due to the imperfections of the real switches a so called parasitic insensitive version of the switched capacitor emulating an ungrounded resistor is in use as depicted in Fig. 18.8a. Again, we will not go into details in this matter. Note, using the trick to change the phases of the switch on the right-hand side of the capacitor (as in Fig. 18.8b), one may produce a negative resistance which is frequently indispensible component in filter design.

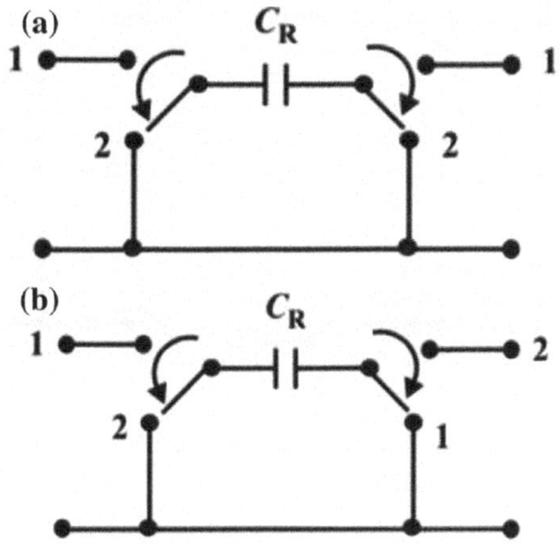

Fig. 18.8 A parasitics insensitive switched capacitor. **a** Positive resistance and **b** negative resistance

One may substitute the resistor by a switched capacitor directly into an existing circuit and that is done frequently. The only problem which is encountered is the fact that some of the switches become redundant such as parallel or series connection of two switches having the same controlling signal. Procedure of switch sharing is applied in complex schematics.

In the next we will introduce the so called sampled data operation and, consequently, the new variable z so enabling the treatment of these circuits as sampled data circuits and use the z-domain transfer functions for filter synthesis. In fact we will show that difference (in place of differential) equations may be used to describe the circuit equilibrium. These equations will be Laplace transformed to introduce the z variable as it was done in (18.5).

Following the developments in [3] we will consider the circuit (lossy integrator) of Fig. 18.9a. In the first phase (ϕ_1 is high) the circuit of Fig. 18.9b is representing the temporary status. For this circuit one may write

$$i_1 = C_1 \frac{dv_1}{dt} = -(C_2 + C_R) \frac{dv_{out}}{dt}. \tag{18.31}$$

After discretization with the Euler Backward rule with a discretization interval of $T_s/2$ it becomes

$$C_1 \cdot [v_1(nT_s) - v_1(n - 1/2)T_s] = -(C_2 + C_R) \cdot [v_{out}(nT_s) - v_{out}(n - 1/2)T_s]. \tag{18.32a}$$

Since, according to Fig. 18.9c, $C_1 \cdot v_1[(n - 1/2)T_s] = 0$ and $C_2 \cdot v_{out}[(n - 1/2)T_s] = 0$, (18.32a) becomes

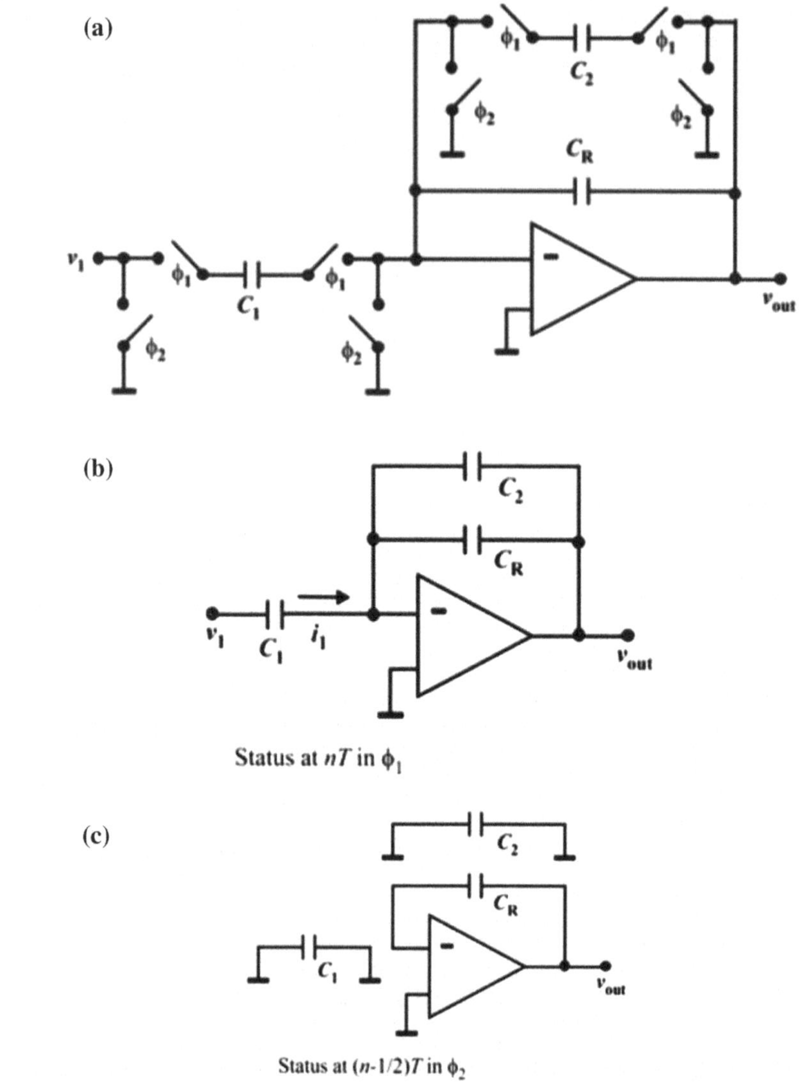

Fig. 18.9 Development of the difference equation. **a** The lossy integrator, **b** phase 1 and **c** phase 2

$$C_1 \cdot v_1(nT_s) = -(C_2 + C_R) \cdot v_{\text{out}}(nT_s) + C_R \cdot v_{\text{out}}(n - 1/2)T_s \tag{18.32b}$$

Further, we take into account that since the operational amplifier was isolated starting with the instant $(n - 1)T_s$ the output voltage remained the same i.e. $v_{\text{out}}[(n - 1/2)T_s] = v_{\text{out}}[(n - 1)T_s]$, which leads to

$$C_1 \cdot v_1(nT_s) = -C_2 \cdot v_{\text{out}}(nT_s) - C_R \cdot \{v_{\text{out}}(nT_s) - v_{\text{out}}[(n-1)T_s]\} \qquad (18.32c)$$

This is a difference equation and it will be Laplace transformed as we did with (18.12) and (18.16). The result is

$$C_R \cdot \left[v_{\text{out}}(n) - v_{\text{out}}(n) \cdot e^{-sT_s}\right] + C_2 \cdot v_{\text{out}}(n) = -C_1 \cdot v_1(n). \qquad (18.33)$$

The corresponding transfer function, after introducing the z-transform, is

$$H_D(z) = \frac{-C_1}{C_R(1 - z^{-1}) + C_2} = -\frac{C_1}{C_2 + C_R} \cdot \frac{z}{z - \frac{C_R}{C_2 + C_R}} \qquad (18.34)$$

This result confirms that the switched capacitor circuits behave as discrete time signal processors and that the theory developed on the transfer function analysis may be implemented directly.

Synthesis of SC circuits with given filter functions, however, is not so straightforward. In fact, due to the complexity of the creation of the transfer function (as seen above) a limited number of solutions are in use mainly restricted to second order cells. In the next paragraph we will further limit our choice to a minimum set of cells which, we believe, are good enough for the task.

18.4 Parallel SC Filter Design

The procedure for parallel synthesis of active SC filters is conceptually the same as the procedure of parallel RC synthesis described in Chap. 16. Figure 18.10 depicts the structure of the system. Here, to repeat, the input signal is considered

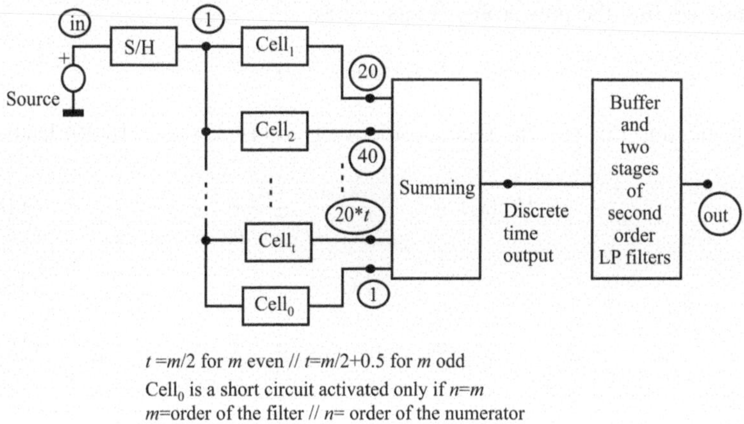

Fig. 18.10 Structure of the filter and node notation

band-limited so no anti-aliasing filter is shown at the input. Detailed information about the cell's schematic will be given in the subsequent paragraphs.

18.4.1 Analogue Prototype Function Decomposition

As for the parallelization of the transfer function in the s-domain nothing is changed with respect to what was given in Chap. 16. We will repeat here the main results related to the ith second order biquad

$$H_{e,i}(s) = G_i \frac{s + a_{0,i}}{s^2 + b_{1,i}s + b_{0,i}} \tag{18.35}$$

and the first order cell

$$H_o(s) = G_o \frac{1}{s + b_{00}}. \tag{18.36}$$

In Fig. 18.10 S/H stands for the sample-and-hold circuit. As in the case of the active RC circuits, in the case when the numerator and the denominator are of the same order (high-pass, all-pass and band-stop filters) additional path (Cell_0) is created from the S/H circuit to the summing with a gain equal to unity.

18.4.2 Analogue Prototype Function Prewarping and s-to-z Transform

Given the digital normalization frequency in the z-domain f_d and the sampling frequency we find the prewarping frequency as

$$f_w = 2 \cdot f_s \cdot \tan(\pi \cdot f_d / f_s). \tag{18.37}$$

Implementing this (for the second order cell) to (18.35) and (18.36) leads to

$$H_{e,i}(s_z) = f_w \cdot \frac{\alpha_{1,i} s_z + \alpha_{0,i}}{s^2 + \beta_{1,i} s_z + \beta_{0,i}} \tag{18.38}$$

$$\alpha_{1,i} = G_i \tag{18.38a}$$

$$\alpha_{0,i} = f_w \cdot a_{0,i} \tag{18.38b}$$

$$\beta_{0,i} = f_w^2 \cdot b_{0,i} \tag{18.38c}$$

$$\beta_{1,i} = f_w \cdot b_{1,i} \tag{18.38d}$$

$$H_o(s) = f_w \cdot G_o \frac{1}{s_z + \beta_{00}} \tag{18.39}$$

$$\beta_{00} = f_w \cdot b_{00} \tag{18.39a}$$

The bilinear transformation is now

$$s_z = \mathbf{2} \cdot \boldsymbol{f_s} \cdot \frac{z - \mathbf{1}}{z + \mathbf{1}} \tag{18.40}$$

In the case of the second order cell, after substitution into (18.38) we get

$$H_e(z) = \frac{\gamma + \varepsilon \cdot z^{-1} + \delta \cdot z^{-2}}{1 + \alpha \cdot z^{-1} + \beta \cdot z^{-2}} \tag{18.41}$$

where

$$D = 4 \cdot f_s^2 + 2 \cdot f_s \cdot \beta_{1,i} + \beta_{0,i} \tag{18.42a}$$

$$\alpha = \left(2 \cdot \beta_{0,i} - 8 \cdot f_s^2\right)/D \tag{18.42b}$$

$$\beta = \left(4 \cdot f_s^2 + \beta_{0,i} - 2 \cdot f_s \cdot \beta_{1,i}\right)/D \tag{18.42c}$$

$$\gamma = f_w \cdot \left(2 \cdot \alpha_{1,i} \cdot f_s + \alpha_{0,i}\right)/D \tag{18.42d}$$

$$\varepsilon = f_w \cdot \left(2 \cdot \alpha_{0,i}\right)/D \tag{18.42e}$$

$$\delta = f_w \cdot \left(\alpha_{0,i} - 2 \cdot \alpha_{1,i} \cdot f_s\right)/D. \tag{18.42f}$$

For the first order cell one gets

$$H_o(z) = \frac{\varepsilon \cdot (z + 1)}{z - \alpha} \tag{18.43}$$

$$D = 2 \cdot f_s + \beta_{00} \tag{18.44a}$$

$$\varepsilon = f_w G_0/D \tag{18.44b}$$

$$\alpha = (2 \cdot f_s - \beta_{00})/D. \tag{18.44c}$$

18.4.3 Switched Capacitor Amplifier

The switched capacitor amplifier is depicted in Fig. 18.11. Figure 18.11a depicts the inverting amplifier whose gain is

$$H_D(z) = -\frac{C_1}{C_2} \tag{18.45}$$

while Fig. 18.11b depict the non-inverting amplifier whose gain is

$$H_D(z) = \frac{C_1}{C_2}. \tag{18.46}$$

18.4.4 Switched Capacitor Inverting Summing Cell

The SC circuit performing summing is depicted in Fig. 18.12. It may be considered as an extension of the single input amplifier. Note, alike the active RC summing

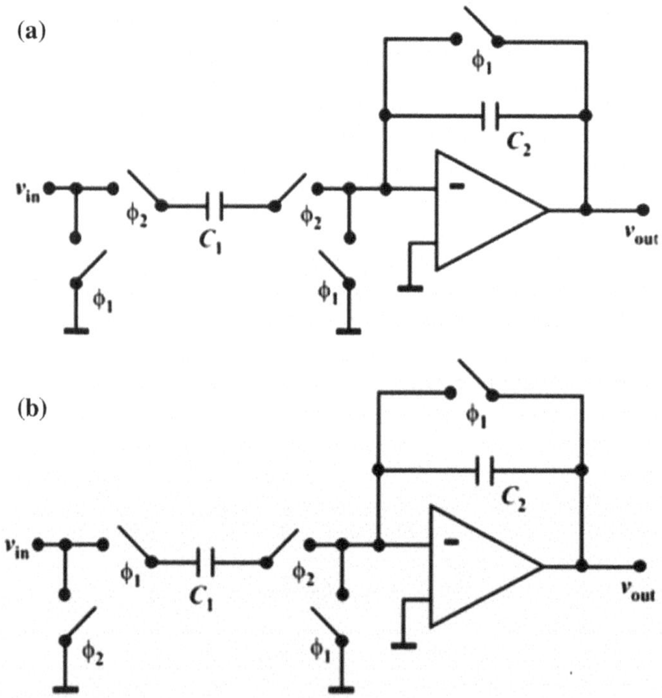

Fig. 18.11 Switched capacitor amplifier. **a** Inverting and **b** non-inverting

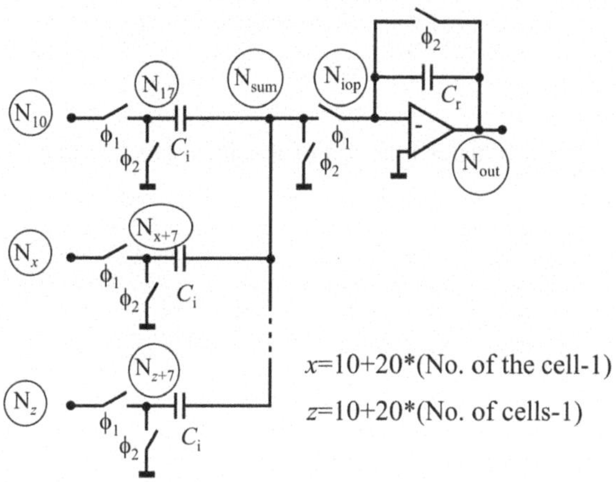

Fig. 18.12 SC summing amplifier and node enumeration in $\mathcal{RM}$ software

circuit, here inversion (if necessary) is performed simply by interchanging ϕ_1 and ϕ_2 in the proper branch on the left hand side of the capacitor as it was done in Fig. 18.11b.

Note, the node enumeration on Fig. 18.12 correspond to the node enumeration on Fig. 18.10.

18.4.5 First Order Bilinear Cell

In this case the schematic of the cell is depicted in Fig. 18.13 [1]. Switch sharing is implemented so it is not easy to recognize its s-domain prototype. By circuit analysis one gets the following transfer function

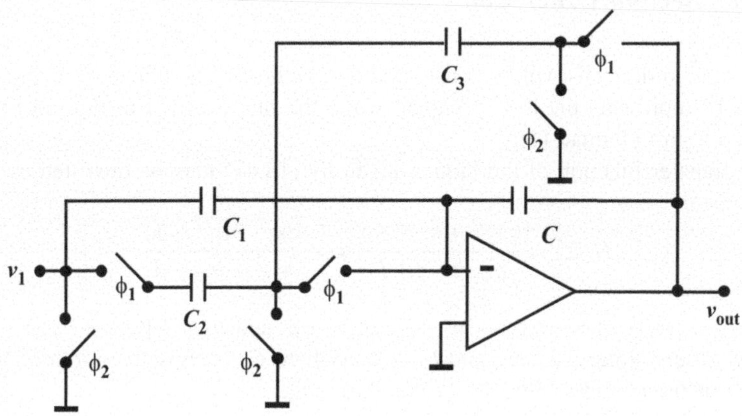

Fig. 18.13 First order SC cell

$$H_D(z) = -\frac{\frac{C_1+C_2}{C+C_3}z - \frac{C_1}{C+C_3}}{z - \frac{C}{C+C_3}} \cdot H_o(z) = \frac{\varepsilon \cdot (z+1)}{z-\alpha} \tag{18.47}$$

For a transfer function of the form (18.43) may be obtained if

$$-\frac{C_1+C_2}{C+C_3} = \varepsilon \tag{18.48a}$$

$$\frac{C_1}{C+C_3} = \varepsilon \tag{18.48b}$$

and

$$\frac{C}{C+C_3} = \alpha. \tag{18.48c}$$

To perform this synthesis, if $\varepsilon > 0$, to satisfy (18.48a) and (18.48b) one needs $|C_1| < |C_2|$ and $C_2 < 0$. Now, by choosing the value of C and $|C_2| = 2 \cdot C_1$, the rest may calculated as

$$C_3 = C \cdot (1/\alpha - 1) \tag{18.49a}$$

and

$$C_1 = \varepsilon \cdot (C + C_3) = \varepsilon \cdot C/\alpha. \tag{18.49b}$$

If $\varepsilon < 0$ one has to reverse its sign and transfer the subtraction to the summing amplifier.

18.4.6 Second Order Cell

Two second order cells will be considered for synthesis. The schematic depicted in Fig. 18.14 represents the low-Q biquad while the one depicted in Fig. 18.15 represents a high-Q biquad [1].

The transfer function of the biquad given by (18.41) may be rewritten as

$$H_e(z) = \frac{\gamma \cdot z^2 + \varepsilon \cdot z + \delta}{z^2 + \alpha \cdot z + \beta} = -\frac{a_2 z^2 + a_1 z + a_0}{b_2 z^2 + b_1 z + 1} \tag{18.50}$$

where $a_0 = -\delta/\beta$, $a_1 = -\varepsilon/\beta$, $a_2 = -\gamma/\beta$, $b_1 = \alpha/\beta$, and $b_2 = 1/\beta$.

Now, after setting $C_1 = C_0$ and $C_2 = C_0$, where C_0 is chosen appropriately, by analysis of the circuit of Fig. 18.14 one may write

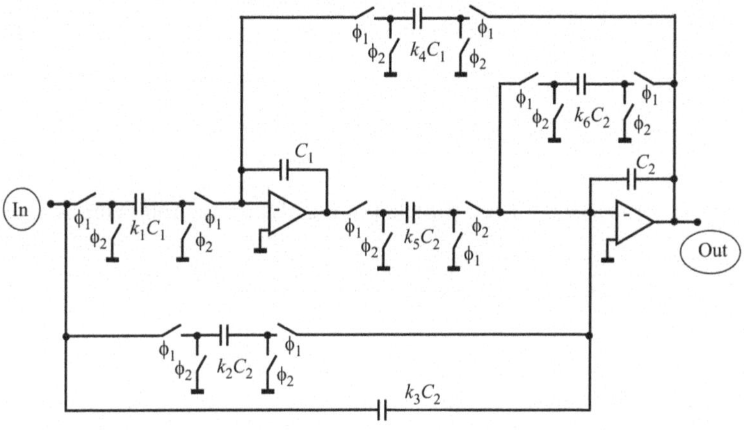

Fig. 18.14 Low Q SC biquad cell

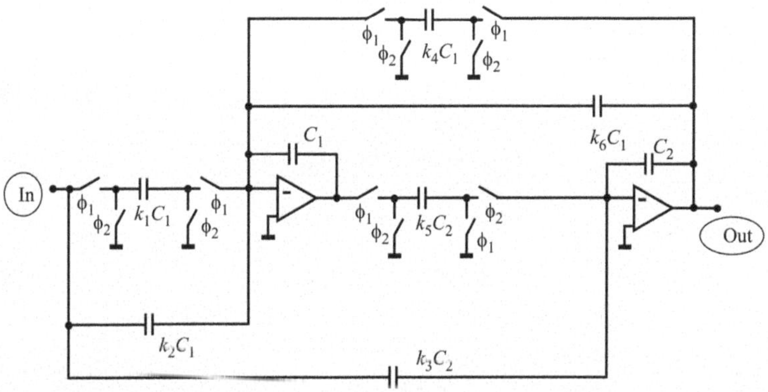

Fig. 18.15 High-Q SC biquad cell

$$H_D(z) = -\frac{(k_2 + k_3) \cdot z^2 + (k_1 k_5 - k_2 - 2k_3) \cdot z + k_3}{(1 + k_6) \cdot z^2 + (k_4 k_5 - k_6 - 2) \cdot z + 1}, \tag{18.51}$$

which, when contrasted to (18.50), gives

$$k_3 = a_0, \tag{18.52a}$$

$$k_2 = a_2 - a_0, \tag{18.52b}$$

$$k_6 = b_2 - 1, \tag{18.52c}$$

$$k_4 = k_5 = \sqrt{b_1 + b_2 + 1}, \tag{18.52d}$$

and

$$k_1 = (a_0 + a_1 + a_2)/k_5. \tag{18.52e}$$

In the case of the High-Q cell depicted in Fig. 18.15, the following is applicable

$$H_e(z) = \frac{\gamma \cdot z^2 + \varepsilon \cdot z + \delta}{z^2 + \alpha \cdot z + \beta} = -\frac{a_2 z^2 + a_1 z + a_0}{z^2 + b_1 z + b_0} \tag{18.53}$$

where $a_0 = -\delta$, $a_1 = -\varepsilon$, $a_2 = -\gamma$, $b_1 = \alpha$, and $b_0 = \beta$.

Now, by circuit analysis

$$H_D(z) = -\frac{k_3 \cdot z^2 + (k_1 k_5 + k_2 k_5 - 2k_3) \cdot z + k_3 - k_2 k_5}{z^2 + (k_4 k_5 + k_5 k_6 - 2) \cdot z + 1 - k_5 k_6} \tag{18.54}$$

where from one may find

$$k_3 = a_2, \tag{18.55a}$$

$$k_4 = k_5 = \sqrt{b_1 + b_0 + 1}, \tag{18.55b}$$

$$k_1 = (a_0 + a_1 + a_2)/k_5, \tag{18.55c}$$

$$k_2 = (a_2 - a_0)/k_5, \tag{18.55d}$$

and

$$k_6 = (1 - b_0)/k_5. \tag{18.55e}$$

18.4.7 The Sample and Hold Circuit

The simplest version of the sample-and-hold circuit is depicted in Fig. 18.16. The sampling period is defined by the sampling frequency which is recognizable from the notation used for the controlling signal ϕ_s.

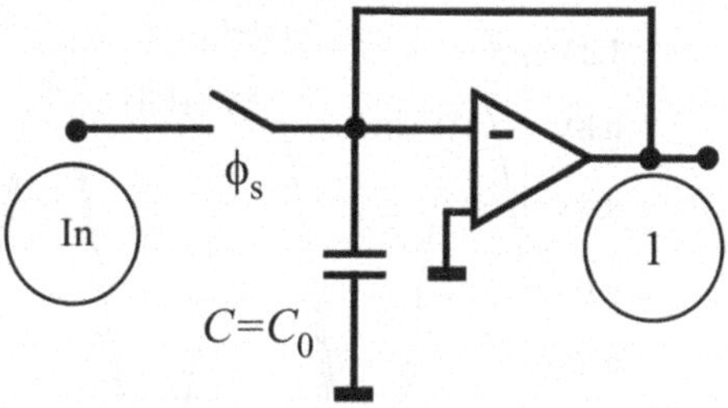

Fig. 18.16 A sample and hold circuit

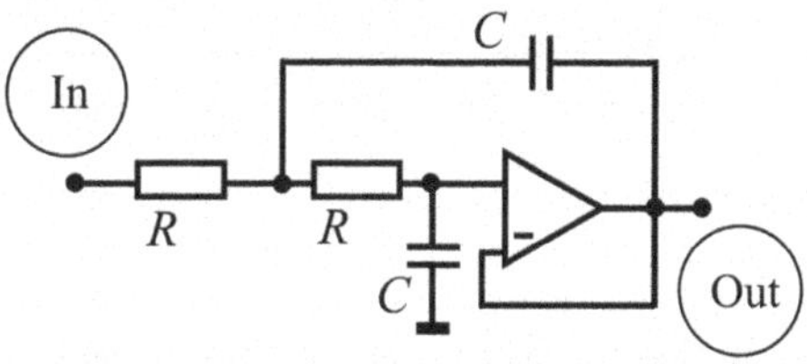

Fig. 18.17 Sallen and Key 2nd order unity gain low-pass cell

18.4.8 Antialiazing Low-Pass Filter

To limit the spectrum of the output signal a low-pass continuous-time filter must be implemented. One among unlimited number of possible solutions is the Sallen and Key [17] cell depicted in Fig. 18.17.

The transfer function of this circuit is given by

$$T(s) = \frac{V_{\text{out}}}{V_{\text{in}}} = \frac{1}{1 + 2s \cdot CR + s^2C^2R^2} \tag{18.56}$$

and its cut-off frequency (preventing aliasing) is

$$f_c = \frac{\sqrt{\sqrt{2} - 1}}{2\pi \cdot RC}. \tag{18.57}$$

So, after choosing the resistance to be R_0, one may find the capacitance as

$$C = \frac{\sqrt{\sqrt{2} - 1}}{2\pi \cdot R_0 f_c}. \tag{18.58}$$

18.5 Design Example

As an example here, a 6th order Bessel filter will be used. The sampling frequency was chosen to be 10 MHz while the required cut-off frequency is 100 kHz. For synthesis the *SC_parallel* program of the *RM* software was used.

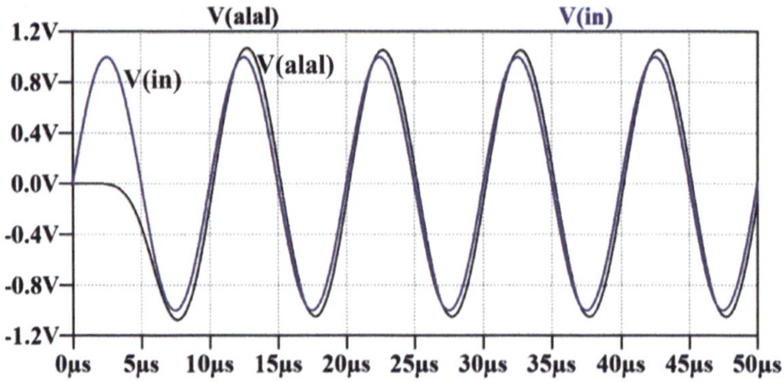

Fig. 18.18 Time domain responses of the 6th order Bessel SC filter at f = 100 kHz

In order to verify the synthesis result, as usual, a SPICE file was automatically created. The difference with respect to the rest of the programs is in that this file is intended to be used exclusively for time domain simulation since real switches are embedded. In order to perform the verification, in this example, we performed multiple time domain simulations the results of two of which are depicted in Figs. 18.18 and 18.19. In the first case both the input (V(in)) and the output (V(alal)) signals are depicted for the signal frequency of 100 kHz. To compare, however, the output signal should be divided by the response at the origin which is done later and depicted in Fig. 18.20. The second case represents the output signal at 600 kHz which is depicted in Fig. 18.19.

Figure 18.20 is constructed in the following way. First, the amplitude characteristic was calculated by the *LP_analysis* program from the transfer function of the 6th order Bessel filter produced by the *Bessel* program. Then, repetitive time domain simulations were performed using the SPICE file generated by the *SC_parallel* program. The frequency of the input signal was varied to allow

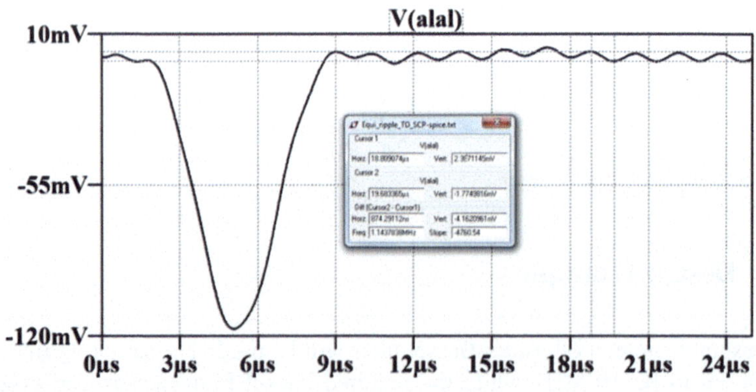

Fig. 18.19 Time response of the 6th order Bessel SC filter at f = 600 kHz

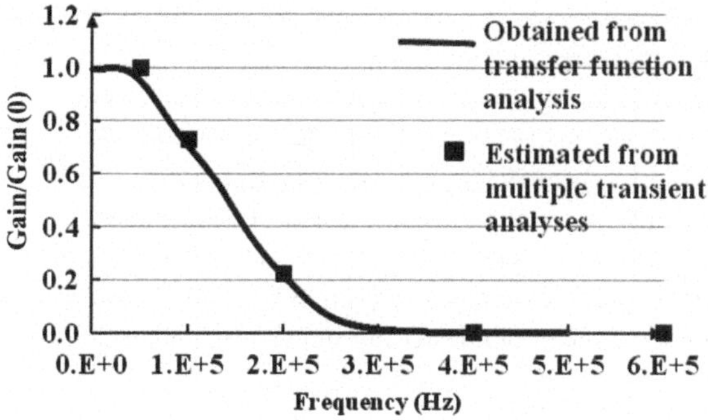

Fig. 18.20 Frequency response of the 6th order Bessel low-pass filter

insight of the amplitude both in the passband and in the stopband. The amplitude of the output signal was estimated after steady state for every signal frequency. It was subsequently divided by the amplitude obtained for a very low frequency to compensate for the gain introduced by the SC circuit. The result obtained in that way is superimposed to the original frequency response as seen in the figure. Acceptable agreement was obtained.

Start.E

SC-Parallel, Program for synthesis of active RC-circuits
PROJECT NAME: Equi_ripple_TD

++

Read in general data

Order of the numerator, n=0 Order of the denominator, m=6 Sampling frequency=1.00000e+007 [Hz] Normalization capacitance=1.00000e-011 [F]

++

poles:
sigma=[1]=-5.4815723e-001 omega=[1]=4.5139051e-001
sigma=[2]=-5.4815723e-001 omega=[2]=-4.5139051e-001
sigma=[3]=-5.2907193e-001 omega=[3]=1.3377439e+000
sigma=[4]=-5.2907193e-001 omega=[4]=-1.3377439e+000
sigma=[5]=-4.4240098e-001 omega=[5]=2.1556661e+000
sigma=[6]=-4.4240098e-001 omega=[6]=-2.1556661e+000

START SOLUTION
Residues in the poles
p1[1]= 5.0604979e-003 q1[1]= -1.5665128e-001
p1[2]= 5.0604979e-003 q1[2]= 1.5665128e-001
p1[3]= -9.2408196e-003 q1[3]= 8.1456379e-002
p1[4]= -9.2408196e-003 q1[4]= -8.1456379e-002
p1[5]= 4.1803217e-003 q1[5]= -1.7623854e-002
p1[6]= 4.1803217e-003 q1[6]= 1.7623854e-002

Residues after multiplication wit 1/gain
p1[1]= 2.5571844e-002 q1[1]= -7.9159444e-001
p1[2]= 2.5571844e-002 q1[2]= 7.9159444e-001
p1[3]= -4.6695957e-002 q1[3]= 4.1161756e-001
p1[4]= -4.6695957e-002 q1[4]= -4.1161756e-001
p1[5]= 2.1124113e-002 q1[5]= -8.9057332e-002
p1[6]= 2.1124113e-002 q1[6]= 8.9057332e-002
Ordered vector of transfer functions to be parallelized T(s)=(g*s+q)/(s**2+a*s+b)

g=	q=	a=	b=
5.11436873e-002	7.42671210e-001	1.09631446e+000	5.04229738e-001
-9.33919142e-002	-1.15068882e+000	1.05814386e+000	2.06947594e+000
4.22482269e-002	4.02646406e-001	8.84801955e-001	4.84261510e+000

Transformation into the z-domain

BILINEAR TRANSFORM
T(z)=(gamma*z**2+epsilon*z+delta)/(ksi*z**2+alfa*z+beta)
cell no.
__beta____________alfa______________ksi________________delta________________
epsilon____gamma
1 4.83263113e-001 -1.00000000e+000 5.17733348e-001 -4.37110405e-004
7.33835097e-004 1.17094550e-003
2 4.85387214e-001 -1.00000000e+000 5.18708829e-001 9.01105276e-004
-1.13875945e-003 -2.03986472e-003
3 4.90835776e-001 -1.00000000e+000 5.18775423e-001 -4.67257876e-004
3.99568680e-004 8.66826556e-004
SYNTHESIS OF PARALLEL ACTIVE SC FILTERS
++
The SPICE simulation will take place at f=5.000000000e+004 Hz
All capacitances in F ,
Nominal capacitance C0=1.000000000e-011
Capacitance in the sample and hold circuit: chold=1.0000000e-011

CELL NO.= 1
The input biquad function parameters after
z-transform:(T(z)=(a2*z^2+a1*z+a0)/(b2*z^2+b1*z+b0))
beta=5.177333483e-001
b0=9.334208709e-001, b1=-1.931496210e+000, b2=1.000000000e+000,
a0=-8.442770899e-004, a1=1.417399709e-003, a2=2.261676799e-003

Carusone-Johns-Martin circuit-Second order-high Q---- 1--------
Capacitances: c0=c1=c2=1.000000000e-011, k1c1=6.461676836e-013,
k2c1=7.079749688e-013, k3c2=2.261676799e-014, k4c1=4.387095626e-013
k5c2=4.387095626e-013, k6c1=1.517612899e-011

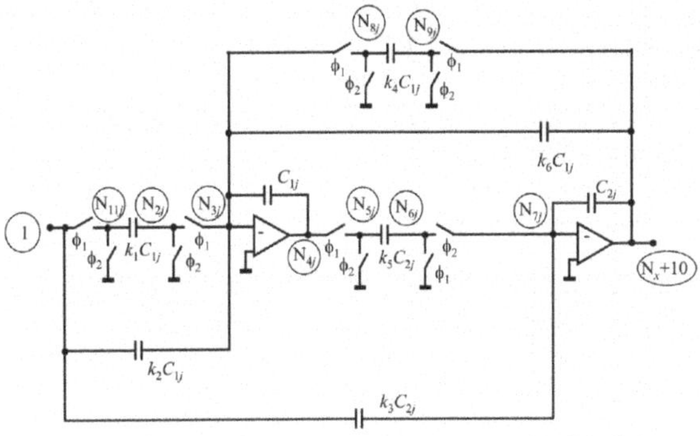

N_x=20*(j-1), for j=1,2,..., No. of parallel cells-1

--------------------End of the 1-th cell description ----------------------------

CELL NO.= 2

The input biquad function parameters after

z-transform:(T(z)=(a2*z^2+a1*z+a0)/(b2*z^2+b1*z+b0))

beta=5.187088285e-001

b0=9.357604642e-001, b1=-1.927863852e+000, b2=1.000000000e+000,

a0=-1.737208288e-003, a1=2.195373176e-003, a2=3.932581464e-003

Carusone-Johns-Martin circuit-Second order-high Q---- 2--------

Capacitances: c0=c1=c2=1.000000000e-011, k1c1=4.941034997e-013,

k2c1=6.380379858e-013, k3c2=3.932581464e-014, k4c1=8.886288712e-013,

k5c2=8.886288712e-013, k6c1=7.229062417e-012

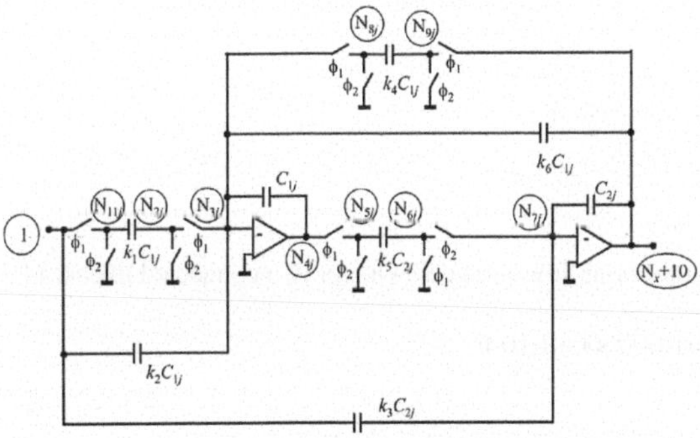

N_x=20*(j-1), for j=1,2,..., No. of parallel cells-1

--------------------End of the 2-th cell description ----------------------------

CELL NO.= 3

The input biquad function parameters after

z-transform:(T(z)=(a2*z^2+a1*z+a0)/(b2*z^2+b1*z+b0))

beta=5.187754225e-001

b0=9.461430804e-001, b1=-1.927616376e+000, b2=1.000000000e+000,

a0=-9.006939336e-004, a1=7.702151305e-004, a2=1.670909064e-003

Carusone-Johns-Martin circuit-Second order-high Q---- 3--------

Capacitances: c0=c1=c2=1.000000000e-011, k1c1=1.131730338e-013, k2c1=1.889317032e-013, k3c2=1.670909064e-014, k4c1=1.361128362e-013, k5c2=1.361128362e-012, k6c1=3.956784762e-012

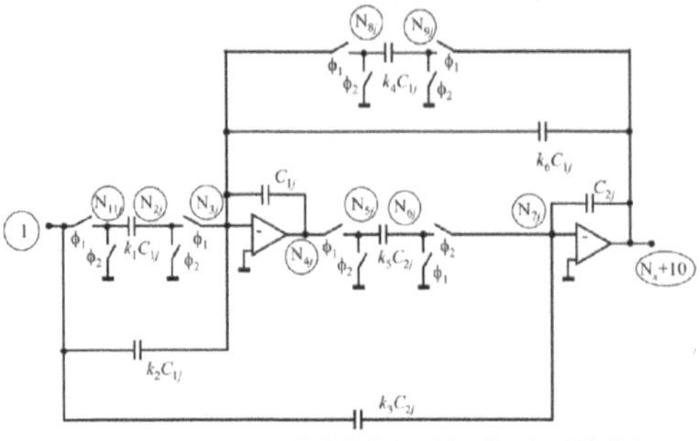

N_x=20*(j-1), for j=1,2,..., *No. of parallel cells*-1

--------------------End of the 3-th cell description ----------------------------

The summing amplifier

Input capacitance Ci=1.000000000e-011 F Feedback capacitance Cr=3.333333333e-012

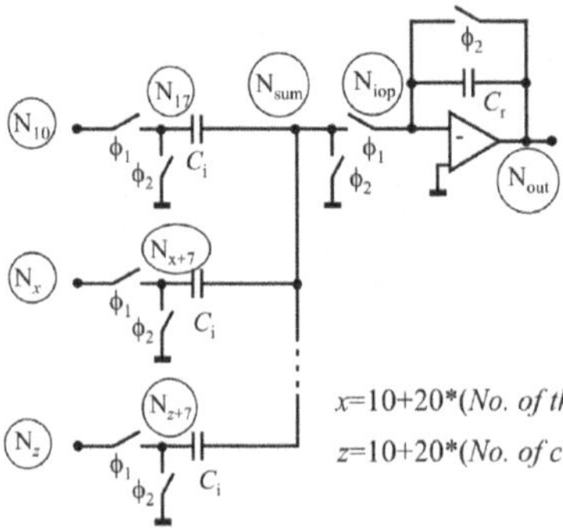

x=10+20*(*No. of the cell*-1)

z=10+20*(*No. of cells*-1)

The antialiazing Sallen and Keysecond order low-pass cell. One of two connected in cascade.

R=1.024312068e+003 Ohm C=1.000000e-010 F

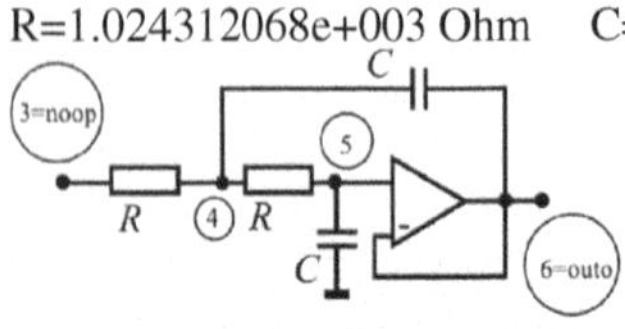

The Sample-and-hold circuit.

C=1.000000e-011 F

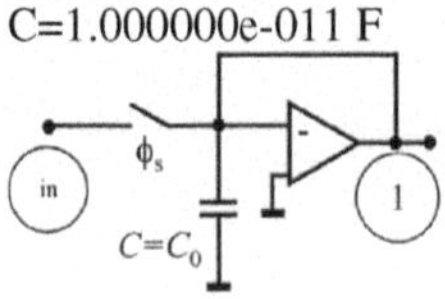

The overall schematic of the filter

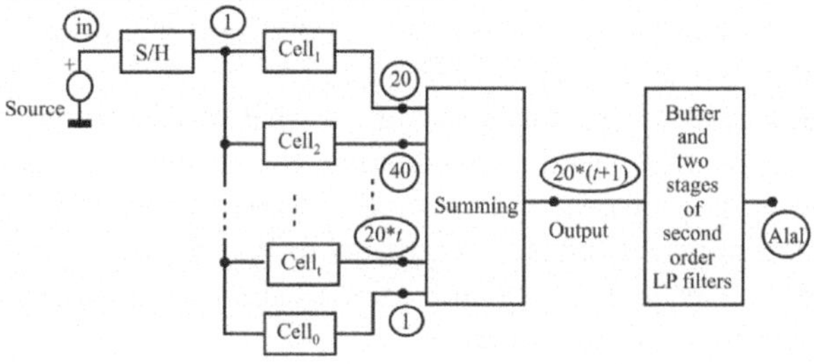

$t = m/2$ for m even // $t = m/2 + 0.5$ for m odd
$Cell_0$ is a short circuit activated only if $n = m$
m=order of the filter // n= order of the numerator

Here ends the synthesis process

End.E

18.6 Developer's Corner

To summarize we will repeat here the steps undertaken for synthesis of a parallel SC filter:

- Synthesis of the continuous time prototype. This will take as much an effort as necessary if complex transfer function requirements are to be fulfilled
- Transforming the transfer function into a sum of partial fractions
- Prewarping the fractions to prepare the entrance into the z-domain
- Performing the bilinear transform
- Synthesis of the building blocks of the parallel SC circuit.

In cases when linearity of the phase or even time domain responses are to be preserved, one may use other formulas (in place of the bilinear transform) which are listed in [11].

As listed in [11], these methods may be categorized in two groups. In the first group are put the ones which implement a specific criterion of approximation such as: the Impulse Response Invariant method, first introduced in [18] and discussed in [19]; the Modified Impulse Response Invariant method, discussed in [20], the Step Response Invariant method (or Zero Order Hold), discussed in [21]; the Magnitude-invariance method, first introduced in [22]; and the Phase-invariance method, first introduced in [23].

There are, however, transformations based on substitution of the complex frequency in the s-domain by an expression being a function of z. In that way, one has the Matched-z Transform method, discussed in [24], and three methods obtained by approximation of the analog integrator by a digital one. These are known as the Backward Euler (backward difference); the Trapezoidal method or the Bilinear Transform method, discussed in [25], and the second order formula introduced in [11].

From the circuit designer point of view, no matter how complex the procedure looks like, it is inherently straightforward the only iterations being encountered in the continuous-time-domain-synthesis when the complexity (order of the filter and number and type of zeros) of the filter is to be defined.

When using a design tool one is to decide as to what should be the sampling frequency. That problem is frequently solved within the existing standards of the design centre and, most probably, is out of the scope of the filter designer. Similarly, the normalization capacitance (suggested to be adopted in the previous text) is dependent on the technology intended to be used for the application. In the design example above $C_0 = 10$ pF was used.

References

1. Carusone TC, Johns DA, Martin KW (2012) Analog integrated circuit design. Wiley, New York
2. Allen PE, Holberg DR (2002) CMOS analog circuit design. Oxford University Press, New York
3. Schaumann R, Van Valkenboug E (2001) Design of analog filters. Oxford University Press, New York
4. Gray PR, Hurst PJ, Lewis SH, Meyer RG (2009) Analysis and design of analog integrated circuits. Wiley Inc, New York
5. Steensgaard-Madsen J (1999) Bootstrapped low-voltage analog switches. In: Proceedings of IEEE international symposium on circuits and systems, vol 2, Orlando, FL, USA, ISCAS '99, pp 29–32
6. Kazim MI (2006) Design of highly linear sampling switches for CMOS track-and-hold circuits. Division of Electronics Systems, Department of Electrical Engineering, Linköpings University, Linköping, Sweden
7. Fayomi CJB, Roberts GW, Sawan M (2005) Low-voltage CMOS analog bootstrapped switch for sample-and-hold circuit: design and chip characterization. In: Proceedings of the IEEE international symposium on circuits and systems, Kobe, Japan, ISCAS'05, pp 2200–2203
8. Oppenheim A, Schafer RW (2009) Discrete time signal processing. Prentice-Hall Signal Processing Series, Pearson
9. Damper RI (1995) Introduction to discrete-time signals and systems. Springer, Netherlands
10. Fleischer PE, Laker KR (1979) A family of active switched capacitor biquad building blocks. Bell Syst Tech J 58(10):2235–2269
11. Mirković D, Petković P, Litovski VB (2014) A second order s-to-z transform and its implementation to IIR filter design. COMPEL: Int J Comput Math Electr Electron Eng 33 (5):1831–1843
12. https://en.wikipedia.org/wiki/Bilinear_transform. Last visited may 2019
13. Sandberg IW, Shichman H (1968) Numerical integration of system of stiff nonlinear differential equations. Bell Syst Tech J 47(4):511–527
14. Gear CW (1971) Numerical initial value problems in ordinary differential equations. Prentice-Hall, Englewood Cliffs, NJ
15. Nagel LW (1975) SPICE2: a computer program to simulate semiconductor circuits. University of California at Berkeley, ERL Memo.-M520, 9 May 1975, Berkeley, CA
16. Litovski V, Zwolinski M (1997) VLSI circuit simulation and optimization. Chapman and Hall, London
17. Sallen RP, Key EL (1995) A practical method of designing RC active filters. IRE Trans Circuit Theory, CT 2:74–85

18. Kaiser JF (1996) Digital filters. In: Kuo FF, Kaiser JF (eds) System analysis by digital computer. Wiley, New York Chap. 7
19. Hongyan C, Lisheng W (2012) Software and hardware implementation of IIR based on matlab & acceldsp. In: Proceedings of the 2nd international conference on computer application and system modeling, Atlantis Press, Paris, France
20. Nelatury SR (2007) Additional correction to the impulse invariance method for the design of IIR digital filters. Digit Signal Proc 17(2):530–540
21. Németh JG, Kollár I (2000) Step-invariant transform from z- to s-domain, a general framework. In: Proceedings of the 17th IEEE instrumentation and measurement technology conference, IMTC/2000, Baltimore, MD, pp 902–907
22. Paarmann LD (1998) Mapping from the s-domain to the z-domain via magnitude-invariance method. Sig Process 69(3):219–228
23. Paarmann LD (2006) Mapping from the s-domain to the z-domain via phase-invariance method. Sig Process 86(2):223–229
24. Shi R, Wang S, Zhao J (2012) An unsplit complex-frequency-shifted PML based on matched z-transform for FDTD modelling of seismic wave equations. J Geophys Eng 9(2):218–229
25. Park W, Park KS, Koh HM (2008) Active control of large structures using a bilinear pole-shifting transform with H_{∞} control method. Eng Struct 30(11):3336–3344

Chapter 19
IIR Digital Filter Synthesis Based on Bilinear Transformation of Analog Prototypes

Digital signal processing means dealing with discretized signals expressed as numbers in a form convenient for electronic implementation of the mathematical operations. Digital filtering is its important part. Two types of digital filters are in use the so called *non-recursive* or finite impulse response (FIR) and the *recursive* or infinite impulse response (IIR) filters. Since the subject of digital filters is very broad in this chapter one of the two types (IIR) was chosen and the rationale of the selection was explained. Parallel architecture is recommended as in most of the previous chapters for analog filters. Bilinear transformation of the corresponding analog counterparts was used. To facilitate the design for some popular polynomial analog selective filters the transfer functions of the biquads are given in appendices. The following issues are considered: the problem of finite number of figures used to represent the coefficients; the problem of stability related to the same number of figures; and the problem of nonlinearity introduced by the bilinear transform and its influence to the frequency characteristic of the digital IIR filter. Specially, digital CMAC and selective linear phase lowpass filters are discussed. The final results are reported for 16- and 32-bit floating point accuracy.

19.1 Introduction

In Chap. 2 we partly discussed the rationale of introduction of digital signal processing as a substitution to the analog. We will here reconsider the difference but related to the filters as such only. Main advantages of the digital filters may be summarized as [1]

- Component tolerances are uncritical
- Component drift and spurious environmental signals have no influence on the system performance
- Accuracy is high

V. Litovski, *Electronic Filters*, Lecture Notes in Electrical Engineering 596,
https://doi.org/10.1007/978-981-32-9852-1_19

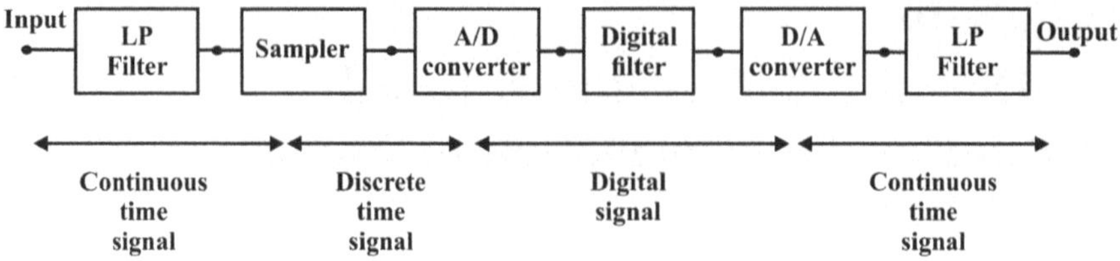

Fig. 19.1 Digital filtering as a system

- Physical size is small
- Reliability is high
- Easeness of adjusting the coefficients available which allows for programmable filters.

These, the ones mentioned in Chap. 2, and many others are the reason for the digital filters to become by far the most dominant in modern electronic systems design.

The structure of a stand alone system which is supposed to perform digital filtering is depicted in Fig. 19.1. Here a continuous time (analog) signal is brought to the input. The first step is limiting of the bandwidth to avoid subsequent aliasing and reduce the thermal noise. Then, sampling is performed to produce discrete time signal. It is used as an input to an analog-to-digital (A/D) converter which is in fact a quantizer with prescribed resolution of the signal levels. In that way a digital signal is produced and is ready for digital filtering. As already mentioned, the output of the A/D converter may produce a single series of pulses or a number of series of pulses, depending on whether the filter will process the result in series or in parallel. The results of the filtering are braught to the input of an digital-to-analog (D/A) converter which restores the continuous time signal and is ready to activate whatever is the target analog device (e.g. a loudspeker).

The subject of this chapter is creation a transfer function of an recursive (IIR) digital filter obtained from an analogue prototype by bilinear transform. That is only the part of the system marked as "Digital signal" will be considered. Parallel realization will be considered only.

After making some comparison related to non-recursive (FIR) and recursive (IIR) filters, implementation issues will be studied related to the bilinear transform. That will be followed by the synthesis procedure and synthesis example.

19.2 FIR Versus IIR Filters

A norecursive (FIR) filter is the one whose output signal is a function of the input signal only:

$$y(l \cdot T_s) = f\{\ldots, x(l \cdot T_s - T_s), x(l \cdot T_s), x(l \cdot T_s + T_s), \ldots\} \tag{19.1a}$$

which is usually expressed as

$$y(l \cdot T_s) = \sum_{i=-\infty}^{\infty} a_i \cdot x(l \cdot T_s - i \cdot T_s). \tag{19.2}$$

If we introduce the limitation that $a_i = 0 \ \forall \ i < 0$, expressing the causallity principle, that $x(l \cdot T_s) = 0$ for $l < 0$, and that the summation goes to N only ($a_i = 0 \ \forall \ i > N$), we may write

$$y(l \cdot T_s) = \sum_{i=0}^{N} a_i \cdot x(l \cdot T_s - i \cdot T_s) \tag{19.3}$$

which expresses the output signal at instant $l \cdot T_s$ as a function $N + 1$ samples of the input signal taken at the present instant ($t_l = l \cdot T_s$) and N previous instants.

In the case of recursive (IIR) filter the output signal is a function of the output state in the previous instants which is expressed as

$$y(l \cdot T_s) = \sum_{i=0}^{N} a_i \cdot x(l \cdot T_s - i \cdot T_s) - \sum_{i=1}^{N} b_i \cdot y(l \cdot T_s - i \cdot T_s). \tag{19.4}$$

Here the output signals is obtained by two linear combinations: one containing the actual and N previous input signal values and another containing previous N output signal values.

Of course by putting $b_i = 0 \ \forall \ i$, into (19.4), the response of the recursive filter reduces into a non-recursive one making the later a special case of the former.

If we define the output signal az $Y(z)$ and the input signal as $X(z)$, after implementation of the z-transform to (19.4) we get the transfer function of a digital recursive filter to be

$$H_D(z) = \frac{Y(z)}{X(z)} = \frac{\sum_{i=0}^{N} a_i \cdot z^{N-i}}{z^N + \sum_{i=1}^{N} b_i \cdot z^{N-i}} \tag{19.5}$$

It may be represented in factored form to be

$$H_D(z) = H_0 \frac{\prod_{i=1}^{N} (z - z_i)}{\prod_{i=1}^{N} (z - p_i)} \tag{19.6}$$

where z_i, and p_i, $i = 1, 2, \ldots, N$, are the complex zeros and poles, respectively. Note, to get the transition from (19.5) into (19.6) one must first rearrange the polynomials in new forms:

$$H_{\mathrm{D}}(z) = \frac{\sum_{i=0}^{N} a_{N-i} \cdot z^i}{1 + \sum_{i=1}^{N} b_{N-i} \cdot z^i}. \tag{19.7}$$

In the nonrecursive case when $b_i = 0 \ \forall \ i$, the denominator of (19.5) reduces to z^{-N} which means that all poles of the transfer function are located in the origin.

From the physical implementation point of view the following are the most important transfer function

- The delay: $H_{\mathrm{D}}(z) = z^{-1}$, or $Y(z) = z^{-1} \cdot X(z)$
- The adder: $Y(z) = \sum_{i=1}^{n} X(z)$, and
- The multiplier by constant: $H_{\mathrm{D}}(z) = \alpha$, or $Y(z) = \alpha \cdot X(z)$.

It is well known from electronic circuit theory that the same transfer function of a filter may have many different realizations. First, one may look for the way imposed requirements are to be fulfilled, which leads to the transfer function synthesis problem. We already saw that almost equal stopband attenuation characteristic may be realized by circuits with largely different passband frequency responses. When solving this problem, in digital filter design, one is to decide first on the choice between FIR and IIR filter functions and then to proceed to the approximation problem. Finally, classification originates from the nature of different circuit structures to have the same transfer function while the same technology and the same signal processing concepts are used. The best example for the latter are the canonical (or state variable) form and the parallel form of active filter structure (which are taken from a larger set of possible configurations). These two are illustrated in Fig. 19.2 for an IIR digital filter [1].

Here electronic (active) digital filters produced as integrated circuits on silicon are dealt with. Hence, before proceeding, one is to decide which of the two basic digital filter solutions, namely IIR or FIR, will be chosen for implementation of the transfer function of the filter and consequently for its physical realization.

For complete decision, however, one has to have in mind the next design step i.e. the synthesis step where from one may state the following as criteria for choice among structures and technologies:

- the ability to fulfill the imposed technical requirements by a given technology e.g. linear phase;
- the possibility to develop a proper transfer function performing best approximation of the requirements e.g. minimum stop-band attenuation and minimum transition region for a given complexity;
- the way of signal processing e.g. software or hardware;
- the reliability of the system produced in that way; and similar.

As an additional set of metrics for decision making one may use: stability; silicon area; power dissipation; overall delay or time elapsed for performance; sensitivity and mapping of the tolerance; and others.

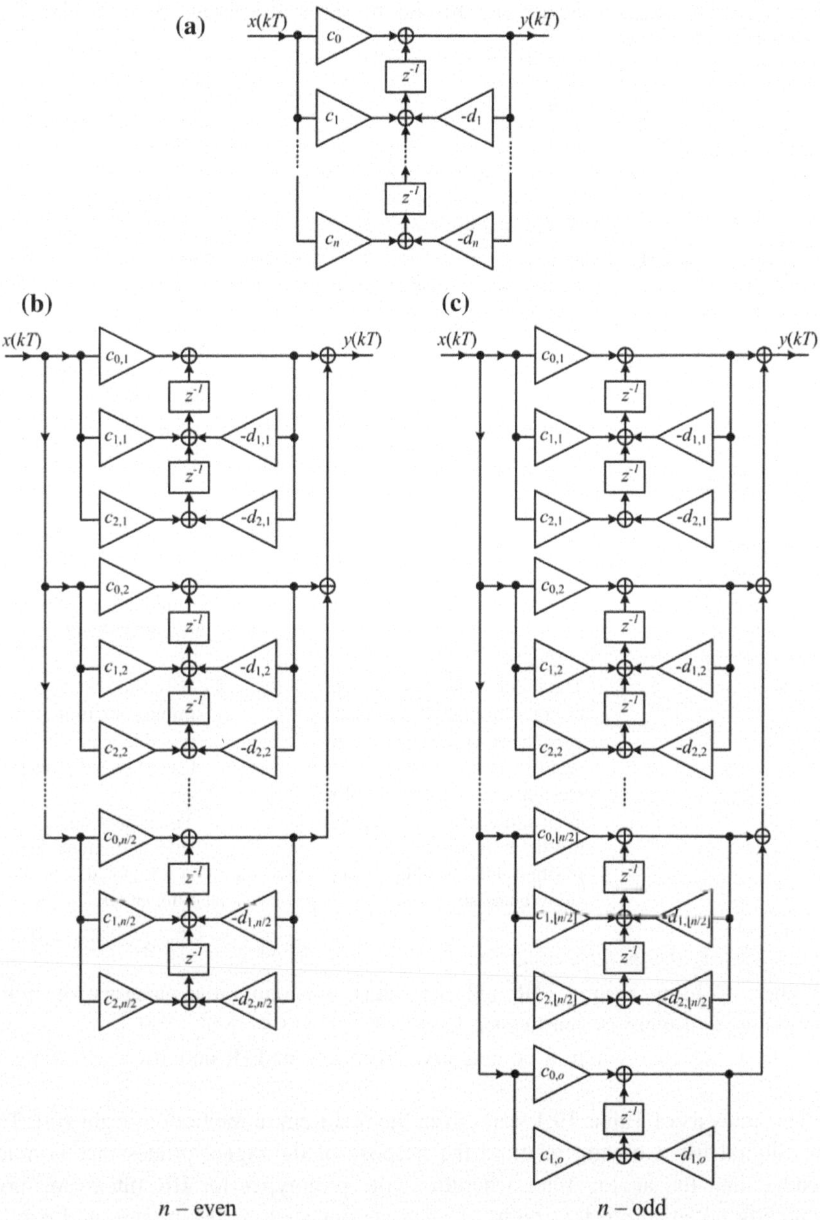

Fig. 19.2 **a** Canonical realization of an nth order IIR filter and **b** and **c** parallel realization forn n even and n odd, respectively

Table 19.1 Comparative overview of FIR and IIR filter realizations (A or D stands for advantages/disadvantages)

Property	A/D	IIR	A/D	FIR
Phase	D	Difficult to control	A	Linear phase always possible
Stability	D	Can be unstable, can have limit cycles. To mitigate parallel synthesis is recommended	A	Always stable No limit cycles
Sensitivity (finite-word length)	D	More To mitigate parallel synthesis is recommended	A	Less
Order	A	Less	D	More Excessive memory requirements
Control of the stop-band attenuation value	A	Continuous	D	Discrete
Power consumption	A	Less	D	More
Latency (amount of processing i.e. computational time)	A	Less	D	More To mitigate pipelining is recommended
Others	A	The order of the filter affects both the minimum stopband attenuation and the transition region Design freedom: to control the stopband attenuation, one may use the transmission zero's position and the filter order simultaneously	A	• Polyphase implementation possible • Can always be made causal • The minimum stopband attenuation is not affected by the filter order

When deciding among FIR and IIR filters, no matter the category of signal processing (software or hardware), based on [2] and several other resources, we came to Table 19.1 where a comparative overview of FIR and IIR realizations is given.

The analysis of Table 19.1 starts from the first item in the bottom-right cell. The term digital filter, having in mind the purpose of the signal processing, is much broader than the analog filter wherefrom the prototypes for IIR filters are most frequently taken. From that point of view actions such as decimation and similar cannot, simply, be part of the argument in the comparison.

Besides, taking an overview, we see that three main advantages of FIR filters are recognizable. The first, related to sensitivity and stability, arises from limited number of significant figures representing the coefficients implementing the filter's hardware. These are much more serious in the case of IIR filters. This problem is partly mitigated by using parallel synthesis as shown in Fig. 19.2. Later, in

Paragraph 19.3, an example illustrates how limiting the number of significant figures propagates the poles of the digital filter function in the complex z-plane and how the filter can become unstable. Note, if a proper number of significant digits is used in conjunction with parallel realization, no stability problems are encountered in IIR filters.

The second "disadvantage" of IIR filter is most frequently related to the cases when linear phase is required for the purpose of proper signal processing. Linear phase transfer function in FIR design is obtained in a simple manner. That is the fact. The fact is, however, that the symmetry of the coefficients needed to achieve linear phase reduces the number of freedoms in amplitude approximation by two. That, and not only that, leads to very hardware intensive solutions as demonstrated in [3]. There, an IIR elliptic filter was designed with an equalizer of the group delay characteristic. The authors claim that for their solution (compared to the FIR counterpart) it: "represents a reduction of 75% in the number of multiplications, a critical operation in digital filtering implementations."

The complexity of selective filters realized as non-recursive come from an additional fact. Namely, a correspondence is easily recognizable between the analog polynomial and the non-recursive digital filters. Similarly, analog filters with finite transmission zeros correspond to the IIR filters. Now, we concluded in Chap. 8 that a transmission zero has much larger influence to the selectivity than a pole. That is because the pole, begin a zero at infinity, defines the assymptotic selectivity while a zero at the imaginary axis may be located near to the cut-off frequency to increase significantly the skirt selectivity which is of prime interest. So, the IIR filter may exhibit the same selectivity as the FIR one but with smaller complexity.

We will turn back now to the linear phase solutions. We want to stress here that the claim that the FIR filter have linear phase per se should not be taken for granted.

To improve the selectivity of a linear phase IIR filter in [4] direct synthesis method was proposed which creates a transfer function with ω-axis zeros.

Similar results were reported in [5] where a linear phase polynomial analog filter was extended by transmission zeros as explained in Paragraph 10.4. The result so obtained was transformed by a bilinear transform to produce an IIR filter.

In both cases linear phase solutions were obtained outperforming the selectivity of the FIR counterpart for the same complexity. We will revisit the results described in Paragraph 10.4 where the results of the bilinear transform will be analized from numerical point of view based on Paragraph 19.3.

It is out of the scope of this book to discuss further but we believe that any amplitude requirement (under linear phase constraint) obtained by FIR filters may be outperformed (in complexity of the final schematic) by these functions.

The third "advantage" of FIR filters is the simple synthesis procedure of the filter function. There is no doubt in that. We, however, think that is not an advantage of prime importance. Furthermore, having in mind the complexity of FIR filters and the optimizations that must be undertaken to mitigate

- the increased silicon area needed for the circuitry and the memory (to store the coefficients and intermediate results of computations) and
- the increased power consumption,

one may easily come to a conclusion that the difficulties are postponed for the much more difficult (from an optimization point of view) succeeding phases of design. Let's, for example, mention the complexity of the optimization needed from the power consumption point of view only [6]. In some cases the damage done by high complexity and large power consumption may be irreparable in the later phases of design and may make the design unacceptable after large time and amount of money was spent.

The latency or the processing time is frequently used when comparing FIR and IIR filter to advantage the first one. That is why we here implement a parallel variant of the IIR filters implementation which reduces the processing time by a factor, for n even, of $n/2$.

In addition, for final decision, it is necessary to take into account the properties of the signal to be filtered as these are very important when deciding which filter to use.

To conclude, there are hypes related to the choice between IIR and FIR solutions and one is to be very careful when taking decisions. Note that if software implementation is sought, a different way of thinking through many of the above considerations is needed. The general conclusions, however, will not differ significantly.

19.3 The Bilinear Transform and Implementation Issues

There are several transformations claiming to preserve some of the original properties of the analog filter function when producing a digital domain counterpart as mentioned in the previous chapter. As listed in [7], these methods may be categorized in two groups. In the first group are put the ones which implement a specific criterion of approximation such as: the Impulse Response Invariant method, first introduced in [8] and discussed in [9]; the Modified Impulse Response Invariant method, discussed in [10], the Step Response Invariant method (or Zero Order Hold), discussed in [11]; the Magnitude-invariance method, first introduced in [12]; and the Phase-invariance method, first introduced in [13].

There are, however, transformations based on substitution of the complex frequency in the s-domain by an expression being a function of z. In that way, one has the Matched-z Transform method, discussed in [14], and three methods obtained by approximation of the analog integrator by a digital one. These are known as the Backward Euler (backward difference); the Trapezoidal method or the Bilinear Transform method, discussed in [15], and the second order formula introduced in [7].

The most popular among all of these is the bilinear transform. Its main properties are simplicity of implementation and good preservation of the properties of the amplitude characteristic of the analog filter. It preserves the stability of the analog prototype. It introduces distortions (reduced by increasing the sampling rate) into the phase (group delay) characteristic which, however, has no importance in many applications.

To preserve completeness we will here partly repeat some of the developments of the previous chapter.

The bilinear transform is implemented via the following transformation into the analog transfer function:

$$s \Rightarrow \frac{2}{T_s} \cdot \frac{z-1}{z+1}, \tag{19.8}$$

where z is the complex variable, and T_s is the sampling rate $T_s = f_c/f_s$, f_s being the sampling frequency and f_c the cut-off frequency. In that way

$$H_A(s) \Rightarrow H_A\left(\frac{2}{T_s} \cdot \frac{z-1}{z+1}\right) = H_D(z), \tag{19.9}$$

is obtained, where H_A stands for the transfer function of the analog filter while H_D stands for that of the digital filter.

In the sequel, we will develop a procedure for implementation of the bilinear transform on analog prototypes and a parallel IIR realization.

In order to obtain parallel realization, one needs to perform partial fraction expansion of the original, analog, transfer function. The procedure begins with factorisation of numerator and denominator polynomials. Assuming $n > m$ the factorized form of the transfer function is:

$$H_A(s) = \frac{\sum_{k=0}^{m} b_k s^k}{\sum_{i=0}^{n} a_i s^i} = \frac{b_m \prod_{k=0}^{m-1}(s - z_k)}{a_n \prod_{i=0}^{n-1}(s - p_i)} = \frac{b_m}{a_n} H(s), \tag{19.10}$$

where, b_m and a_n are the highest order numerator and denominator coefficients, z_k and p_i are the zeros and poles of transfer function, respectively. It should be emphasised that if $m = n$ polynomial long division must be carried out first i.e.:

$$H_A(s) = \frac{b_n}{a_n}\left(1 + \frac{\prod_{k=0}^{n-2}(s - q_k)}{\prod_{i=0}^{n-1}(s - p_i)}\right) = \frac{b_n}{a_n}(1 + H(s)), \tag{19.11a}$$

where q_k are the roots of the polynomial

$$P(s) = \sum_{k=0}^{n-1}\left(\frac{b_k}{b_n} - \frac{a_k}{a_n}\right) s^k \tag{19.11b}$$

Next, partial fraction expansion of the function $H(s)$ is to be performed to get

$$H(s) = \sum_{i=0}^{n-1} \frac{r_i}{(s - p_i)^1} \tag{19.11c}$$

which is valid for the case when all poles are simple. In (19.11c) r_i, $i = 0, 1, 2, \ldots, n - 1$ stands for the residue.

Here by carefully combining complex-conjugate pole pairs with its associated residues the following form emerges

$$H(s) = \begin{cases} \sum_{i=0}^{n/2} H_e(s), & n - \text{even} \\ H_o(s) + \sum_{i=1}^{\lfloor n/2 \rfloor} H_e(s), & n - \text{odd} \end{cases} \tag{19.11d}$$

In (19.11d) index e is used for the complex pair of poles while o means simple real pole. In (19.11d) we use

$$H_e(s) = G_i \frac{s + b_{0,i}}{s^2 + a_{1,i}s + a_{0,i}} \tag{19.12}$$

with $G_i = 2\mathrm{re}\{r_i\}$, $b_{0,i} = -\left(\mathrm{re}\{p_i\} + \frac{\mathrm{im}\{r_i\}\mathrm{im}\{p_i\}}{\mathrm{re}\{r_i\}}\right)$, $a_{1,i} = -2\mathrm{re}\{p_i\}$, $a_{i0} = |p_i|^2$,

and

$$H_o(s) = G_o \frac{1}{s + a_o}, \tag{19.13}$$

with $G_o = r_o$, and $a_o = -p_o$. In the above p_i stands for the ith pole, r_i for the residue in the ith pole, "re" for "real part" and "im" for "imaginary part".

Implementing bilinear transform one gets (With reference to Figs. 19.2 and 19.3.)

$$H_e(z) = \frac{c_{0,i} + c_{1,i}z^{-1} + c_{2,i}z^{-2}}{1 + d_{1,i}z^{-2} + d_{1,i}z^{-2}} \tag{19.14}$$

$$H_o(z) = \frac{c_{0,o} + c_{1,o}z^{-1}}{1 + d_{1,o}z^{-1}} \tag{19.15}$$

With reference to (19.12) and (19.13), Tables 19.7, 19.8 and 19.9 (found in Appendix) give complete information about the transfer functions of the CMAC filters for orders $n = 3$ to $n = 10$. The coefficients needed for computation of (19.14) and (19.15) may be calculated after T_s is specified.

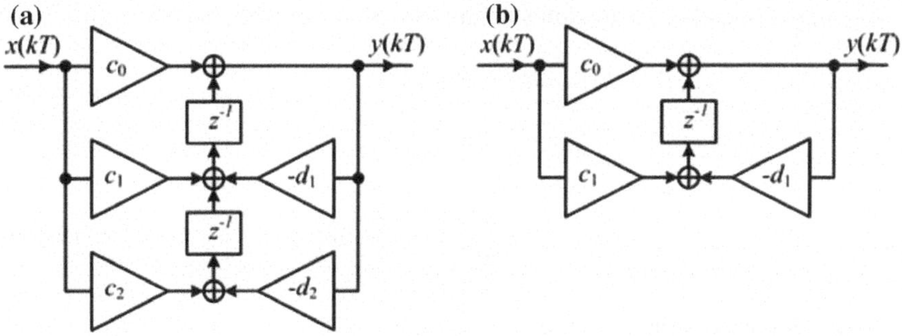

Fig. 19.3 Second-order, **a** and first-order, **b** section

It is important to have in mind that these tables are developed under the condition of $\varepsilon^2 = 1$, or, $a_{\max} \approx 3$ dB.

19.3.1 Implementation and Discussion

Among the issues related to the implementation of the synthesis method described above we will consider three. First the problem of finite number of figures used to represent the coefficients will be considered. Then, the problem of stability related to the same number of figures and finally the problem of nonlinearity introduced by the bilinear transform and the influence to the frequency characteristic of the digital IIR filter.

We will first consider the problem of finite number of figures used for representing the coefficients within the hardware realization of the filters. This is an important issue in both implementations: fixed and floating point arithmetic. Namely truncation or rounding is encountered no matter which implementation is used. The change of the coefficient values will lead to migration of the poles and zeroes of the transfer function so changing its presumed properties while in, some cases, may jeopardize the stability of the system.

To check for the changes in the pole (or zero) positions due to increments of the coefficients of the transfer function one may invoke the Rouché's Theorem [16]. Let the denominator polynomial of the transfer faction be

$$P_n(z) = \sum_{i=0}^{n} a_i z^i \quad \text{with} \quad a_n = 1. \tag{20.16}$$

The upper bound of the modulus of any root of this polynomial was shown to be

$$\bar{R} = \max(1, \alpha), \tag{19.17a}$$

where

$$\alpha = \frac{1}{|a_n|}\sum_{i=0}^{n-1}|a_i| = \sum_{i=0}^{n-1}|a_i|. \tag{19.17b}$$

Now, let consider the situation when all the coefficients (but a_n, which is fixed) are incremented by Δa_i. In general, the sign of the increment may be positive or negative since both, truncation (negative increment for positive coefficients) or rounding (negative or positive increment) are included. In this case α becomes

$$\alpha(\mathbf{a} + \Delta\mathbf{a}) = \sum_{i=0}^{n-1}|a_i + \Delta a_i|. \tag{19.17c}$$

Here $\mathbf{a}$ and $\Delta\mathbf{a}$ are vectors of coefficients and their increments. Assuming none of the coefficients is equal to zero, after some simple manipulation the increment of α is transformed into

$$\Delta\alpha = \alpha(\mathbf{a} + \Delta\mathbf{a}) - \alpha(\mathbf{a}) = \sum_{i=0}^{n-1}|a_i| \cdot \left(\left|1 + \frac{\Delta a_i}{a_i}\right| - 1\right). \tag{19.18}$$

The worst case arises (i.e. the increment is maximal) when $(\Delta a_i / a_i) > 0 \; \forall \, i = 0, 1, \ldots, n - 1$. In that case

$$(\Delta\alpha)_{\max} = \sum_{i=0}^{n-1}|\Delta a_i|. \tag{19.19}$$

The maximal increment of the radius of the circle encompassing all roots of the polynomial rises from two reasons. One is the amount of the increments themselves and the other is their number. That means: the higher the order of the polynomial the larger the circle encompassing the roots.

This result will be demonstrated by an example.

We will compare the effects of the reduction of the number of significant figures on the two realizations depicted in Fig. 19.2. Namely, three cases will be exercised: coefficients represented in 10 significant figures (as in Tables 19.7 and 19.8); the number of significant figures reduced to 4; and the number of figures reduced to 2. For both realizations, the polynomials generated in that way were solved and new positions of the poles were obtained. The results are illustrated in Fig. 19.4.

Figure 19.4 depicts the original position of the poles for the LSM filter of $n = 7$ in the s- and z-domains. We will refer to these values as the exact pole coordinates.

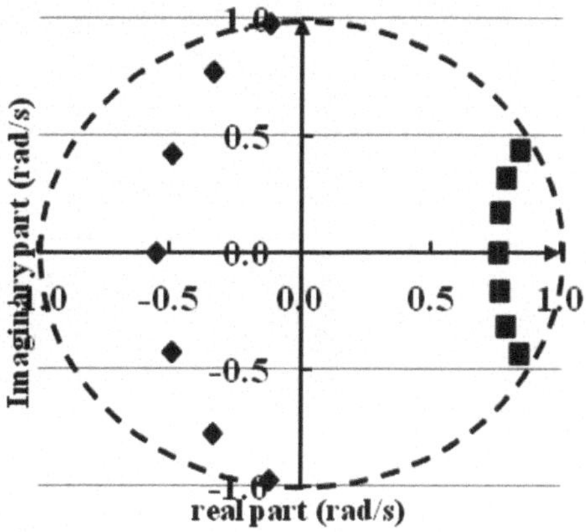

Fig. 19.4 Pole locations of the 7th order LSM with amax = 3 dB. ◆ analog (s-plane) and ■ digital (z-plane) transformed by bilinear transform with $T_s = 0.5$

As can be seen all mapped poles are confined within the unit circle. Hence, the IIR filter is stable.

Table 19.2 contains the exact pole coordinates of the 7th order LSM with $a_{max} = 3$ dB.

Figure 19.5a represents the position of the poles of the two configurations of Fig. 19.2 when the number of decimal significant digits is truncated to 4. That would approximately correspond to 14 bit representation. As can be seen the stability is preserved for parallel realizations while for canonical is already threatened which means that, in this case, the circuit is not acceptable. Figure 19.5b depicts the position of the poles of the two configurations of Fig. 19.2 when the number of decimal significant digits is truncated to 2. That would approximately correspond to 7 bit representation. This results with even worse behavior of the realization in Fig. 19.2a since three poles are out of the unit circle. The parallel realization, however, resists and stays stable.

Table 19.3 contains the pole coordinates corresponding to Fig. 19.5.

Figure 19.6a depicts all three situations related to the parallel synthesis. One can recognize that not only the stability is preserved but very small migration of the poles is obtained while even a very coarse truncation scheme was implemented. A similar conclusion is drawn after the last two columns of Tables 19.2 and 19.3 are compared. Namely, for the case when only two significant decimal figures are

Table 19.2 Exact coordinates of the poles in the s- and z-domain (Re stands for the real part and Im stands for the imaginary part)

s-domain		*z*-domain	
Re{*s*}	Im{*s*}	Re{*z*}	Im{*z*}
−0.5510897460	0.0	0.7578207520	0.0
−0.1179475625	±0.9751626241	0.8395567140	±0.4356216113
−0.3342221750	±0.7735798237	0.7887919250	±0.3192668225
−0.4935853895	±0.4252967357	0.7645095574	±0.1670025358

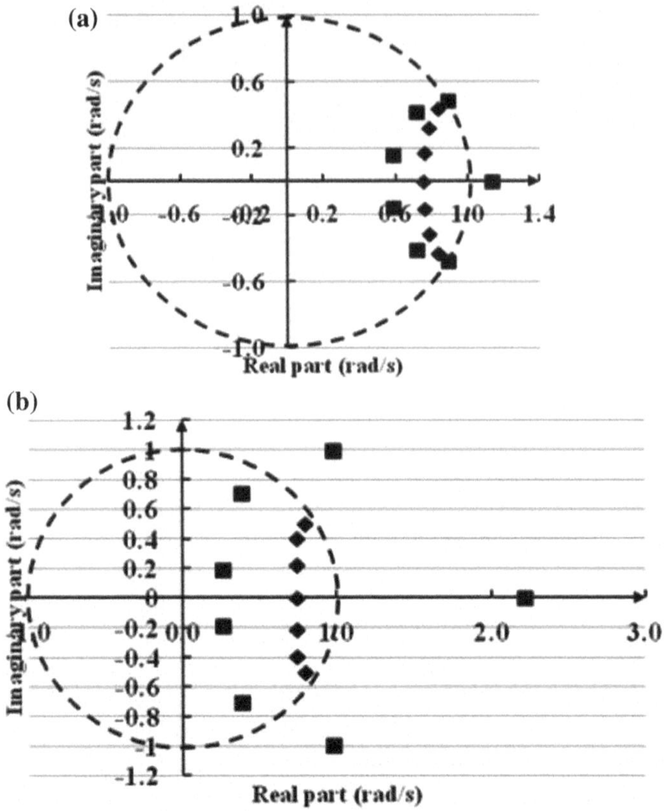

Fig. 19.5 Pole location in the z-plane for implementation as a canonical (■) and as parallel connection of sections (◆); **a** four decimal significant figures and **b** two decimal significant figures

Table 19.3 Coordinates of the poles in the z-domain (reduced precision)

	For canonical realization (Fig. 19.2a)		For parallel realization (Fig. 19.2c)	
	Re	Im	Re	Im
Coefficients of denominator truncated to four significant figures	1.1360207906	0	0.7578	0
	0.8939351215	±0.4816186107	0.8395	±0.4357060362
	0.7208726759	±0.4140006562	0.7885	±0.3199496054
	0.5886818073	±0.1575365844	0.7645	±0.1668524798
Coefficients of denominator truncated to two significant figures	2.2161297265	0	0.7500	0
	0.9837501045	±0.9948786645	0.8000	±0.5000000000
	0.3926899052	±0.7054768233	0.7500	±0.3968626967
	0.2654951271	±0.1886575446	0.7500	±0.2179449472

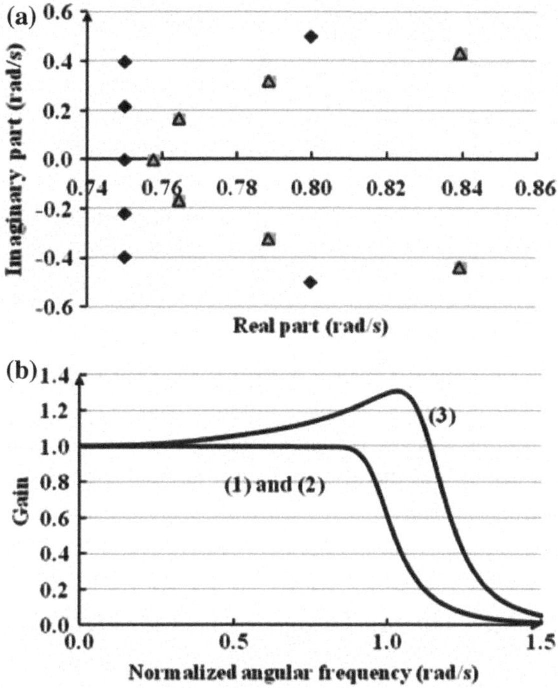

Fig. 19.6 **a** Position of the poles and **b** attenuation characteristics for implementation as parallel structure: (1, ■) full precision, (2, Δ) four decimal significant figures, (3, ◆) two decimal significant figures

preserved i.e. approximately 7 bit arithmetic is used, the changes of the pole coordinates are clearly noticeable for the canonical realization (e.g. the maximal migration of the real pole from 0.7578207520 to 2.2161297265) while not as clear for the parallel (the same pole is migrating to 0.75). As the best confirmation of the theory developed above and expressed by (19.19) one may use this maximal migration of the real pole since it represents the minimal radius of the disc encompassing all poles of the system.

Therefore, the new radius of the circle encompassing the poles is now $R_{c7bits} = 2.2161297265$. That is to be compared with the radius of the circle encompassing the exact pole coordinates which is $R_{exact} = 0.9458444186$. This gives increase of about 1.2703 in radius value.

Note, the worst case increase of the radius (with reference to Table 19.4), as predicted with (19.19), for the canonical realization using 7-bit arithmetic is

$$(\Delta\alpha)_{max} = \sum_{i=0}^{n-1} |\Delta a_i| = 1.9563. \tag{19.20}$$

As can be seen, the prediction was quite good.

Figure 19.6b represents the magnitude characteristics of the functions whose poles are given in Fig. 19.3. As expected, while keeping the stability, in the case of two significant digital figures, significant deterioration of the amplitude

Table 19.4 Denominator coefficients of the 7th order LSM filter z-domain transfer function for canonical realization

	Original	Two sig. fig.	$\|\Delta a_i\|$
a_7	1	1.0	0
a_6	−5.5435371447	−5.5	0.0435371447
a_5	13.4863201902	13	0.4863201903
a_4	−18.6187031540	−18	0.6187031541
a_3	15.7256233042	15	0.7256233043
a_2	−8.1149764659	−8.1	0.0149764659
a_1	2.3665384182	2.3	0.0665384182
a_0	−0.3006277069	0.3	0.0006277070

characteristic is observed. In the case of truncation to four significant decimal figures the effects on the amplitude characteristic are not clearly noticeable.

In this way the common opinion that use of second order sections will preserve the stability of the system even if large change of the coefficients are encountered, was confirmed.

19.3.2 *On the Design of CMAC IIR Filters*

We will use the context here to explain how an optimum solution for the design of CMAC IIR filters may be reached.

Namely, for full freedom of implementation of the CMAC characteristic, one needs to cater for values of $a_{\max}$ different than 3 dB (here denoted as a_x) to be allowed. That may be accomplished in two ways.

Usually the change of the maximum pass-band attenuation is accomplished by proper change of ε^2 in (6.1). This approach will be here referred to as "Variant A". Implementation of this approach means complete restart of the synthesis process.

First, one needs to create (or acquire) $L_n(\omega^2)$ for the given class of approximation.

Then, one needs to get the poles of the transfer function (and calculate the corresponding coefficients of the cells) and finally to proceed to the transformation into the digital domain (after proper pre-warping). For complete implementation of Variant A, replicas of Tables 19.7, 19.8 and 19.9 should be created for every conceived value of a_x.

If such a procedure was implemented for the 7th order LSM filter for $a_x = 0.1$ dB (To which corresponds $\varepsilon^2 = 10^{a_{\max}/10} - 1 = 0.02329$.), one would get the new coefficients of the cells as given in Table 19.5. The corresponding attenuation characteristic is depicted in Fig. 19.7 and denoted by "A". Note this would be the only way to accommodate to different values of $a_{\max}$ if non-monotonic pass-band amplitude characteristic (e.g. Chebyshev) was to be used.

Alternatively, one may rescale the original function. That may be done if, by analysis of the attenuation characteristic, the frequency is found in the pass-band at which the prescribed attenuation a_x is reached.

Table 19.5 Coefficients of the LSM cells for $a_{min} = 0.1$ dB, $n = 7$

	Cell no.	a_1	a_0	b_0	G
Variant A	I	1	0.8382296053	1	2.2609376252
	II	0.3681575640	1.3596015121	0.7323259440	0.4606771994
	III	1.0365060084	1.1149293509	4.0922660128	−0.4753827606
	IV	1.5071404049	0.8265477896	0.0914022622	−2.2462320640
Variant B	I	1	0.6256965514	1	0.5528172113
	II	0.2678307250	1.2437822669	1.5837820003	0.0867650311
	III	0.7589386803	0.9154209518	−22.3082096556	0.0192998639
	IV	1.1208144528	0.5472231928	0.0934529761	−0.6588821063

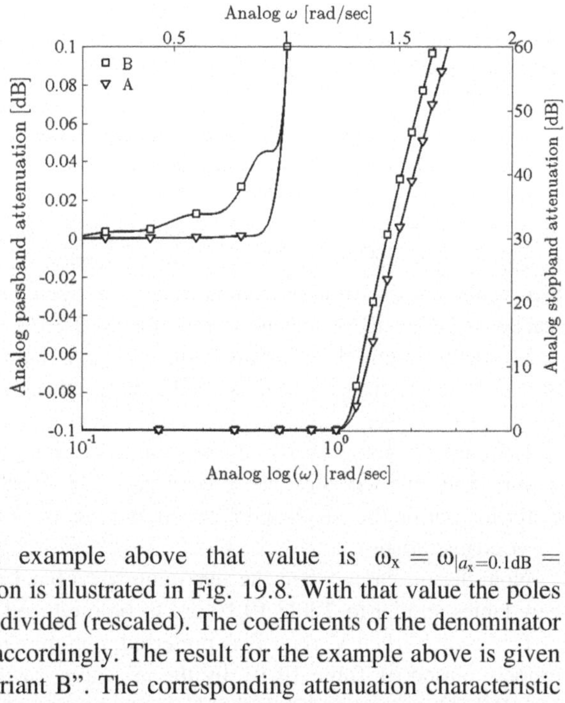

Fig. 19.7 Attenuation characteristics for the 7th order LSM with $a_{max} = 0.1$ dB

In the 7th order LSM example above that value is $\omega_x = \omega_{|a_x=0.1\text{dB}} = 0.881291$ rad/s. This operation is illustrated in Fig. 19.8. With that value the poles of the prototype function are divided (rescaled). The coefficients of the denominator polynomial are recomputed accordingly. The result for the example above is given in Fig. 19.7, denoted as "Variant B". The corresponding attenuation characteristic is depicted in Fig. 19.7 and denoted as B. As can be seen the attenuation characteristics obtained in two different ways of "rescaling" are not identical.

Variant B leads to a better attenuation characteristic in the stop-band thanks to the rise of the pass-band distortions being allowed. That may be exploited to reduce the filter order in some situations.

For implementation of Variant B one may produce a table with the values of ω_x for different values of a_x in advance. One such attempt is presented in Table 19.10 for $n = 6$ and for all four main classes of CMAC. The top part of Table 19.10, denoted as "Pass-band ω_x", is to be used. To use these numbers in conjunction with Tables 19.7 and 19.8, one has to divide a_1 and a_0 with ω_x and ω_x^2, respectively.

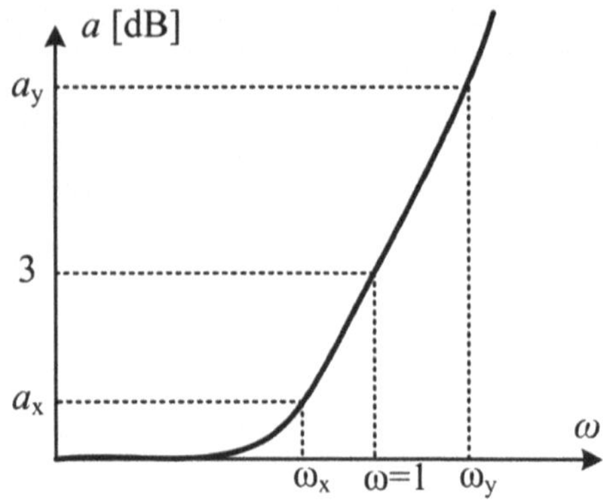

Fig. 19.8 Definition of the quantities in Table 19.10

Note, if such table is available the design process becomes much simpler since no polynomial solution is to be involved anew.

To complete the design process one must know the value of the filter order, *n*. While, for the Butterworth filters there are expressions to estimate *n*, for the rest of the filter classes such information is not given in advance. To overcome this, tables or nomograms may be created similar to the bottom part of Table 19.10 (denoted by "Stop-band ω_y"). With reference to Fig. 19.8, here, the values of the frequency at which the 3 dB (Variant A) version of the attenuation characteristic reaches the named value of attenuation a_y are given. For example, the Halpern filter reaches $a_y = 70$ dB in the stop-band at $\omega_y = 2.7033$.

So, if the required transition region width is larger than 1.7033, this function ($n = 6$, Halpern) may be used as a solution (or check for $n = 5$). If not, one is to go for $n = 7$ or higher.

Looking to Table 19.10, if the width of the transition region is the main requirement, one will always use the Halpern characteristic. In the cases when low distortion within the pass-band is required, however, one will rethink and go for the LSM characteristic.

Note that in cases when the maximum pass-band attenuation is less than 3 dB new tables similar to Table 19.10 are to be produced if Variant A is adopted.

In the case when Variant B is preferred, one may rescale the bottom part of the proper column of Table 19.10, by the ω_x value for the given a_x. For example, if $a_x = 0.2$ dB is required for the Halpern filter ($\omega_x = 0.2184$), and rescaling is implemented (Variant B), the new function will reach 50 dB in the stop-band at the frequency $\omega_{y\ new} = \omega_y/\omega_x = 1.9145/0.2184 = 8.766$ rad/s.

That is a large number.

To summarize, in cases when the maximum pass-band attenuation is lower than 3 dB, one should carefully choose between Variant A and Variant B since, one can see from Fig. 19.7, for the LSM characteristic Variant B is preferred, while for the Halpern filter one should choose Variant A.

19.3.3 Linear Phase Selective IIR Filters

The issue of synthesis linear phase recursive filters was one which is used most frequently in comparisons of FIR and IIR solutions. We already mentioned that by using the method proposed in Chap. 10 one may produce an analog prototype which, after transformation into an IIR filter may produce a solution which, from complexity point of view, is at least favorably comparable with any FIR counterpart with the same selectivity. In this paragraph we will exemplify how such a solution may be obtained and we will try to establish how is the bilinear transform influencing the frequency response of the digital filter obtained as a result.

Consider an 8th order Bessel filter in the s-domain whose amplitude characteristic was "corrected" by three zeros at the positive half of the ω-axis so that the minimum stopband attenuation is 40 dB. Its attenuation characteristic is depicted in Fig. 19.9, marked as analog. Its group delay characteristic is depicted in Fig. 19.10. Its cut-off frequency is chosen to be f_c = 100 kHz. Its zero-pole values are given in Table 19.6. It was transformed by the bilinear transform with f_s = 10 MHz. In that way T_s = 0.01. Prewarping was used but it has not much of significance since T_s is so small.

The attenuation characteristic of the resulting digital filter is depicted together with the analog in Fig. 19.9. One may see that for such a high sampling frequency the bilinear transform performs excellent. Practically there is no difference between the prototype and the transformed filter characteristic. When group delay is considered, however, as can be seen from Fig. 19.10, even with such a high sampling frequency and prewarping performed, distorsion are noteceable, though small and practically negligible. This is a sign that if smaller sampling frequency is used the distortions will become noticeable and not acceptable.

To check for that, the prototype was transformed anew with f_s = 2 MHz which increases T_s to a value of 0.05. The results are depicted in Fig. 19.11. Figure 19.11a depicts the attenuation in the transition region. One may see that the resulting digital filter has higher selectivity. Next, Fig. 19.11b represents the passband attenuation which is in fact not affected. Finally, Fig. 19.11c depicts the group delay. One may recognize serious distortions with a pick being almost 10% above the nominal value. A more detailed analysis of this issues may be found in [5].

Concluding this paragraph one may say that there are two alternatives in front of the designer of linear phase selective recursive filters. The first is described here and comprises use of high sampling frequency to avoid distortions of the group delay. The second is to use other transformation from the list of references at the end of this chapter.

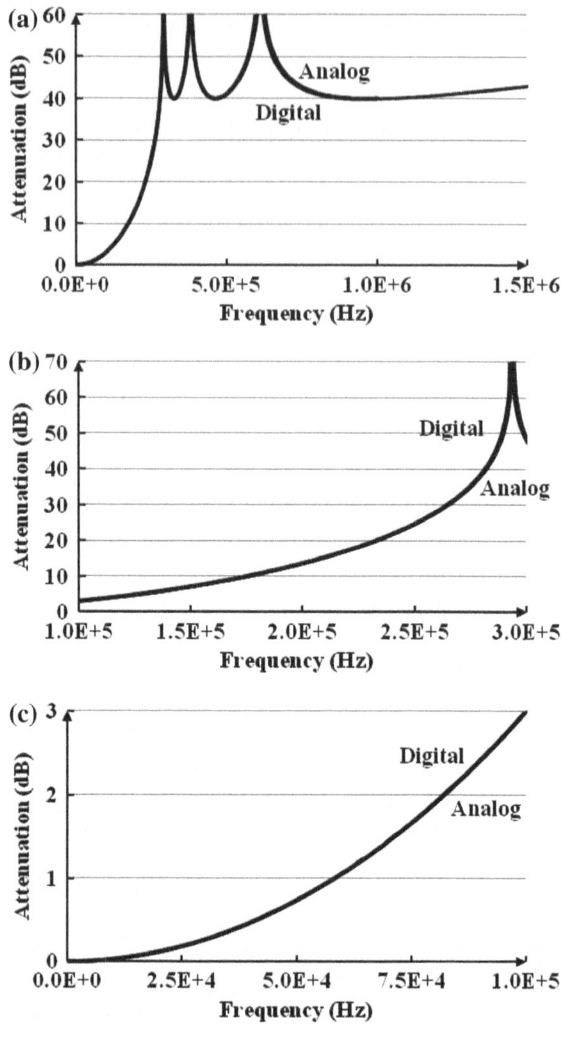

Fig. 19.9 Attenuation characteristics of the analog and the digital filter approximating constant group delay in maximally flat manner. Order of the filter is 8 and the number of transmission zeros is 6. **a** Overall attenuation, **b** transition region, and **c** passband attenuation. $T_s = 0.01$

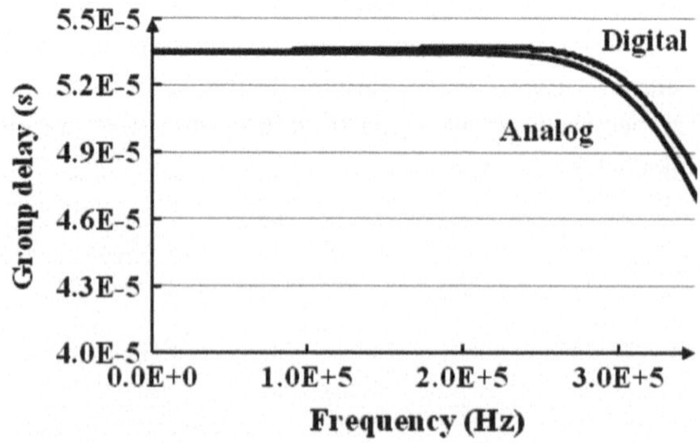

Fig. 19.10 Group delay characteristics of the analog and the digital filter approximating constant group delay in maximally flat manner. Order of the filter is 8 and the number of transmission zeros is 6. $T_s = 0.01$

Table 19.6 Zero-pole coordinates of the prototype linear phase selective analog filter

Zeros	
Real part	Imag. part
0.000000000e+000	2.938369488e+000
0.000000000e+000	−2.938369488e+000
0.000000000e+000	3.834516583e+000
0.000000000e+000	−3.834516583e+000
0.000000000e+000	6.173399911e+000
0.000000000e+000	−6.173399911e+000
Poles	
Real part	Imag. part
−2.896632789e+000	−4.497515574e−001
−2.896632789e+000	4.497515574e−001
−2.698070859e+000	−1.356165583e+000
−2.698070859e+000	1.356165583e+000
−1.471664584e+000	−3.293722825e+000
−1.471664584e+000	3.293722825e+000
−2.264421587e+000	−2.288346396e+000
−2.264421587e+000	2.288346396e+000

19.4 Design Example

Here, as an example, we will use the other version of selective filters exhibiting linear phase: the phase corrected selective filters.

The example was created as follows. First a 7th order L-filter was synthesized. To its characteristic function maximum number of transmission zeros were added to get a selective rational L_Z function. That was transformed into a band-pass function with passband width of 20%. To end the developments in the s domain a 2nd order corrector was synthesized to produce a 2% error of the group delay in the passband. Finally, the bilinear transform was implemented with $T_s = 0.1$ and $T_s = 0.2$. The attenuation and group delay characteristics of the filters so obtained are depicted in Fig. 19.12.

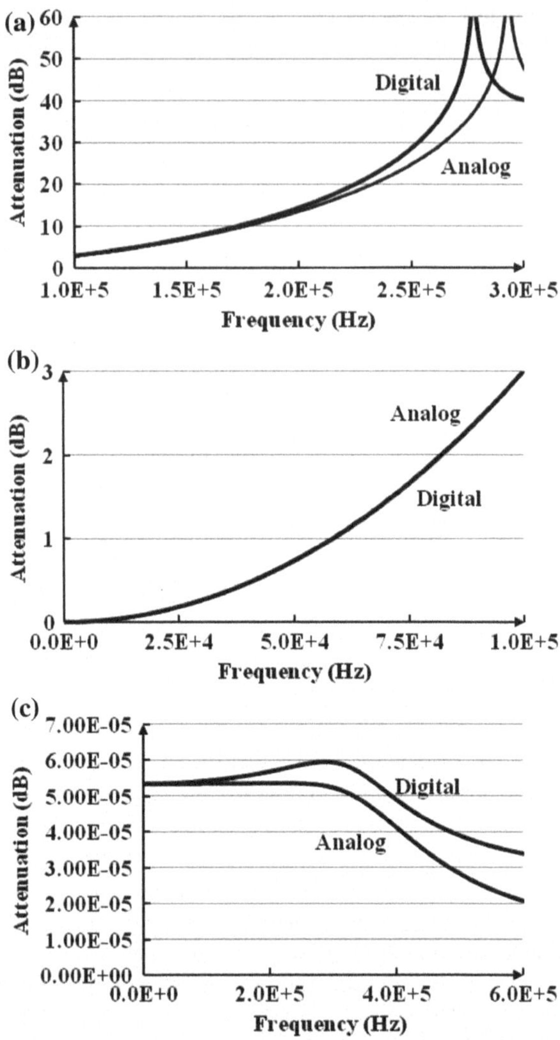

Fig. 19.11 Attenuation and group delay characteristics of the analog and the digital filter approximating constant group delay in maximally flat manner. Order of the filter is 8 and the number of transmission zeros is 6. **a** Transition region, **b** passband attenuation, and **c** group delay. $T_s = 0.05$

A glimpse to the results given reveals that the transformed digital filters tend to have narrower passband with the rise of T_s. Similarly, the frequency band in which the group delay behaves as a constant is getting narrower with a tendency to be further distorted. That would mean that for a proper bandpass digital design using the bilinear transform and a selective analog filter with corrected group delay one

Fig. 19.12 **a** Overall attenuation, **b** passband attenuation, and **c** passband group delay of the 14th order L_Z bandpass filter with a 2nd order corrector for analog, and two digital solutions based on the bilinear transform

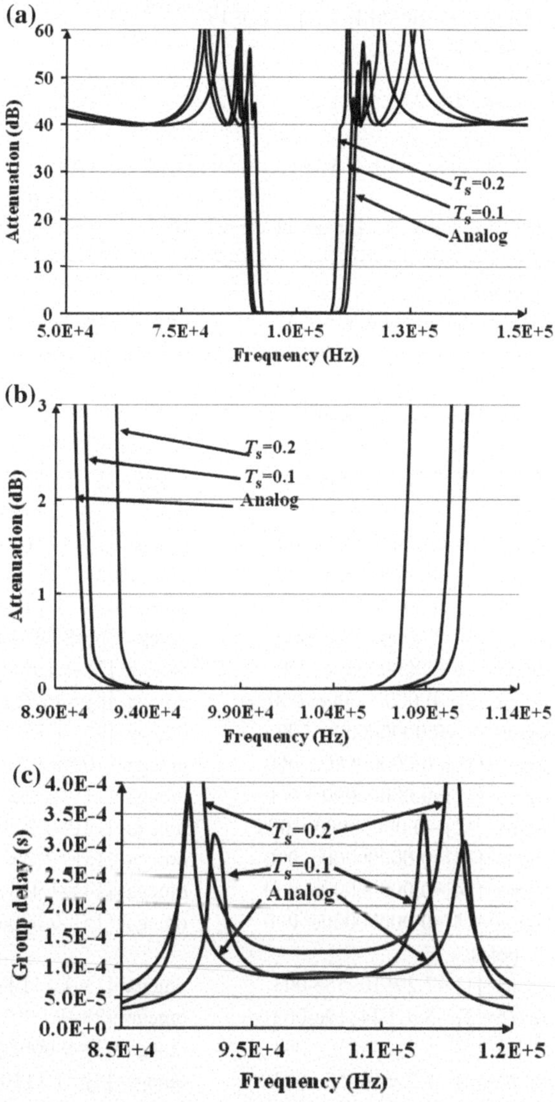

will need several iterations in which the relative passband width will be varied until satisfactory results are obtained. Of course, the sampling frequency may be used as an additional degree of freedom during such an optimization. Finally, if not satisfied, one may try another transform as listed in the references below.

Below is the edited report of the *IIR_parallel* program of the *RM* software.

Start.E

```
Program: IIR_PARALLEL

SYNTHESIS OF IIR PARALLEL DIGITAL FILTER
++++++++++++++++++++++++++++++++++++++++++++++++++++++++++++++++++++++++++++++
 PROJECT NAME: Linear_phase_BP_IIR
------------------------------------------------------------------------------
 Input data
Order of the numerator: n=15           Order of the denominator: m=16
Central frequency, f0=1.000000e+005    sampling frequency, fsample=1.000000e+006

 Original transfer function
    Zeros:
sigma=[1]=7.57859386e-002              omega=[1]=1.00888476e+000
sigma=[2]=7.57859386e-002              omega=[2]=-1.00888476e+000
sigma=[3]=0.00000000e+000              omega=[3]=1.13039221e+000
sigma=[4]=-0.00000000e+000             omega=[4]=-8.84648701e-001
sigma=[5]=0.00000000e+000              omega=[5]=-1.13039221e+000
sigma=[6]=0.00000000e+000              omega=[6]=8.84648701e-001
sigma=[7]=0.00000000e+000              omega=[7]=1.15311260e+000
sigma=[8]=-0.00000000e+000             omega=[8]=-8.67217996e-001
sigma=[9]=0.00000000e+000              omega=[9]=-1.15311260e+000
sigma=[10]=0.00000000e+000             omega=[10]=8.67217996e-001
sigma=[11]=0.00000000e+000             omega=[11]=1.26214789e+000
sigma=[12]=-0.00000000e+000            omega=[12]=-7.92300180e-001
sigma=[13]=0.00000000e+000             omega=[13]=-1.26214789e+000
sigma=[14]=0.00000000e+000             omega=[14]=7.92300180e-001
sigma=[15]=0.00000000e+000             omega=[15]=0.00000000e+000
    poles:
sigma=[1]=-7.29610955e-003             omega=[1]=1.11103081e+000
sigma=[2]=-5.91044819e-003             omega=[2]=-9.00026237e-001
sigma=[3]=-5.91044819e-003             omega=[3]=9.00026237e-001
sigma=[4]=-7.29610955e-003             omega=[4]=-1.11103081e+000
sigma=[5]=-2.87161739e-002             omega=[5]=1.10392653e+000
sigma=[6]=-2.35479151e-002             omega=[6]=-9.05244838e-001
sigma=[7]=-2.35479151e-002             omega=[7]=9.05244838e-001
sigma=[8]=-2.87161739e-002             omega=[8]=-1.10392653e+000
sigma=[9]=-9.87157760e-002             omega=[9]=9.95115669e-001
sigma=[10]=-9.87157760e-002            omega=[10]=-9.95115669e-001
sigma=[11]=-7.09346789e-002            omega=[11]=1.07577461e+000
```

```
sigma=[12]=-6.10283839e-002        omega=[12]=-9.25538638e-001
sigma=[13]=-6.10283839e-002        omega=[13]=9.25538638e-001
sigma=[14]=-7.09346789e-002        omega=[14]=-1.07577461e+000
sigma=[15]=-7.57859386e-002        omega=[15]=1.00888476e+000
sigma=[16]=-7.57859386e-002        omega=[16]=-1.00888476e+000
--------------------------------------------------------------------
Transformations in the s-domain

Residues in the poles
p1[ 1 ]= 2.47283482e-001    q1[ 1 ]= 6.15390127e-001
p1[ 2 ]= 3.00340883e-001    q1[ 2 ]= 4.34259478e-001
p1[ 3 ]= 3.00340883e-001    q1[ 3 ]= -4.34259478e-001
p1[ 4 ]= 2.47283482e-001    q1[ 4 ]= -6.15390127e-001
p1[ 5 ]= -3.84493707e+000   q1[ 5 ]= -1.19139402e-001
p1[ 6 ]= -2.85007766e+000   q1[ 6 ]= 6.77633018e-001
p1[ 7 ]= -2.85007766e+000   q1[ 7 ]= -6.77633018e-001
p1[ 8 ]= -3.84493707e+000   q1[ 8 ]= 1.19139402e-001
p1[ 9 ]= 7.72838506e+001    q1[ 9 ]= -2.28726761e+001
p1[ 10 ]= 7.72838506e+001   q1[ 10 ]= 2.28726761e+001
p1[ 11 ]= 9.36797118e+000   q1[ 11 ]= -1.76873781e+001
p1[ 12 ]= 2.53355193e+000   q1[ 12 ]= -1.33165032e+001
p1[ 13 ]= 2.53355193e+000   q1[ 13 ]= 1.33165032e+001
p1[ 14 ]= 9.36797118e+000   q1[ 14 ]= 1.76873781e+001
p1[ 15 ]= -8.25379834e+001           q1[ 15 ]= 2.80947280e+001
p1[ 16 ]= -8.25379834e+001           q1[ 16 ]= -2.80947280e+001

Gain at nominal frequency (0, 1, or infinty)=7.31678261e+001      ipb=1     (m-n) is
equal:0

Residues after multiplication with 1/gain
p1[ 1 ]= 3.37967513e-003    q1[ 1 ]= 8.41066572e-003
p1[ 2 ]= 4.10482174e-003    q1[ 2 ]= 5.93511522e-003
p1[ 3 ]= 4.10482174e-003    q1[ 3 ]= -5.93511522e-003
p1[ 4 ]= 3.37967513e-003    q1[ 4 ]= -8.41066572e-003
p1[ 5 ]= -5.25495601e-002   q1[ 5 ]= -1.62830315e-003
p1[ 6 ]= -3.89526081e-002   q1[ 6 ]= 9.26135235e-003
p1[ 7 ]= -3.89526081e-002   q1[ 7 ]= -9.26135235e-003
p1[ 8 ]= -5.25495601e-002   q1[ 8 ]= 1.62830315e-003
p1[ 9 ]= 1.05625457e+000    q1[ 9 ]= -3.12605654e-001
p1[ 10 ]= 1.05625457e+000   q1[ 10 ]= 3.12605654e-001
p1[ 11 ]= 1.28034024e-001   q1[ 11 ]= -2.41737100e-001
p1[ 12 ]= 3.46265848e-002   q1[ 12 ]= -1.81999437e-001
p1[ 13 ]= 3.46265848e-002   q1[ 13 ]= 1.81999437e-001
p1[ 14 ]= 1.28034024e-001   q1[ 14 ]= 2.41737100e-001
```

p1[15]= -1.12806390e+000 q1[15]= 3.83976531e-001
p1[16]= -1.12806390e+000 q1[16]= -3.83976531e-001

Ordered vector of analog transfer functions to be parallelized
T(s)=(g*s+q)/(s**2+a*s+b)

g	q	a	b
6.759350261e-003	-1.863970048e-002	1.459221910e-002	1.234442688e+000
8.209643471e-003	1.073204150e-002	1.182089638e-002	8.100821608e-001
-1.050991202e-001	5.770094915e-004	5.743234781e-002	1.219478404e+000
-7.790521619e-002	1.493307741e-002	4.709583016e-002	8.200227216e-001
2.112509138e+000	8.306955474e-001	1.974315521e-001	1.000000000e+000
2.560680473e-001	5.382733758e-001	1.418693578e-001	1.162322749e+000
6.925316957e-002	-3.326686133e-001	1.220567677e-001	8.603462340e-001
-2.256127803e+000	-9.457589029e-001	1.515718773e-001	1.023591966e+000

++

Transformations in the z-domain

BILINEAR TRANSFORM

T(z)=(gamma*z**2+epsilon*z+delta)/(ksi*z**2+alfa*z+beta)

NORMALIZED EXACT COEFFICIENTS

cell no.	beta	alfa	ksi	delta
	epsilon	gamma		
1	6.471270571e-001	-1.000000000e+000	6.525788543e-001	-2.394046497e-003
	-2.262733004e-003	1.313134932e-004		
2	5.914207748e-001	-1.000000000e+000	5.956208155e-001	-8.389836516e-004
	1.238971926e-003	2.077955578e-003		
3	6.370587715e-001	-1.000000000e+000	6.584771558e-001	1.963239350e-002
	6.991802172e-005	-1.956247548e-002		
4	5.864008213e-001	-1.000000000e+000	6.031534985e-001	1.471898665e-002
	1.725945669e-003	-1.299304098e-002		
5	5.821733837e-001	-1.000000000e+000	6.538945931e-001	-3.346817774e-001
	9.805030863e-002	4.327320861e-001		
6	6.136016213e-001	-1.000000000e+000	6.661454000e-001	-1.503180829e-002
	6.477562864e-002	7.980743693e-002		
7	5.780929634e-001	-1.000000000e+000	6.217136411e-001	-3.168957729e-002
	-3.862943978e-002	-6.939862488e-003		
8	5.935482882e-001	-1.000000000e+000	6.487637516e-001	3.549658079e-001
	-1.119434191e-001	-4.669092270e-001		

NORMALIZED 16 bit COEFFICIENTS

cell no. beta alfa ksi delta
epsilon gamma

```
1   6.471252441e-001   -1.000000000e+000   6.525573730e-001   -2.380371094e-003
-2.258300781e-003   1.220703125e-004
2   5.914001465e-001   -1.000000000e+000   5.956115723e-001   -8.239746094e-004
1.220703125e-003   2.075195313e-003
3   6.370544434e-001   -1.000000000e+000   6.584472656e-001   1.962280273e-002
6.103515625e-005   -1.956176758e-002
4   5.863952637e-001   -1.000000000e+000   6.031494141e-001   1.470947266e-002
1.708984375e-003   -1.296997070e-002
5   5.821533203e-001   -1.000000000e+000   6.538696289e-001   -3.346557617e-001
9.802246094e-002   4.327087402e-001
6   6.135864258e-001   -1.000000000e+000   6.661376953e-001   -1.501464844e-002
6.475830078e-002   7.980346680e-002
7   5.780639648e-001   -1.000000000e+000   6.217041016e-001   -3.167724609e-002
-3.860473633e-002   -6.927490234e-003
8   5.935363770e-001   -1.000000000e+000   6.487426758e-001   3.549499512e-001
-1.119384766e-001   -4.668884277e-001

NORMALIZED 32 bit COEFFICIENTS
cell no.  beta              alfa               ksi                delta
   epsilon              gamma
1   6.471270570e-001   -1.000000000e+000   6.525788540e-001   -2.394046169e-003
-2.262732945e-003   1.313132234e-004
2   5.914207748e-001   -1.000000000e+000   5.956208152e-001   -8.389833383e-004
1.238971483e-003   2.077955287e-003
3   6.370587712e-001   -1.000000000e+000   6.584771555e-001   1.963239349e-002
6.991764531e-005   -1.956247538e-002
4   5.864008209e-001   -1.000000000e+000   6.031534984e-001   1.471898658e-002
1.725945622e-003   -1.299304096e-002
5   5.821733833e-001   -1.000000000e+000   6.538945930e-001   -3.346817773e-001
9.805030841e-002   4.327320857e-001
6   6.136016212e-001   -1.000000000e+000   6.661453997e-001   -1.503180806e-002
6.477562850e-002   7.980743656e-002
7   5.780929630e-001   -1.000000000e+000   6.217136411e-001   -3.168957727e-002
-3.862943966e-002   -6.939862389e-003
8   5.935482881e-001   -1.000000000e+000   6.487637511e-001   3.549658079e-001
-1.119434186e-001   -4.669092270e-001
```

End.E

19.5 Developer's Corner

The example given in the previous paragraph is probably a best demonstration why tables and nomograms are still extensively in use for filter design. Namely, the number of steps needed to come to the final solution is large and every each of them represents a challenge. To avoid that, designers prefer to sacrifice quality by using as standard as solution as possible.

What we claim is that it is possible to create any shape of the final characteristic and one is not to hesitate to go for it. One only needs programming skills and the advice given throughout this book. One is not to forget [17, 18].

Appendix

See Tables 19.7, 19.8, 19.9 and 19.10.

Table 19.7 Coefficients of the cells of the parallel realization of Butterworth and Papoulis critical monotonic low-pass filters in the *s*-domain ($a_2 = 1$ ($b_1 = 1$) for second (first) order cells and $b_2 = 0$ for all cells)

n	Cell no.	Butterworth (B)			Papoulis (L)		
		a_1	a_0	b_0	a_1	a_0	b_0
3	I	1	1	1	1	0.6203082849	1
	II	1	1	0	0.6903917151	0.9307442993	0.0700834301
4	I	0.7653672194	1	0.7653672194	1.0995792894	0.4308098003	1.9118630981
	II	1.8477587806	1	1.8477587806	0.4633207106	0.9476323420	1.2756045193
5	I	1	1	1	1	0.4680710942	1
	II	1.6180340252	1	0.0000001317	0.3071340598	0.9608864565	−1.4132095572
	III	0.6180339748	1	3.2360678683	0.7762948460	0.4971652024	0.0360607687
6	I	1.4142151072	1	0.7071090985	0.8777336926	0.2502090573	1.2366018523
	II	1.931849679	1	1.4142110600	0.6180711616	0.5829658180	0.8578725685
	III	0.5176382137	1	−1.4142129778	0.2302951458	0.9695466448	0.2893476789
7	I	1	1	1	1	0.3820989548	1
	II	1.8019355769	1	−0.0000200688	0.1724393813	0.9762400047	14.8125340076
	III	0.445041614	1	−0.0000004674	0.4749524633	0.6623099563	−1.4276050748
	IV	1.246981809	1	1.6038825628	0.6984092005	0.3059759366	0.0849687232
8	I	1.1111393411	1	−3.6245565722	0.7343271929	0.1675321898	0.9792360679
	II	1.9615718304	1	1.2727606091	0.6007456553	0.3829961151	0.7441819695
	III	0.3901810003	1	0.8504310221	0.3884180817	0.7177426430	0.2912036234
	IV	1.6629388282	1	0.7209589468	0.1379090701	0.9809337153	−0.3301999028
9	I	1	1	1	1	0.3256519872	1
	II	1.8793736817	1	−0.0002223787	0.1107524513	0.9844603612	1.5344757725
	III	0.3472963815	1	5.7587720018	0.3135345509	0.7666049295	9.8384542102
	IV	0.9999967257	1	−0.0000098228	0.4977032803	0.4635299526	−0.9503328861
	V	1.5321032111	1	1.3054500093	0.6186577302	0.2089995013	0.1145312707

(continued)

Table 19.7 (continued)

n	Cell no.	Butterworth (B)			Papoulis (L)		
		a_1	a_0	b_0	a_1	a_0	b_0
10	I	1.7820096531	1	0.7434857812	0.6344118300	0.1217691252	0.8361740517
	II	0.9079809304	1	0.7434958874	0.5548133009	0.2702436256	0.6781545908
	III	0.3128686266	1	−1.2030038515	0.4283437543	0.5282533454	0.3525623940
	IV	1.9753764521	1	1.2030014647	0.2650385298	0.8012475974	−0.2051855948
	V	1.4142173378	1	−66217.1264399413	0.0918018850	0.9869322390	−1.1302164506

Table 19.8 Coefficients of the cells of the parallel realization of Halpern and LSM critical monotonic low-pass filters in the *s*-domain ($a2 = 1$ ($b1 = 1$) for second (first) order cells and $b2 = 0$ for all cells)

n	Cell no.	Halpern (H)			LSM		
		a_1	a_0	b_0	a_1	a_0	b_0
3	I	1	0.4702478436	1	1	0.7958988355	1
	II	0.6367521564	1.0632686715	0.1665043127	0.8153011645	0.9281027526	0.0194023290
4	I	0.8944625355	0.3143884544	2.4466636219	1.3772131318	0.6148237093	1.7777273416
	II	0.4139374645	1.0602599916	1.9661385509	0.5676868682	0.9390504811	0.9682010780
5	I	1	0.3282851703	1	1	0.6482864897	1
	II	0.2685842245	1.0572314618	−0.757205088	0.3984099499	0.9480204612	−4.375417099
	III	0.6284306051	0.4802053253	0.0996537476	1.0437035604	0.6131549502	0.0051829652
6	I	0.6617351889	0.1605479062	1.6082146540	1.1840327597	0.4128235590	1.2022258690
	II	0.4995965403	0.5933304958	1.2385017508	0.8448171966	0.6639747508	0.7223013471
	III	0.1993682707	1.0497882884	0.7138490124	0.3058500437	0.9573866056	−0.115686059
7	I	1	0.2574597373	1	1	0.5510897460	1
	II	0.1483753562	1.0434494931	−2.6107189660	0.2358951250	0.9648537709	1.3949350022
	III	0.3858424129	0.6954210571	−0.6065981603	0.6684443499	0.7101302059	−19.64822335
	IV	0.5252224936	0.2676335438	0.1192759812	0.9871707791	0.4245038502	0.0823098301
8	I	0.5304223592	0.0995163613	1.2978876890	1.0318940112	0.3010195783	0.9673497876
	II	0.4539728649	0.3654510662	1.0875926592	0.8556487737	0.4750725507	0.6512773766
	III	0.3158674364	0.7567369297	0.6675958151	0.5544582466	0.7506293161	−0.044374113
	IV	0.1184373395	1.0381602388	0.1631480199	0.1914989685	0.9707353681	−1.704969690
9	I	1	0.2139715266	1	1	0.4824732273	1
	II	0.0947590447	1.0331292237	17.7450272126	0.1562054203	0.9753746968	0.2404567516
	III	0.2570961950	0.8091515505	−1.9813784437	0.4561969335	0.7863440743	1.2311153294
	IV	0.3790969096	0.4683539787	−0.3484663478	0.7222177021	0.5266260241	−10.95628962
	V	0.4446763241	0.1705226892	0.1247991336	0.9012067168	0.3136118329	0.1375269472

(continued)

Table 19.8 (continued)

n	Cell no.	Halpern (H)			LSM		
		a_1	a_0	b_0	a_1	a_0	b_0
10	I	0.4454570681	0.0685710396	1.0948206554	0.9166602950	0.2321592186	0.8301260924
	II	0.4012051766	0.2457418738	0.9527933042	0.8111224547	0.3568113268	0.6049824304
	III	0.3089548571	0.5393899795	0.7021883258	0.6281627374	0.5753122253	0.0836489006
	IV	0.2159139308	0.8437755380	0.3190843324	0.3890472425	0.8149796901	−1.452015087
	V	0.0796689673	1.0287391826	−0.2135508298	0.1310072703	0.9787969491	4.5505889595

Table 19.9 Cell's gains of the parallel realization of critical monotonic low-pass filters in the s-domain

n	Cell no.	Butterworth (B)	Papoulis (L)	Halpern (H)	LSM
		G	G	G	G
3	I	1	0.6507013320	0.5076292764	0.8093655101
	II	−1	−0.6507013320	−0.5076292764	−0.8093655101
4	I	−0.9238800780	0.3234425432	0.1686937496	0.5375188797
	II	0.9238800780	−0.3234425432	−0.1686937496	−0.5375188797
5	I	1.8944273525	0.6114897244	0.4055185021	0.9514555431
	II	−1.6180341569	0.1698023137	0.1635836040	0.0994002520
	III	−0.2763931956	−0.7812920381	−0.5691021061	−1.0508557951
6	I	−3.0472259343	0.3706559579	0.1508760739	0.7964360226
	II	2.6389775882	−0.6148223552	−0.2904240863	−1.1712122529
	III	0.4082483461	0.2441663972	0.1395480125	0.3747762303
7	I	4.3119060349	0.6167129225	0.3545488355	1.1842138663
	II	−2.9839132723	0.0124834590	−0.0430683868	0.1858631583
	III	0.7369748670	0.2583847482	0.2625637728	0.0413430805
	IV	−2.0649676297	−0.8875811298	−0.5740442215	−1.4114201052
8	I	0.6935110900	0.3774798517	0.1268605481	1.0218205613
	II	7.1357913788	−0.7528469870	−0.2927226517	−1.7832499664
	III	0.5879388222	0.5156456831	0.2628447381	0.8827853623
	IV	−8.4172412911	−0.1402785478	−0.0969826345	−0.1213559572
9	I	10.7204116638	0.6331191284	0.3215415390	1.4724703065
	II	−6.1526839179	−0.0678860920	−0.0044949903	−0.2068562996
	III	0.1157654862	0.0377157435	−0.0963164797	0.5255373963
	IV	3.2742337540	0.3817890029	0.3386674330	0.0959483109
	V	−7.9577269861	−0.9847377829	−0.5593975020	−1.8870997141
10	I	−22.9393185497	0.3730208554	0.1104025225	1.2419942101
	II	3.7977123674	−0.8194721420	−0.2766519222	−2.4113618506
	III	−0.3717469536	0.7230379003	0.3087634168	1.5451351412
	IV	19.5132218984	−0.3423648113	−0.2076174012	−0.3374220626
	V	0.0001312376	0.0657781976	0.0651033841	−0.0383454382

Table 19.10 Sampled attenuation characteristics of the critical monotonic filters for $n = 6$

	a(dB)	B	L	H	LSM
Pass-band ω_x	0.1	0.7313	0.2899	0.1792	0.8448
	0.2	0.7755	0.4032	0.2184	0.8728
	0.5	0.8395	0.7600	0.2925	0.9085
	1.0	0.8938	0.9407	0.3904	0.9386
Stop-band ω_y	30.0	1.7788	1.4417	1.4115	1.5336
	50.0	2.6111	1.9915	1.9145	2.1639
	70.0	3.8327	2.8389	2.7033	3.1173
	100.0	6.8156	4.9601	4.6951	5.4818

References

1. Antoniou A (1983) Digital filters: analysis and design. McGraw-Hill Inc, New York
2. Litwin L (2000) FIR and IIR digital filters. The effects of finite bit precision. IEEE Potentials 19(4):28–31
3. Quélhas MF, Petraglia A, Petraglia MR (2004) Efficient group delay equalization of discrete-time IIR filters. In: Proceedings of the XII European signal processing conference, EUSIPCO-2004, Vienna, Austria, vol 1, pp 125–128
4. Mollowa GS, Unbehauen R (1998) Design of recursive filters with constant group delay and chebyshev stopband attenuation. In: Proceedings No. 3 of the IEEE international conference on acoustics, speech and signal processing, ICASSP '98, Seattle, WA, Art. No. 681673, pp 1257–1260
5. Mirković D, Litovski IV, Litovski VB (2015) On the synthesis and realization of selective linear phase IIR filters. In: Proceedings of 2nd international conference on electrical, electronic and computing engineering, IcETRAN 2015, Silver Lake, Serbia, ELI1.3.1-6
6. Rabey J (2009) Low power design essentials. Springer Science+Business Media, LLC, New York
7. Mirković D, Petković P, Litovski V (2014) A second order s-to-z transform and its implementation to IIR filter design. COMPEL Int J Comput Math Electr Electron Eng 33 (5):1831–1843
8. Kaiser JF (1996) Digital filters. In: Kuo FF, Kaiser JF (eds) (1996) System analysis by digital computer. Wiley, New York
9. Hongyan C, Lisheng W (2012) Software and hardware implementation of IIR based on Matlab&Acceldsp. In: Proceedings of the 2nd international conference on computer application and system modeling. Atlantis Press, Paris, France, pp 1411–1415
10. Nelatury SR (2007) Additional correction to the impulse invariance method for the design of IIR digital filters. Digit Signal Proc 17(2):530–540
11. Németh JG, Kollár I. (2000) Step-invariant transform from z- to s-domain, A general framework. In: Proceedings of the 17th IEEE instrumentation and measurement technology conference, IMTC/2000, Baltimore, MD, vol 2, pp 902–907
12. Paarmann LD (1998) Mapping from the s-domain to the z-domain via magnitude-invariance method. Sig Process 69(3):219–228
13. Paarmann LD (2006) Mapping from the s-domain to the z-domain via phase-invariance method. Sig Process 86(2):223–229
14. Shi R, Wang S, Zhao J (2012) An unsplit complex-frequency-shifted PML based on matched z-transform for FDTD modelling of seismic wave equations. J Geophys Eng 9(2):218–229
15. Park W, Park KS, Koh HM (2008) Active control of large structures using a bilinear pole-shifting transform with H_∞ control method. Eng Struct 30(11):3336–3344
16. Krantz SG (1999) Rouché's theorem, §5.3.1 in "Handbook of Complex Variables". Birkhäuser Publication Co., Boston, MA, p 74
17. https://www.youtube.com/channel/UCF_Ipw_YD2gwrRpJDUJJULw/playlists?disable_polymer=1
18. http://skysupervisor.com/projects/RM%20software/RM%20software.htm

Index

V. Litovski, *Electronic Filters*, Lecture Notes in Electrical Engineering 596,
https://doi.org/10.1007/978-981-32-9852-1

The manufacturer's authorised representative in the EU is Springer Nature Customer Service Centre GmbH, Europaplatz 3, 69115 Heidelberg, Germany. If you have any concerns regarding our products, please contact ProductSafety@springernature.com

Printed and bound by CPI Group (UK) Ltd, Croydon, CR0 4YY
25/09/2026
02222884-0011